Numerical Issues in
Statistical Computing
for the Social Scientist

Numerical Issues in Statistical Computing for the Social Scientist

MICAH ALTMAN

JEFF GILL

MICHAEL P. McDONALD

WILEY-INTERSCIENCE

A JOHN WILEY & SONS, INC., PUBLICATION

Library of Congress Cataloging-in-Publication Data:

Altman, Micah.
    Numerical issues in statistical computing for the social scientist / Micah Altman, Jeff Gill, Michael P. McDonald.
        p. cm.—(Wiley series in probability and statistics)
    Includes bibliographical references and index.
    ISBN 0-471-23633-0 (acid-free paper)
    1. Statistics–Data processing. 2. Social sciences–Statistical methods–Data processing.
I. Gill, Jeff. II. McDonald, Michael P., 1967–III. Title. IV. Series.

    QA276.4.A398 2004
    519.5–dc21   2003053470

Printed in the United States of America.

10 9 8 7 6 5 4 3 2 1

Contents

Preface

Overview

This book is intended to serve multiple purposes. In one sense it is a pure research book in the traditional manner: new principles, new algorithms, and new solutions. But perhaps more generally it is a guidebook like those used by naturalists to identify wild species. Our "species" are various methods of estimation requiring advanced statistical computing: maximum likelihood, Markov chain Monte Carlo, ecological inference, nonparametrics, and so on. Only a few are wild; most are reasonably domesticated.

A great many empirical researchers in the social sciences take computational factors for granted: "For the social scientist, software is a tool, not an end in itself" (MacKie-Mason 1992). Although an extensive literature exists on statistical computing in statistics, applied mathematics, and embedded within various natural science fields, there is currently no such guide tailored to the needs of the social sciences. Although an abundance of package-specific literature and a small amount of work at the basic, introductory level exists, a text is lacking that provides social scientists with modern tools, tricks, and advice, yet remains accessible through explanation and example.

The overall purpose of this work is to address what we see as a serious deficiency in statistical work in the social and behavioral sciences, broadly defined. Quantitative researchers in these fields rely on statistical and mathematical computation as much as any of their colleagues in the natural sciences, yet there is less appreciation for the problems and issues in numerical computation. This book seeks to rectify this discrepancy by providing a rich set of interrelated chapters on important aspects of social science statistical computing that will guide empirical social scientists past the traps and mines of modern statistical computing.

The lack of a bridging work between standard statistical texts, which, at most, touch on numerical computing issues, and the comprehensive work in statistical computing has hindered research in a number of social science fields. There are two pathologies that can result. In one instance, the statistical computing process fails and the user gives up and finds less sophisticated means of answering research questions. Alternatively, something disastrous happens during the numerical calculations, yet seemingly reasonable output results. This is much worse, because there are no indications that something has failed, and incorrect statistical output becomes a component of the larger project.

Fortunately, many of the most common problems are easy to describe and easier still to avoid. We focus here to a great extent on problems that can occur in maximum likelihood estimation and nonlinear regression because these are, with the exception of simple linear models, the methods most widely used by social scientists. False convergence, numerical instability, and problematic likelihood surfaces can be diagnosed without much agony by most interested social scientists if they have specific advice about how to do so. Straightforward computational techniques such as data rescaling, changes of starting values, function reparameterization, and proper use of analytic derivatives can then be used to reduce or eliminate many numerical problems. Other important and recent statistical approaches that we discuss are ecological inference, logistic regression, Markov chain Monte Carlo, and spatial analysis.

Starters

In this book we introduce the basic principles of numerical computation, outlines the optimization process, and provides specific tools to assess the sensitivity of the subsequent results to problems with these data or model. The reader is not required to have an extensive background in mathematical statistics, advanced matrix algebra, or computer science. In general, the reader should have at least a year of statistics training, including maximum likelihood estimation, modest matrix algebra, and some basic calculus. In addition, rudimentary programming knowledge in a statistical package or compiled language is required to understand and implement the ideas herein.

Some excellent sources for addressing these preliminaries can be found in the following sources.

- *Introductory statistics.* A basic introductory statistics course, along the lines of such texts as: Moore and McCabe's *Introduction to the Practice of Statistics* (2002), Moore's *The Basic Practice of Statistics* (1999), *Basic Statistics for the Social and Behavioral Sciences* by Diekhoff (1996), Blalock's wellworn *Social Statistics* (1979), Freedman et al.'s *Statistics* (1997), Amemiya's *Introduction to Statistics and Econometrics* (1994), or *Statistics for the Social Sciences* by Sirkin (1999).

- *Elementary matrix algebra.* Some knowledge of matrix algebra, roughly at the level of Greene's (2003) introductory appendix, or the first half of the undergraduate texts by either Axler (1997) or Anton and Rorres (2000). It will not be necessary for readers to have an extensive knowledge of linear algebra or experience with detailed calculations. Instead, knowledge of the structure of matrices, matrix and vector manipulation, and essential symbology will be assumed. Having said that, two wonderful reference books that we advise owning are the theory book by Lax (1997), and the aptly entitled book by Harville (1997), *Matrix Algebra from a Statisticians Perspective*.

- *Basic calculus.* Elementary knowledge of calculus is important. Helpful, basic, and inexpensive basic texts include Kleppner and Ramsey (1985), Bleau (1994), Thompson and Gardner (1998), and for a very basic introduction, see Downing (1996). Although we avoid extensive derivations, this material is occasionally helpful.

Programming

Although knowledge of programming is not required, most readers of this book are, or should be, programmers. We do not mean necessarily in the sense of generating hundreds of lines of FORTRAN code between seminars. By *programming* we mean working with *statistical* languages: writing likelihood functions in Gauss, R, or perhaps even Stata, coding solutions in WinBUGS, or manipulating procedures in SAS. If all available social science statistical solutions were available as point-and-click solutions in SPSS, there would not be very many truly interesting models in print.

There are two, essentially diametric views on programming among academic practitioners in the social sciences. One is emblemized by a well-known quote from Hoare (1969, p. 576): "Computer programming is an exact science in that all the properties of a program and all the consequences of executing it in any given environment can, in principle, be found out from the text of the program itself by means of purely deductive reasoning." A second is by Knuth (1973): "It can be an aesthetic experience much like composing poetry or music." Our perspective on programming agrees with both experts; programming is a rigorous and exacting process, but it should also be creative and fun. It is a rewarding activity because practitioners can almost instantly see the fruits of their labor. We give extensive guidance here about the practice of statistical programming because it is important for doing advanced work and for generating high-quality work.

Layout of Book and Course Organization

There are two basic sections to this book. The first comprises four chapters and focuses on general issues and concerns in statistical computing. The goal in this section is to review important aspects of numerical maximum likelihood and related estimation procedures while identifying specific problems. The second section is a series of six chapters outlining specific problems that center on problems that originate in different disciplines but are not necessarily contained within. Given the extensive methodological cross-fertilization that occurs in the social sciences, these chapters should have more than a narrow appeal. The last chapter provides a summary of recommendations from previous chapters and an extended discussion of methods for ensuring the general replicability of one's research.

The book is organized as a single-semester assignment accompanying text. Obviously, this means that some topics are treated with less detail than in a fully developed mathematical statistics text that would be assigned in a one-year statistics department course. However, there is a sufficient set of references to lead interested readers into more detailed works.

A general format is followed within each chapter in this work, despite widely varying topics. A specific motivation is given for the material, followed by a detailed exposition of the tool (mode finding, EI, logit estimation, MCMC, etc.). The main numerical estimation issues are outlined along with various means of avoiding specific common problems. Each point is illustrated using data that social scientists care about and can relate to. This last point is not trivial; a great many books in overlapping areas focus on examples from biostatistics, and the result is often to reduce reader interest and perceived applicability in the social sciences. Therefore, every example is taken from the social and behavioral sciences, including: economics, marketing, psychology, public policy, sociology, political science, and anthropology.

Many researchers in quantitative social science will simply read this book from beginning to end. Researchers who are already familiar with the basics of statistical computation may wish to skim the first several chapters and pay particular attention to Chapters 4, 5, 6, and 11, as well as chapters specific to the methods being investigated.

Because of the diversity of topics and difficulty levels, we have taken pains to ensure that large sections of the book are approachable by other audiences. For those who do not have the time or training to read the entire book, we recommend the following:

- *Undergraduates* in courses on statistics or research methodology, will find a gentle introduction to statistical computation and its importance in Section 1.1 and Chapter 2. These may be read without prerequisites.
- *Graduate students* doing any type of quantitative research will wish to read the introductory chapters as well, and will find Chapters 3 and 11 useful and approachable. Graduate students using more advanced statistical models should also read Chapters 5 and 8, although these require more some mathematical background.
- *Practitioners* may prefer to skip the introduction, and start with Chapters 3, 4, and 11, as well as other chapters specific to the methods they are using (e.g., nonlinear models, MCMC, ecological inference, spatial methods).

However, we hope readers will enjoy the entire work. This is intended to be a research work as well as a reference work, so presumably experienced researchers in this area will still find some interesting new points and views within.

Web Site

Accompanying this book is a Web site: <http://www.hmdc.harvard.edu/ numerical_issues/>. This site contains links to many relevant resources, including bibliographies, discussion-lists, benchmark data, high-precision libraries, and optimization software.

In addition, the Web site includes links to all of the code and data used in this book and not otherwise described in detail, in order to assist other scholars in carrying out similar analyses on other datasets.

Debts and Dedications

We would like to thank the support of our host institutions: Harvard University, University of Florida, and George Mason University. All three of us have worked in and enjoyed the Harvard–MIT Data Center as a research area, as a provider of data, and as an intellectual environment to test these ideas. We also unjammed the printer a lot and debugged the e-mail system on occasion while there. We thank Gary King for supporting our presence at the center.

The list of people to thank in this effort is vast. We would certainly be remiss without mentioning Chris Achen, Bob Anderson, Attic Access, Neal Beck, Janet Box–Steffensmeier, Barry Burden, Dieter Burrell, George Casella, Suzie De Boef, Scott Desposato, Karen Ferree, John Fox, Charles Franklin, Hank Heitowit, Michael Herron, Jim Hobert, James Honaker, Simon Jackman, Bill Jacoby, David James, Dean Lacey, Andrew Martin, Michael Martinez, Rogerio Mattos, Ken McCue, Ken Meier, Kylie Mills, Chris Mooney, Jonathan Nagler, Kevin Quinn, Ken Shotts, Kevin Smith, Wendy Tam Cho, Alvaro Veiga, William Wei, Guy Whitten, Jason Wittenberg, Dan Wood, and Chris Zorn (prison rodeo consultant to the project). A special thanks go to our contributing authors, Paul Allison, Gary King, James LeSage, and Bruce McCullough, for their excellent work, tireless rewriting efforts, and general patience with the three of us. Special thanks also go to our editor, Steve Quigley, as well, since this project would not exist without his inspiration, prodding, and general guidance.

Significant portions of this book, especially Chapters 2, 3, and 11, are based in part upon research supported by National Science Foundation Award No. 11S-987 47 47.

This project was typeset using LaTeX and associated tools from the TeX world on a Linux cluster housed at the Harvard–MIT Data Center. We used the John Wiley & Sons LaTeX style file with the default computer modern font. All of this produced very nice layouts with only moderate effort on our part.

CHAPTER 1

Introduction: Consequences of Numerical Inaccuracy

1.1 IMPORTANCE OF UNDERSTANDING COMPUTATIONAL STATISTICS

How much pollution is bad for you? Well-known research conducted from 1987 to 1994 linked small-particle air pollution to health problems in 90 U.S. cities. These findings were considered reliable and were influential in shaping public policy. Recently, when the same scientists attempted to replicate their own findings, they produced different results with the same data—results that showed a much weaker link between air pollution and health problems. "[The researchers] re-examined the original figures and found that the problem lay with how they used off-the-shelf statistical software to identify telltale patterns that are somewhat akin to ripples from a particular rock tossed into a wavy sea. Instead of adjusting the program to the circumstances that they were studying, they used standard default settings for some calculations. That move apparently introduced a bias in the results, the team says in the papers on the Web" (Revkin, 2002).

Problems with numerical applications are practically as old as computers: In 1962, the *Mariner I* spacecraft, intended as the first probe to visit another planet, was destroyed as a result of the incorrect coding of a mathematical formula (Neumann 1995), and five years later, Longley (1967) reported on pervasive errors in the accuracy of statistical programs' implementation of linear regression. Unreliable software is sometimes even expected and tolerated by experienced researchers. Consider this report on the investigation of a high-profile incident of academic fraud, involving the falsification of data purporting to support the discovery of the world's heaviest element at Lawrence Berkeley lab: "The initial suspect was the analysis software, nicknamed Goosy, a somewhat temperamental computer program known on occasion to randomly corrupt data. Over the years, users had developed tricks for dealing with Goosy's irregularities, as one might correct a wobbling image on a TV set by slapping the side of the cabinet" (Johnson 2002).

In recent years, many of the most widely publicized examples of scientific application failures related to software have been in the fields of space exploration

Numerical Issues in Statistical Computing for the Social Scientist, by Micah Altman, Jeff Gill, and Michael P. McDonald
ISBN 0-471-23633-0   Copyright © 2004 John Wiley & Sons, Inc.

and rocket technology. Rounding errors in numerical calculations were blamed for the failure of the Patriot missile defense to protect an army barracks in Dhahran from a Scud missile attack in 1991 during Operation Desert Storm (Higham 2002). The next year, the space shuttle had difficulties in an attempted rendezvous with *Intelsat 6* because of a round-off error in the routines that the shuttle computers used to compute distance (Neumann 1995). In 1999, two Mars-bound spacecraft were lost, due (at least in part) to software errors—one involving failure to check the units as navigational inputs (Carreau 2000). Numerical software bugs have even affected our understanding of the basic structure of the universe: highly publicized findings suggesting the existence of unknown forms of matter in the universe, in violation of the "standard model," were later traced to numerical errors, such as failure to treat properly the sign of certain calculations (Glanz 2002; Hayakawa and Kinoshita 2001).

The other sciences, and the social sciences in particular, have had their share of less publicized numerical problems: Krug et al. (1988) retracted a study analyzing suicide rates following natural disasters that was originally published in the *Journal of the American Medical Association*, one of the world's most prestigious medical journals, because their software erroneously counted some deaths twice, undermining their conclusions (see Powell et al. 1999). Leimer and Lesnoy (1982) trace Feldstein's (1974) erroneous conclusion that the introduction of Social Security reduced personal savings by 50% to the existence of a simple software bug. Dewald et al. (1986), in replicating noted empirical results appearing in the *Journal of Money, Credit and Banking*, discovered a number of serious bugs in the original authors' analyses programs. Our research and that of others has exposed errors in articles recently published in political and social science journals that can be traced to numerical inaccuracies in statistical software (Altman and McDonald 2003; McCullough and Vinod 2003; Stokes 2003).

Unfortunately, numerical errors in published social science analyses can be revealed only through replication of the research. Given the difficulty and rarity of replication in the social sciences (Dewald et al. 1986; Feigenbaum and Levy 1993), the numerical problems reported earlier are probably the tip of the iceberg. One is forced to wonder how much of the critical and foundational findings in a number of fields are actually based on suspect statistical computing.

There are two primary sources of potential error in numerical algorithms programmed on computers: that numbers cannot be perfectly represented within the limited binary world of computers, and that some algorithms are not guaranteed to produce the desired solution.

First, small computational inaccuracies occur at the precision level of all statistical software when digits beyond the storage capacity of the computer must be rounded or truncated. Researchers may be tempted to dismiss this threat to validity because measurement error (miscoding of data, survey sampling error, etc.) is almost certainly an order of magnitude greater for most social science applications. But these small errors may propagate and magnify in unexpected ways in the many calculations underpinning statistical algorithms, producing wildly erroneous results on their own, or exacerbating the effects of measurement error.

Second, computational procedures may be subtly biased in ways that are hard to detect and are sometimes not guaranteed to produce a correct solution. Random number generators may be subtly biased: random numbers are generated by computers through non-random, deterministic processes that mimic a sequence of random numbers but are not genuinely random. Optimization algorithms, such as maximum likelihood estimation, are not guaranteed to find the solution in the presence of multiple local optima: Optimization algorithms are notably susceptible to numeric inaccuracies, and resulting coefficients may be far from their true values, posing a serious threat to the internal validity of hypothesized relationships linking concepts in the theoretical model.

An understanding of the limits of statistical software can help researchers avoid estimation errors. For typical estimation, such as ordinary least squares regression, well-designed off-the-shelf statistical software will generally produce reliable estimates. For complex algorithms, our knowledge of model building has outpaced our knowledge of computational statistics. We hope that researchers contemplating complex models will find this book a valuable tool to aid in making robust inference within the limits of computational statistics.

Awareness of the limits of computational statistics may further aid in model testing. Social scientists are sometimes faced with iterative models that fail to converge, software that produces nonsensical results, Hessians that cannot be inverted, and other problems associated with estimation. Normally, this would cause researchers to abandon the model or embark on the often difficult and expensive process of gathering more data. An understanding of computational issues can offer a more immediately available solution—such as use of more accurate computations, changing algorithmic parameters of the software, or appropriate rescaling of the data.

1.2 BRIEF HISTORY: DUHEM TO THE TWENTY-FIRST CENTURY

The reliability of scientific inference depends on one's tools. As early as 1906, French physicist and philosopher of science Pierre Duhem noted that every scientific inference is conditioned implicitly on a constellation of background hypotheses, including that the instruments are functioning correctly (Duhem 1991, Sec. IV.2). The foremost of the instruments used by modern applied statisticians is the computer.

In the early part of the twentieth century the definition of a *computer* to statisticians was quite different from what it is today. In antiquated statistics journals one can read where authors surprisingly mention "handing the problem over to my computer." Given the current vernacular, it is easy to miss what is going on here. Statisticians at the time employed as "computers" *people* who specialized in performing repetitive arithmetic. Many articles published in leading statistics journals of the time addressed methods by which these calculations could be made less drudgingly repetitive because it was noticed that as tedium increases linearly, careless mistakes increase exponentially (or thereabouts). Another rather prescient development of the time given our purpose here was the attention paid to creating self-checking procedures where "the computer" would at regular intervals have a clever means to check calculations against some summary value as a way of detecting errors (cf. Kelley and McNemar 1929). One of the reasons that Fisher's normal tables (and therefore the artificial 0.01 and 0.05 significance thresholds) were used so widely was that the task of manually calculating normal integrals was time consuming and tedious. Computation, it turns out, played an important role in scholarship even before the task was handed over to machines.

In 1943, Hotelling and others called attention to the accumulation of errors in the solutions for inverting matrices in the method of least squares (Hotelling 1943) and other matrix manipulation (Turing 1948). Soon after development of the mainframe computer, programmed regression algorithms were criticized for dramatic inaccuracies (Longley 1967). Inevitably, we improve our software, and just as inevitably we make our statistical methods more ambitious. Approximately every 10 years thereafter, each new generation of statistical software has been similarly faulted (e.g., Wampler 1980; Simon and LeSage 1988).

One of the most important statistical developments of the twentieth century was the advent of *simulation* on computers. While the first simulations were done *manually* by Buffon, Gosset, and others, it was not until the development of machine-repeated calculations and electronic storage that simulation became prevalent. In their pioneering postwar work, von Neumann and Ulam termed this sort of work *Monte Carlo simulation*, presumably because it reminded them of long-run observed odds that determine casino income (Metropolis and Ulam 1949; Von Neumann 1951). The work was conducted with some urgency in the 1950s because of the military advantage of simulating nuclear weapon designs. One of the primary calculations performed by von Neumann and his colleagues was a complex set of equations related to the speed of radiation diffusion of fissile materials. This was a perfect application of the Monte Carlo method because it avoided both daunting analytical work and dangerous empirical work. During this same era, Metropolis et al. (1953) showed that a new version of Monte Carlo simulation based on Markov chains could model the movement of atomic particles in a box when analytical calculations are impossible.

Most statistical computing tasks today are sufficiently routinized that many scholars pay little attention to implementation details such as default settings, methods of randomness, and alternative estimation techniques. The vast majority of statistical software users blissfully point-and-click their way through machine implementations of noncomplex procedures such as least squares regression, cross-tabulation, and distributional summaries. However, an increasing number of social scientists regularly use more complex and more demanding computing methods, such as Monte Carlo simulation, nonlinear estimation procedures, queueing models, Bayesian stochastic simulation, and nonparametric estimation. Accompanying these tools is a general concern about the possibility of knowingly or unknowingly producing invalid results.

In a startling article, McCullough and Vinod (1999) find that econometric software packages can still produce "horrendously inaccurate" results (p. 635) and that inaccuracies in many of these packages have gone largely unnoticed (pp. 635–37). Moreover, they argue that given these inaccuracies, past inferences are in question and future work must document and archive statistical software alongside statistical models to enable replication (pp. 660–62).

In contrast, when most social scientists write about quantitative analysis, they tend not to discuss issues of accuracy in the implementation of statistical models and algorithms. Few of our textbooks, even those geared toward the most sophisticated and computationally intensive techniques, mention issues of implementation accuracy and numerical stability. Acton (1996), on the other hand, gives a frightening list of potential problems: "loss of significant digits, iterative instabilities, degenerative inefficiencies in algorithms, and convergence to extraneous roots of previously docile equations."

When social science methodology textbooks and review articles in social science do discuss accuracy in computer-intensive quantitative analysis, they are relatively sanguine about the issues of accurate implementation:

- On finding maximum likelihood: "Good algorithms find the correct solution regardless of starting values. ... The computer programs for most standard ML estimators automatically compute good starting values." And on accuracy: "Since neither accuracy nor precision is sacrificed with numerical methods they are sometimes used even when analytical (or partially analytical) solutions are possible" (King 1989, pp. 72–73).
- On the error of approximation in Monte Carlo analysis: "First, one may simply run ever more trials, and approach the infinity limit ever more closely" (Mooney 1997, p. 100).
- In the most widely assigned econometric text, Greene (2003) provides an entire appendix on computer implementation issues but also understates in referring to numerical optimization procedures: "Ideally, the iterative procedure should terminate when the gradient is zero. In practice, this step will not be possible, primarily because of accumulated rounding error in the computation of the function and its derivatives" (p. 943).

However, statisticians have been sounding alarms over numerical computing issues for some time:

- Grillenzoni worries that when confronted with the task of calculating the gradient of a complex likelihood, software for solving nonlinear least squares and maximum likelihood estimation, can have "serious numerical problems; often they do not converge or yield inadmissible results" (Grillenzoni 1990, p. 504).
- Chambers notes that "even a reliable method may perform poorly if not careful checked for special cases, rounding error, etc. are not made" (Chambers 1973, p. 9).
- "[M]any numerical optimization routines find local optima and may not find global optima; optimization routines can, particularly for higher dimensions,

'get lost' in subspaces or in flat spots of the function being optimized" (Hodges 1987, p. 268).

- Beaton et al. examine the famous Longley data problem and determine: "[T]he computationally accurate solution to this regression problem—even when computed using 40 decimal digits of accuracy—may be a very poor estimate of regression coefficients in the following sense: small errors beyond the last decimal place in the data can result solutions more different than those computed by Longley with his less preferred programs" (Beaton et al. 1976, p. 158). Note that these concerns apply to a *linear model!*

- The BUGS and WinBUGS documentation puts this warning on page 1 of the documentation: "**Beware—Gibbs sampling can be dangerous!**"

A clear discrepancy exists between theoreticians and applied researchers: The extent to which one should worry about numerical issues in statistical computing is unclear and even debatable. This is the issue we address here, bridging the knowledge gap difference between empirically driven social scientists and more theoretically minded computer scientists and statisticians.

1.3 MOTIVATING EXAMPLE: RARE EVENTS COUNTS MODELS

It is well known that binary rare events data are difficult to model reliably because the results often greatly underestimate the probability of occurrence (King and Zeng 2001a). It is true also that rare events counts data are difficult to model because like binary response models and all other generalized linear models (GLMs), the statistical properties of the estimations are conditional on the mean of the outcome variable. Furthermore, the infrequently observed counts are often not temporally distributed uniformly throughout the sample space, thus produce clusters that need to be accounted for (Symons et al. 1983).

Considerable attention is being given to model specification for binary count data in the presence of overdispersion (variance exceeding the mean, thus violating the Poisson assumption) in political science (King 1989; Achen 1996; King and Signorino 1996; Amato 1996; Londregan 1996), economics (Hausman et al. 1984; Cameron and Trivedi 1986, 1990; Lee 1986, Gurmu 1991), and of course, statistics (McCullagh and Nelder 1989). However, little has been noted about the numerical computing and estimation problems that can occur with other rare events counts data.

Consider the following data from the 2000 U.S. census and North Carolina public records. Each case represents one of 100 North Carolina counties, and we use only the following subset of the variables.

- **Suicides by Children.** This is (obviously) a rare event on a countywide basis and refers almost strictly to teenage children in the United States.
- **Number of Residents in Poverty.** Poverty is associated directly with other social ills and can lower the quality of education, social interaction, and opportunity of children.

- **Number of Children Brought Before Juvenile Court.** This measures the number of first-time child offenders brought before a judge or magistrate in a juvenile court for each of these counties.

Obviously, this problem has much greater scope as both a sociological question and a public policy issue, but the point here is to demonstrate numerical computing problems with a simple but *real* data problem. For replication purposes these data are given in their entirety in Table 1.1.

For these we specified a simple Poisson generalized linear model with a log link function:

$$\underbrace{g^{-1}(\boldsymbol{\theta})}_{100 \times 1} = g^{-1}(\mathbf{X}\boldsymbol{\beta}) = \exp[\mathbf{X}\boldsymbol{\beta}]$$

$$= \exp[1\beta_0 + \text{POV}\beta_1 + \text{JUV}\beta_2]$$

$$= E[\mathbf{Y}] = E[\text{SUI}]$$

in standard GLM notation (Gill 2000). This basic approach is run on five commonly used statistical packages and the results are summarized in Table 1.2. Although there is some general agreement among R, S-Plus, Gauss, and Stata, SAS (Solaris v8) produces estimates substantively different from the other four.[1] Although we may have some confidence that the results from the four programs in agreement are the "correct" results, we cannot know for sure, since we are, after all, estimating unknown quantities. We are left with the troubling situation that the results are dependent on the statistical program used to generate statistical estimates.

Even among the four programs in agreement, there are small discrepancies among their results that should give pause to researchers who interpret t-statistics strictly as providing a measure of "statistical significance." A difference in the way Stata handles data input explains some of the small discrepancy between Stata's results and R and S-Plus. Unless specified, Stata reads in data as single precision, whereas the other programs read data as double precision. When we provide the proper commands to read in data into Stata as double precision, the estimates from the program lie between the estimates of R and S-Plus. This does not account for the difference in the estimates generated by Gauss, a program that reads in data as double precision, which are in line with Stata's single-precision estimates.

This example highlights some of the important themes to come. Clearly, inconsistent results indicate that there are some sources of inaccuracy from these data. All numerical computations have limited accuracy, and it is possible for particular characteristics of the data at hand to exacerbate these effects; this is the focus of Chapter 2. The questions addressed there are: What are the sources of inaccuracy associated with specific algorithmic choices? How may even a small error propagate into a large error that changes substantive results?

---

[1]Note that SAS issued warning messages during the estimation, but the final results were not accompanied by any warning of failure.

Table 1.1 North Carolina 2000 Data by Counties

County	Suicide	Poverty	Juvenile/ Court	County	Suicide	Poverty	Juvenile/ Court
Alamance	0	14,519	47	Johnston	1	15,612	45
Alexander	0	2,856	70	Jones	0	1,754	81
Alleghany	0	1,836	26	Lee	0	6,299	87
Anson	0	4,499	49	Lenoir	0	9,900	17
Ashe	0	3,292	56	Lincoln	0	5,868	14
Avery	0	2,627	58	Macon	0	4,890	70
Beaufort	0	8,767	71	Madison	0	3,756	58
Bertie	0	4,644	26	Martin	0	3,024	74
Bladen	0	6,778	66	McDowell	1	5,170	86
Brunswick	1	9,216	19	Mecklenburg	0	63,982	1
Buncombe	0	23,522	52	Mitchell	1	2,165	50
Burke	0	9,539	33	Montgomery	0	4,131	69
Cabarrus	0	9,305	36	Moore	0	8,524	25
Caldwell	0	8,283	29	Nash	1	11,714	22
Camden	0	695	60	New Hanover	0	21,003	62
Carteret	0	6,354	13	Northampton	1	4,704	54
Caswell	0	3,384	67	Onslow	1	19,396	42
Catawba	0	12,893	51	Orange	0	16,670	6
Chatham	0	4,785	79	Pamlico	0	1,979	26
Cherokee	0	3,718	68	Pasquotank	2	6,421	74
Chowan	0	2,557	46	Pender	0	5,587	10
Clay	0	1,000	20	Perquimans	0	2,035	35
Cleveland	0	12,806	41	Person	0	4,275	82
Columbus	0	12,428	2	Pitt	0	27,161	27
Craven	1	11,978	12	Polk	0	1,851	20
Cumberland	2	38,779	73	Randolph	1	11,871	42
Currituck	0	1,946	61	Richmond	0	9,127	9
Dare	0	2,397	75	Robeson	1	28,121	64
Davidson	1	14,872	55	Rockingham	0	11,767	4
Davie	0	2,996	72	Rowan	0	13,816	44
Duplin	0	9,518	69	Rutherford	0	8,743	32
Durham	2	29,924	53	Sampson	1	10,588	71
Edgecombe	0	10,899	34	Scotland	0	7,416	18
Forsyth	1	33,667	57	Stanly	0	6,217	83
Franklin	1	5,955	84	Stokes	0	4,069	16
Gaston	0	20,750	59	Surry	0	8,831	24
Gates	0	1,788	15	Swain	1	2,373	56
Graham	0	1,559	37	Transylvania	0	2,787	78
Granville	0	5,674	85	Tyrrell	0	967	11
Greene	0	3,833	40	Union	1	10,018	38
Guilford	1	44,631	77	Vance	0	8,806	7
Halifax	1	13,711	8	Wake	5	48,972	80
Harnett	0	13,563	39	Warren	0	3,875	48
Haywood	1	6,214	21	Washington	0	2,992	43
Henderson	0	8,650	30	Watauga	0	7,642	63
Hertford	1	4,136	56	Wayne	0	15,639	42
Hoke	0	5,955	76	Wilkes	0	7,810	23
Hyde	0	897	81	Wilson	0	13,656	31
Iredell	1	10,058	28	Yadkin	2	3,635	3
Jackson	0	5,001	5	Yancey	0	2,808	65

Table 1.2 Rare Events Counts Models in Statistical Packages

		R	S-Plus	SAS	Gauss	Stata
Intercept	Coef.	−3.13628	−3.13678	0.20650	−3.13703	−3.13703
	Std. err.	0.75473	0.75844	0.49168	0.76368	0.76367
	t-stat.	−4.15550	−4.13585	0.41999	−4.10788	−4.10785
Poverty/1000	Coef.	0.05264	0.05263	−1.372e-04	0.05263	0.05269
	Std. err.	0.00978	0.00979	1.2833-04	0.00982	0.00982
	t-stat.	5.38241	5.37136	−1.06908	5.35881	5.36558
Juvenile	Coef.	0.36167	0.36180	−0.09387	0.36187	0.36187
	Std. err.	0.18056	0.18164	0.12841	0.18319	0.18319
	t-stat.	2.00301	1.99180	−0.73108	1.97541	1.97531

In this example we used different software environments, some of which required direct user specification of the likelihood function, the others merely necessitating menu direction. As seen, different packages sometimes yield different results. In this book we also demonstrate how different routines within the same package, different version numbers, or even different parameter settings can alter the quality and integrity of results. We do not wish to imply that researchers who do their own programming are doing better or worse work, but that the more responsibility one takes when model building, the more one must be aware of issues regarding the software being used and the general numerical problems that might occur. Accordingly, in Chapter 3 we demonstrate how proven benchmarks can be used to assess the accuracy of particular software solutions and discuss strategies for consumers of statistical software to help them identify and avoid numeric inaccuracies in their software.

Part of the problem with the example just given is attributable to these data. In Chapter 4 we investigate various data-originated problems and provide some solutions that would help with problems, as we have just seen. One method of evaluation that we discuss is to check results on multiple platforms, a practice that helped us identify a programming error in the Gauss code for our example in Table 1.2.

In Chapter 5 we discuss some numerical problems that result from implementing Markov chain Monte Carlo algorithms on digital computers. These concerns can be quite complicated, but the foundational issues are essentially like those shown here: numerical treatment within low-level algorithmic implementation. In Chapter 6 we look at the problem of a non-invertible Hessian matrix, a serious problem that can occur not just because of collinearity, but also because of problems in computation or data. We propose some solutions, including a new approach based on generalizing the inversion process followed by importance sampling simulation.

In Chapter 7 we investigate a complicated modeling scenario with important theoretical concerns: ecological inference, which is susceptible to numerical inaccuracies. In Chapter 8 Bruce McCullough gives guidelines for estimating general nonlinear models in economics. In Chapter 10 Paul Allison discusses numerical issues in logistical regression. Many related issues are exacerbated with spatial data, the topic of Chapter 9 by James LeSage. Finally, in Chapter 11 we provide a summary of recommendations and an extended discussion of methods for ensuring replicable research.

1.4 PREVIEW OF FINDINGS

In this book we introduce principles of numerical computation, outline the optimization process, and provide tools for assessing the sensitivity of subsequent results to problems that exist in these data or with the model. Throughout, there are real examples and replications of published social science research and innovations in numerical methods.

Although we intend readers to find this book useful as a reference work and software guide, we also present a number of new research findings. Our purpose is not just to present a collection of recommendations from different methodological literatures. Here we actively supplement useful and known strategies with unique findings.

Replication and verification is not a new idea (even in the social sciences), but this work provides the first replications of several well-known articles in political science that show where optimization and implementation problems affect published results. We hope that this will bolster the idea that political science and other social sciences should seek to recertify accepted results.

Two new methodological developments in the social sciences originate with software solutions to historically difficult problems. Markov chain Monte Carlo has revolutionized Bayesian estimation, and a new focus on sophisticated software solutions has similarly reinvigorated the study of ecological inference. In this volume we give the first look at numerical accuracy of MCMC algorithms from pseudo-random number generation and the first detailed evaluation of numerical periodicity and convergence.

Benchmarks are useful tools to assess the accuracy and reliability of computer software. We provide the first comprehensive packaged method for establishing standard benchmarks for social science data input/output accuracy. This is a neglected area, but it turns out that the transmission of data across applications can degrade the quality of these data, even in a way that affects estimation. We also introduce the first procedure for using *cyclical redundancy checks* to assess the success of data input rather than merely checking file transfer. We discuss a number of existing benchmarks to test numerical algorithms and to provide a new set of standard benchmark tests for distributional accuracy of statistical packages.

Although the negative of the Hessian (the matrix of second derivatives of the posterior with respect to the parameters) must be positive definite and hence invertible in order to compute the variance matrix, invertible Hessians do not exist for some combinations of datasets and models, causing statistical procedures to fail. When a Hessian is non-invertible purely because of an interaction between the model and the data (and not because of rounding and other numerical errors), this means that the desired variance matrix does not exist; the likelihood function may still contain considerable information about the questions of interest. As such, discarding data and analyses with this valuable information, even if the information cannot be summarized as usual, is an inefficient and potentially biased procedure. In Chapter 6 Gill and King provide a new method for applying generalized inverses to Hessian problems that can provide results even in circumstances where it is not usually possible to invert the Hessian and obtain coefficient standard errors.

Ecological inference, the problem of inferring individual behavior from aggregate data, was (and perhaps still is) arguably once the longest-standing unsolved problem in modern quantitative social science. When in 1997 King provided a new method that incorporated both the statistical information in Goodman's regression and the deterministic information in Duncan and Davis's bounds, he garnered tremendous acclaim as well as persistent criticism. In this book we report the first comparison of the numerical properties of competing approaches to the ecological inference problem. The results illuminate the trade-offs among correctness, complexity, and numerical sensitivity.

More important than this list of new ideas, which we hope the reader will explore, this is the first general theoretical book on statistical computing that is focused purely on the social sciences. As social scientists ourselves, we recognize that our data analysis and estimation processes can differ substantially from those described in a number of (even excellent) texts.

All too often new ideas in statistics are presented with examples from biology. There is nothing wrong with this, and clearly the points are made more clearly when the author actually cares about the data being used. However, we as social scientists often *do not* care about the model's implications for lizards, beetles, bats, coal miners, anchovy larvae, alligators, rats, salmon, seeds, bones, mice, kidneys, fruit flies, barley, pigs, fertilizers, carrots, and pine trees. These are actual examples taken from some of our favorite statistical texts. Not that there is anything wrong with studying lizards, beetles, bats, coal miners, anchovy larvae, alligators, rats, salmon, seeds, bones, mice, kidneys, fruit flies, barley, pigs, fertilizers, carrots, and pine trees, but we would rather study various aspects of human social behavior. This is a book for those who agree.

CHAPTER 2

Sources of Inaccuracy in Statistical Computation

2.1 INTRODUCTION

Statistical computations run on computers contain inevitable error, introduced as a consequence of translating pencil-and-paper numbers into the binary language of computers. Further error may arise from the limitations of algorithms, such as pseudo-random number generators (PRNG) and nonlinear optimization algorithms. In this chapter we provide a detailed treatment of the sources of inaccuracy in statistical computing. We begin with a revealing example, then define basic terminology, and discuss in more detail bugs, round-off and truncation errors in computer arithmetic, limitations of random number generation, and limitations of optimization.

2.1.1 Revealing Example: Computing the Coefficient Standard Deviation

Not all inaccuracies occur by accident. A Microsoft technical note[1] states, in effect, that some functions in Excel (v5.0–v2002) are inaccurate *by design*. The standard deviation, kurtosis, binomial distributions, and linear and logistic regression functions produce incorrect results when intermediate calculations, calculations that are hidden from the user to construct a final calculation, yield large values. Calculation of the standard deviation by Microsoft Excel is a telling example of a software design choice that produces inaccurate results. In typical statistics texts, the standard deviation of a population is defined as

$$s = \sqrt{\frac{\sum_{i=1}^{n}(x_i - \widehat{x})}{n-1}}. \qquad (2.1)$$

Mathematical expressions do not necessarily imply a unique computational method, as sometimes transformations of the expression yield faster and more

---

[1] Microsoft Knowledge base article Q158071.

Numerical Issues in Statistical Computing for the Social Scientist, by Micah Altman, Jeff Gill, and Michael P. McDonald
ISBN 0-471-23633-0   Copyright © 2004 John Wiley & Sons, Inc.

Table 2.1   Reported Standard Deviations for Columns of Data in `Excel`

		Significant Digits				
	2	7	8	9	10	15
	1	1000001	10000001	100000001	1000000001	100000000000001
	2	1000002	10000002	100000002	1000000002	100000000000002
	1	1000001	10000001	100000001	1000000001	100000000000001
	2	1000002	10000002	100000002	1000000002	100000000000002
	1	1000001	10000001	100000001	1000000001	100000000000001
	2	1000002	10000002	100000002	1000000002	100000000000002
	1	1000001	10000001	100000001	1000000001	100000000000001
	2	1000002	10000002	100000002	1000000002	100000000000002
	1	1000001	10000001	100000001	1000000001	100000000000001
	2	1000002	10000002	100000002	1000000002	100000000000002
Reported:	0.50	0.50	0.51	0.00	12.80	11,86,328.32
Correct:	0.50	0.50	0.50	0.50	0.50	0.50

tractable programming. In this case, the textbook formula is not the fastest way to calculate the standard deviation since it requires one pass through data to compute the mean and a second pass to compute the difference terms. For large datasets, a numerically naive but mathematically equivalent formula that computes the standard deviation in a single pass is given by

$$\sqrt{\frac{n\sum_{i=1}^{n} x_i^2 - \left(\sum_{i=1}^{n} x\right)^2}{n(n-1)}}. \tag{2.2}$$

Microsoft `Excel` uses the single-pass formula, which is prone to severe rounding errors when $n\sum_{i=1}^{n} x_i^2 - (\sum_{i=1}^{n} x)^2$ requires subtracting two large numbers. As a consequence, `Excel` reports the standard deviation incorrectly when the number of significant digits in a column of numbers is large. Table 2.1 illustrates this. Each column of 10 numbers in Table 2.1 has a standard deviation of precisely $1/2$, yet the standard deviation reported by `Excel` ranges from zero to over 1 million.[2]

2.1.2   Some Preliminary Conclusions

The inaccuracies in Excel are neither isolated nor harmless. Excel is one of the most popular software packages for business statistics and simulation, and the solver functions are used particularly heavily (Fylstra et al. 1998). Excel exhibits

---

[2]Table 2.1 is an example of a statistics benchmark test, where the performance of a program is gauged by how well it reproduces a known answer. The one presented here is an extension of Simon and LeSage (1988).

similar inaccuracies in its nonlinear solver functions, statistical distribution, and linear models (McCullough and Wilson 1999, 2002). Excel is not alone in its algorithm choice; we entered the numbers in Table 2.1 into a variety of statistical software programs and found that some, but not all, produced errors similar in magnitude.

The standard deviation is a simple formula, and the limitations of alternative implementations is well known; Wilkinson and Dallal (1977) pointed out failures in the variance calculations in statistical packages almost three decades ago. In our example, the inaccuracy of Excel's standard deviation function is a direct result of the algorithm choice, not a limitation of the precision of its underlying arithmetic operators. Excel's fundamental numerical operations are as accurate as those of most other packages that perform the standard deviation calculation correctly. By implementing the textbook equation within Excel, using the "average" function, we were able to obtain the correct standard deviations for all the cases shown in Table 2.1. Excel's designers might argue that they made the correct choice in choosing a more time-efficient calculation over one that is more accurate in some circumstances; and in a program used to analyze massive datasets, serious thought would need to go into these trade-offs. However, given the uses to which Excel is normally put and the fact that internal limits in Excel prohibit analysis of truly large datasets, the one-pass algorithm offers no real performance advantage.

In this case, the textbook formula is more accurate than the algorithm used by Excel. However, we do not claim that the textbook formula here is the most robust method to calculate the standard deviation. Numerical stability could be improved in this formula in a number of ways, such as by sorting the differences before summation. Nor do we claim that textbook formulas are in general always numerically robust; quite the opposite is true (see Higham 2002, pp. 10–14). However, there are other one-pass algorithms for the standard deviation that are nearly as fast and much more accurate than the one that Excel uses. So even when considering performance when used with massive datasets, no good justification exists for choosing the algorithm used in Excel.

An important concern is that Excel produces incorrect results without warning, allowing users unwittingly to accept erroneous results. In this example, even moderately sophisticated users would not have much basis for caution. A standard deviation is requested for a small column of numbers, all of which are similarly scaled, and each of which is well within the documented precision and magnitude used by the statistical package, yet Excel reports severely inaccurate results. Because numeric inaccuracies can occur in intermediate calculations that programs obscure from the user, and since such inaccuracies may be undocumented, users who do not understand the potential sources of inaccuracy in statistical computing have no way of knowing when results received from statistical packages and other programs are accurate.

The intentional, and unnecessary, inaccuracy of Excel underscores the fact that trust in software and its developers must be earned, not assumed. However, there are limits to the internal operation of computers that ultimately affect all algorithms, no matter how carefully programmed. Areas that commonly cause inaccuracies in computational algorithms include floating point arithmetic, random number generation, and nonlinear optimization algorithms. In the remainder of this chapter we discuss the various sources of such potential inaccuracy.

2.2 FUNDAMENTAL THEORETICAL CONCEPTS

A number of concepts are fundamental to the discussion of accuracy in statistical computation. Because of the multiplicity of disciplines that the subject touches on, laying out some terminology is useful.

2.2.1 Accuracy and Precision

For the purposes of analyzing the numerical properties of computations, we must distinguish between precision and accuracy. *Accuracy* (almost) always refers to the absolute or relative error of an approximate quantity. In contrast, *precision* has several different meanings, even in scientific literature, depending on the context. When referring to measurement, precision refers to the degree of agreement among a set of measurements of the same quantity—the number of digits (possibly in binary) that are the same across repeated measurements. However, on occasion, it is also used simply to refer to the number of digits reported in an estimate. Other meanings exist that are not relevant to our discussion; for example, Bayesian statisticians use the word *precision* to describe the inverse of the variance. In the context of floating point arithmetic and related numerical analysis, *precision* has an alternative meaning: the accuracy with which basic arithmetic operations are performed or quantities are stored in memory.

2.2.2 Problems, Algorithms, and Implementations

An *algorithm* is a set of instructions, written in an abstract computer language that when executed solves a specified *problem*. The problem is defined by the complete set of *instances* that may form the input and the properties the *solution* must have. For example, the algorithmic *problem* of computing the maximum of a set of values is defined as follows:

- **Problem:** Find the maximum of a set of values.
- **Input:** A sequence of n keys $k_1, \ldots, k_n$ of fixed size.
- **Solution:** The key k^*, where $k^* \geq k_i$ for all $i \in n$.

An algorithm is said to solve a problem if and only if it can be applied to any instance of that problem and is guaranteed to produce a correct solution to that instance. An example algorithm for solving the problem described here is to enter the values in an array S of size n and sort as follows:

```
MaxSort(S)
    for i = 1 to {n}-1 {
        for j ={i+1} to 2 {
            if A[j] < A[j-1] {
                t = A[j-1]
                A[j-1]=A[j]
                A[j]=A[j-1]
            }
        }
    }
    return (A[n]);
```

This algorithm, called a *bubble sort*, is proven to produce the correct solution for all instances of the problem for all possible input sequences. The proof of correctness is the fundamental distinction between algorithms and *heuristic algorithms*, or simply *heuristics*, procedures that are useful when dealing with difficult problems but do not provide guarantees about the properties of the solution provided. Heuristics may be distinguished from approximations and randomized algorithms. An approximation algorithm produces a solution within some known relative or absolute error of the optimal solution. A randomized algorithm produces a correct solution with some known probability of success. The behavior of approximation and randomized algorithms, unlike heuristics, is formally *provable* across all problem instances.

Correctness is a separate property from *efficiency*. The bubble sort is one of the least efficient common sorting methods. Moreover, for the purpose of finding the maximum, scanning is more efficient than sorting since it requires provably fewer operations:

```
MaxScan(S)
    m = 1
    for i = 2 to n {
        if A[m] < A[i] {
            m=i
        }
    }
    return (A[m]);
```

Note that an algorithm is defined independent of its *implementation* (or *program*), and we use pseudocode here to give the specific steps without defining a particular software implementation. The same algorithm may be expressed using different computer languages, different encoding schemes for variables and parameters, different accuracy and precision in calculations, and run on different types of hardware. An implementation is a particular instantiation of the algorithm in a real computer environment.

Algorithms are designed and analyzed independent of the particular hardware and software used to execute them. Standard proofs of the correctness of particular algorithms assume, in effect, that arithmetic operations are of infinite precision. (This is not the same as assuming that the inputs are of infinite length.) To illustrate this point, consider the following algorithm for computing the average of n of numbers:

```
SumSimple(S)
    x=0
    for i = 1 to n {
        x = x + S[i]
    }
    x = x/n;
    return(x);
```

Although this is a correct and efficient algorithm, it does not lend itself particularly well to accurate implementation in current standard computer languages. For example, suppose that the input $S[i]$ was produced by the function $S[i] = i^{-2}$; then x grows in magnitude at every iteration while S[i] shrinks. In cases like this, implementations of the SumSimple algorithm exhibit significant round-off error, because small numbers added to large numbers tend to "drop off the end" of the addition operator's precision and fail to contribute to the sum (see Higham 2002, pp. 14–17). (For a precise explanation of the mechanics of rounding error, see Section 2.4.2.1.)

Altering the algorithm to sort S before summing reduces rounding error and leads to more accurate results. (This is generally true, not true only for the previous example.) Applying this concept, we can create a "wrapper" algorithm, given by

```
SumAccurate(S)
    Sort(S)
    return(SumSimple(S));
```

Concerning the overall purpose of accuracy in computer implementations, Wilkinson (1994) claims that "[t]he results matter more than the algorithm." There is certainly some truth in this simplification. Implementation matters. A particular algorithm may have been chosen for asymptotic performance that is irrelevant to the current data analysis, may not lend itself easily to accurate implementation, or may elide crucial details regarding the handling of numerical errors or boundary details. Both algorithm and implementation must be considered when evaluating accuracy.

To summarize, an algorithm is a procedure for solving a well-defined problem. An algorithm is correct when given an instance of a problem, it can be proved to produce an output with well-defined characteristics. An algorithm may be correct but still lead to inaccurate implementations. Furthermore, in choosing and implementing algorithms to solve a particular problem, there are often trade-offs between accuracy and efficiency. In the next section we discuss the role of algorithms and implementations in inference.

2.3   ACCURACY AND CORRECT INFERENCE

Ideally, social scientists would like to take data, y, that represent phenomena of interest, M, and infer the process that produced it: $p(M|y)$. This *inverse probability* model of inference is, unfortunately, impossible. A weaker version, where priors are assumed over parameters of a given model (or across several models of different functional forms), is the foundation of Bayesian statistics (Gill 2002), which gives the desired form of the conditional probability statement at the "cost" of requiring a *prior* distribution on M.

Broadly defined, a statistical estimate is a mapping between

$$\{\text{data, model, priors, inference method}\} \Rightarrow \{\text{estimates}\},$$

or symbolically,

$$\{\mathbf{X}, M, \pi, IM\} \Rightarrow e.$$

For example, under the likelihood model,[3] we assume a parameterized statistical model M' of the social system that generates our data and hold it fixed, we assume noninformative priors. According to this inference model, the best point estimate of the parameters is

$$B^* = \text{Max}_{\forall B} L(B|M', y) \tag{2.3}$$

where $L(B|M', y) \propto P(B|M', y)$. Or, in terms of the more general mapping,

$$\{y, M', \text{maximum likelihood inference}\} \Rightarrow \{B^*\}.$$

While the process of maximum likelihood estimation has a defined stochastic component, other sources of error are often ignored. There is always potential error in the collection and coding of social science—people lie about their opinions or incorrectly remember responses to surveys, votes are tallied for the wrong candidate, census takers miss a household, and so on. In theory, some sources of error could be dealt with formally in the model but frequently are dealt with outside the model. Although we rarely model measurement error explicitly in these cases, we have pragmatic strategies for dealing with them: We look for outliers, clean the data, and enforce rigorous data collection procedures.

Other sources of error go almost entirely unacknowledged. That error can be introduced in the act of estimation is known but rarely addressed, even informally. Particularly for estimates that are too complex to calculate analytically, using only pencil and paper, we must consider how computation may affect results. In

---

[3]There is a great deal to recommend the Bayesian perspective, but most researchers settle for the more limited but easily understood model of inference: *maximum likelihood estimation* (see King 1989).

such cases, if the output from the computer is not known to be equivalent to e, one must consider the possibility that the estimate is inaccurate. Moreover, the output may depend on the algorithm chosen to perform the estimation, parameters given to that algorithm, the accuracy and correctness of the implementation of that algorithm, and implementation-specific parameters. Including these factors results in a more complex mapping[4]:

$\{\mathbf{X}, M, \pi, IM$, algorithm, algorithm parameters, implementation,

implementation parameters$\} \Rightarrow$ output.

By *algorithm* we intend to encompass choices made in creating output that are not part of the statistical description of the model and which are independent of a particular computer program or language: This includes the choice of mathematical approximations for elements of the model (e.g., the use of Taylor series expansion to approximate a distribution) and the method used to find estimates (e.g., nonlinear optimization algorithm). Implementation is meant to capture all remaining aspects of the program, including bugs, the precision of data storage, and arithmetic operations (e.g., using floating point double precision). We discuss both algorithmic and implementation choices at length in the following sections of this chapter.

Ignoring the subtle difference between output and estimates may often be harmless. However, as we saw at the beginning of this chapter, the two may be very different. Features of both the algorithm and its implementation may affect the resulting output: An algorithm used in estimation, even if implemented correctly and with infinite accuracy, may produce output only approximating the estimates, where the closeness of the approximation depends on particular algorithmic parameters. An algorithm may be proved to work properly only on a subset of the possible data and models. Furthermore, implementations of a particular algorithm may be incorrect or inaccurate, or be conditioned on implementation-specific parameters.

The *accuracy* of the output actually presented to the user is thus the dissimilarity (using a well-behaved dissimilarity measure) between estimates and output[5]:

$$\text{accuracy} = \text{distance} = D(e, \text{output}). \tag{2.4}$$

The choice of an appropriate dissimilarity measure depends on the form of the estimates and the purpose for which those estimates are used. For output that is a single scalar value, we might choose *log relative error* (LRE) as an informative

---

[4]Renfro (1997) suggests a division of problem into four parts: specification, estimator choice, estimator computation, and estimator evaluation. Our approach is more formal and precise, but is roughly compatible.

[5]Since *accurate* is often used loosely in other contexts, it is important to distinguish between computational accuracy, as discussed earlier, and correct inference. A perfectly accurate computer program can still lead one to incorrect results if the model being estimated is misspecified.

measure, which can be interpreted roughly as the number of numerically "correct" digits in the output:

$$\text{LRE} = -\log_{10}\left|\frac{\text{output} - e}{e}\right|. \tag{2.5}$$

When $e = 0$, LRE is defined as the *log absolute error* (LAE), given by

$$\text{LRE} = -\log_{10}|\text{output} - e|. \tag{2.6}$$

A number of measures of other measures of dissimilarity and distance are commonly used in statistical computation and computational statistics (see Chapter 4; Higham 2002, Chap. 6; and Gentle 2002, Sec. 5.4).

Accuracy alone is often not enough to ensure correct inferences, because of the possibility of model misspecification, the ubiquity of unmodeled measurement error in the data, and of rounding error in implementations (Chan et al. 1983). Where noise is present in the data or its storage representation and not explicitly modeled, correct inference requires the output to be *stable*. Or as Wilkinson puts it: "Accuracy is not important. What matters is how an algorithm handles inaccuracy..." (Wilkinson 1994).

A stable program gives "almost the right answer for almost the same data" (Higham 2002, p. 7). More formally, we can define stability in terms of the distance of the estimate from the output when a small amount of noise is added to the data:

$$S = D(e, \text{output}') \quad \text{where} \quad \text{output}' = \text{output}(\dots, Y', \dots), Y' \equiv Y + \Delta Y. \tag{2.7}$$

Results are said to be stable where S is sufficiently small.

Note that unstable output could be caused by sensitivity in the algorithm, implementation, or model. Any error, from any source, may lead to incorrect inferences if the output is not stable.

Users of statistical computations must cope with errors and inaccuracies in implementation and limitations in algorithms. Problems in implementations include mistakes in programming and inaccuracies in computer arithmetic. Problems in algorithms include approximation errors in the formula for calculating a statistical distribution, differences between the sequences produced by pseudo-random number generators and true random sequences, and the inability of nonlinear optimization algorithms to guarantee that the solution found is a global one. We examine each of these in turn.

2.3.1 Brief Digression: Why Statistical Inference Is Harder in Practice Than It Appears

Standard social science methods texts that are oriented toward regression, such as Hanushek and Jackson (1977), Gujarati (1995), Neter et al. (1996), Fox (1997),

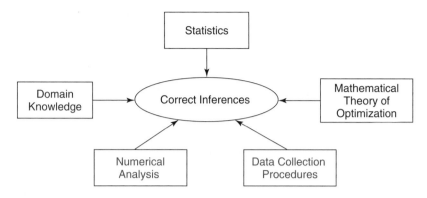

Fig. 2.1 Expertise implicated in correct inference.

Harrell (2001), Greene (2003), and Montgomery et al. (2001), discuss how to choose a theoretically appropriate statistical model, derive the likelihood function for it, and draw inferences from the resulting parameter estimates. This is a necessary simplification, but it hides much of what is needed for correct inference. It is tacitly understood that domain knowledge is needed to select an appropriate model, and it has begun to be recognized that knowledge of data collection is necessary to understand whether data actually correspond to the variables of the model.

What is not often recognized by social scientists [with notable exceptions, such as Renfro (1997) and McCullough and Vinod (2000)] is the sophistication of statistical software and the number of algorithmic and implementation choices contained therein that can affect one's estimates.[6]

In this book we show that knowledge of numerical analysis and optimization theory may be required not only to choose algorithms effectively and implement them correctly, but even to use them properly. Thus, correct inference sometimes requires a combination of expertise in the substantive domain, statistics, computer algorithms, and numerical analysis (Figure 2.1).

2.4 SOURCES OF IMPLEMENTATION ERRORS

Implementation errors and inaccuracies are possible when computing practically any quantity of interest. In this section we review the primary sources of errors and inaccuracy. Our intent in this chapter is not to make *positive* recommendations. We save such recommendations for subsequent chapters. We believe, as Acton (1970: p. 24) has stated eloquently in a similar if more limited context:

[6]This fact is recognized by researchers in optimization. For example, Maros (2002) writes on the wide gap between the simplicity of the simplex algorithm in its early incarnations and the sophistication of current implementations of it. He argues that advances in optimization software have come about not simply through algorithms, but through an integration of algorithmic analysis with software engineering principles, numerical analysis of software, and the design of computer hardware.

"The topic is complicated by the great variety of information that may be known about the statistical structure of the function [problem] .... At the same time, people ... are usually seeing a quick uncritical solution to a problem that can be treated only with considerable thought. To offer them even a hint of a panacea ... is to permit them to draw with that famous straw firmly grasped." Thus, we reserve positive recommendations for subsequent chapters, which deal with more specific computational and statistical problems.

2.4.1 Bugs, Errors, and Annoyances

Any computer program of reasonable complexity is sure to have some programming errors, and there is always some possibility that these errors will affect results. More formally, we define *bugs* to be mistakes in the implementation of an algorithm—failure to instruct the computer to perform the operations as specified by a particular algorithm.

No statistical software is known to work properly with certainty. In limited circumstances it is theoretically possible to prove software correct, but to our knowledge no statistical software package has been proven correct using formal methods. Until recently, in fact, such formal methods were widely viewed by practitioners as being completely impractical (Clarke et al. 1996), and despite increasing use in secure and safety critical environments, usage remains costly and restrictive. In practice, statistical software will be tested but not proven correct. As Dahl et al. (1972) write: "Program testing can be used to show the presence of bugs, but never to show their absence."

As we saw in Chapter 1, bugs are a recurring phenomenon in scientific applications, and as we will see in Chapter 3, serious bugs are discovered with regularity in statistical packages. In addition, evidence suggests that even experienced programmers are apt to create bugs and to be overconfident of the correctness of their results. Bugs in mathematically oriented programs may be particularly difficult to detect, since incorrect code may still return plausible results rather than causing a total failure of the program. For example, in a 1987 experiment by Brown and Gould [following a survey by Creeth (1985)], experienced spreadsheet programmers were given standardized tasks and allowed to check their results. Although the programmers were quite confident of their program correctness, nearly half of the results were wrong. In dozens of subsequent independent experiments and audits, it was not uncommon to find more than half of the spreadsheets reviewed to be in error (Panko 1998). Although we suspect that statistical programs have a much lower error rate, the example illustrates that caution is warranted.

Although the purpose of this book is to discuss more subtle inaccuracies in statistical computing, one should be aware of the potential threat to inference posed by bugs. Since it is unlikely that identical mistakes will be made in different implementations, one straightforward method of testing for bugs is to reproduce results using multiple independent implementations of the same algorithm (see Chapter 4). Although this method provides no explicit information about the bug itself, it is useful for identifying cases where a bug has potentially contaminated a statistical analysis. [See Kit and Finzi (1995) for a practical introduction to software testing.] This approach was used by the National Institute of Standards and Technology (NIST) in the development of their statistical accuracy benchmarks (see Chapter 3).

2.4.2 Computer Arithmetic

Knuth (1998, p. 229) aptly summarizes the motivation of this section: "There's a credibility gap: We don't know how much of the computer's answers to believe. Novice computer users solve this problem by implicitly trusting in the computer as an infallible authority; they tend to believe that all digits of a printed answer are significant. Disillusioned computer users have just the opposite approach; they are constantly afraid that their answers are almost meaningless."

Researchers are often surprised to find that computers do not calculate numbers "exactly." Because no machine has infinite precision for storing intermediate calculations, cancellation and rounding are commonplace in computations. The central issues in computational numerical analysis are how to minimize errors in calculations and how to estimate the magnitude of inevitable errors.

Statistical computing environments generally place the user at a level far removed from these considerations, yet the manner in which numbers are handled at the lowest possible level affects the accuracy of the statistical calculations. An understanding of this process starts with studying the basics of data storage and manipulation at the hardware level. We provide an overview of the topic here; for additional views, see Knuth (1998) and Overton (2001). Cooper (1972) further points out that there are pitfalls in the way that data are stored and organized on computers, and we touch on these issues in Chapter 3.

Nearly all statistical software programs use floating point arithmetic, which represents numbers as a fixed-length sequence of ones and zeros, or bits (b), with a single bit indicating the sign. Surprisingly, the details of how floating point numbers are stored and operated on differ across computing platforms. Many, however, follow ANSI/IEEE Standard 754-1985 (also known, informally, as IEEE floating point) (Overton 2001). The IEEE standard imposes considerable consistency across platforms: It defines how floating point operations will be executed, ensures that underflow and overflow can be detected (rather than occurring silently or causing other aberrations, such as rollover), defines rounding mode, and provides standard representations for the result of arithmetic exceptions, such as division by zero. We recommend using IEEE floating point, and we assume that this standard is used in the examples below, although most of the discussion holds regardless of it.

Binary arithmetic on integers operates according to the "normal" rules of arithmetic as long as the integers are not too large. For a designer-chosen value of the parameter b, the threshold is $> 2^{b-1} - 1$, which can cause an undetected *overflow* error. For most programs running on microcomputers, $b = 32$, so the number 2,147,483,648 $(1 + 2^{32-1} - 1)$ would overflow and may actually roll over to -1.

A floating point numbering system is a subset of the real number system where elements have the form $y = \pm m \times \beta^{e-t}$. Each numbering system is characterized by the following quantities: a base (β), an exponent range $(e, e_{\min} \leq e \leq e_{\max})$, a precision (t), a sign, and a mantissa $(m, 0 \leq m \leq \beta^t - 1)$. For a particular y, only m, e, and a sign are stored.

In IEEE floating point, which uses base 2, one can think of each number as being represented by a single bit for the sign, a sequence of t bits for the mantissa, and an exponent of length e bits. As noted, *guard* bits are used during floating point operations to detect exceptions and to permit proper rounding, but they are not stored once an operation is completed. Under this definition, some numbers (e.g., the number 2) can be represented by more than one pattern of bits. This is dealt with by normalizing the mantissa.[7] Under IEEE standard double-precision floating point, $\beta = 2$, $t = 53$, and $-1021 \leq e \leq 1024$ (for single precision, $t = 24$ and $-125 \leq e \leq 128$).

Some numbers cannot be exactly represented using this scheme. An example is the number 0.1, which has an infinitely repeating binary representation using this technique. The infinitely repeating floating point transformation of 0.1 must be represented somehow, leading to either *rounding* or *truncation* errors at the last stored bit.

The relative error of representation is bounded by *machine epsilon* (or machine precision), ϵ_M. Machine epsilon is defined as the distance between 1.0 and the next floating point number. It is equal to $\boldsymbol{\beta}^{1-t}$ and is sometimes confused with other things, such as rounding error, the smallest quantity that can be represented, or with the smallest floating point number, which when added to 1.0 produces a different result.[8]

Floating point arithmetic is not exact even if the operands happen to be represented exactly. For example, when floating point numbers are added or subtracted, their exponents are first normalized: The mantissa of the smaller number is divided in two while increasing its exponent until the two operands have the same exponent. This division may cause low-order bits in the mantissa of the smaller number to be lost.

Operations in floating point representation are susceptible not only to rounding and overflow, but to *underflow* as well—when a number is smaller than the smallest value capable of being represented by the computer. As Knuth (1998, p. 229) points out, a consequence of the inaccuracies in floating point arithmetic is that the associative law sometimes breaks down:

$$(a \oplus b) \oplus c \neq a \oplus (b \oplus c), \tag{2.8}$$

where $\oplus$ denotes the standard arithmetic operators. Furthermore, as Higham (2002, Sec. 2.10) notes, limits on precision can interfere with the mathematical

---

[7]The remaining denormalized bit patterns are then used to represent *subnormal* numbers between the machine epsilon and 1.0, with reduced precision (see Higham 2002, p. 37).

[8]The smallest quantity that when added to 1.0 produces a different result is actually smaller than the machine precision because of subnormal numbers. In other words, the smallest number that can be added does not have full precision.

properties of functions. While many elementary functions can be calculated efficiently to arbitrary degrees of precision, the necessity of reporting the final results at a fixed precision yields inexact results that may not have all of the mathematical properties of the true function. The requirements of preserving symmetries, mathematical relations and identities, and correct rounding to the destination precision can conflict, even for elementary functions.

Floating point underflow drives the general principle that one should not test two floating point numbers for equality in a program, but instead, test that the difference is less than a tolerance value (Higham 2002, p. 493):

$$\text{tol} : \ |x - y| \le \text{tol}. \tag{2.9}$$

Manipulating numbers in floating point arithmetic, such as adding the squares of a large and a small number, may propagate or accumulate errors, which may in turn produce answers that are wildly different from the truth. If two nearly equal numbers are subtracted, *cancellation* may occur, leaving only the accumulated rounding error as a result, perhaps to be further multiplied and propagated in other calculations (Higham 2002, p. 9).

2.4.2.1 *Floating Point Arithmetic Example*
Consider the following calculation:

$$i = 1000000000 + 2 - 0.1 - 1000000000.$$

(We assume here that operations are executed from left to right, as appears, but see below concerning the reproducibility of floating point calculations for further complications.) How does this floating point operation work using (single-precision) floating point arithmetic?[9]

First, the operand 100000000 must be represented in floating point. We do this by taking the binary expansion

$$(100000000)_{10} = (111011100110101100101000000000)_2, \tag{2.10}$$

normalizing it,

$$(1.11011100110101100101000000000)_2 \times 2^{29}, \tag{2.11}$$

and truncating to 23 points following the decimal:

$$(1.11011100110101100101000)_2 \times 2^{29}.$$

---

[9]The number of bits used to store each floating point number is sometimes referred to as *precision*. A *single-precision floating point number* is typically stored using 32 bits. A *double-precision number* uses double the storage of a single-precision number.

From this we can see that 100000000 has an *exact* representation in floating point. No rounding occurred in the representation. Similarly, the number 2 is represented exactly by

$$(1.00000000000000000000000)_2 \times 2^1.$$

The number 0.1 is more tricky, because it has an infinite binary expansion:

$$(0.1)_{10} = (0.0001100110011\ldots)_2.$$

Expanded to 24 significant digits, normalized, and rounded in binary (the last digit in the mantissa would be zero without rounding), we obtain

$$(0.1)_{10} \cong (1.10011001100110011001101)_2 \times 2^{-4}. \tag{2.12}$$

Next, we add and subtract pairs of numbers. The general method for adding and subtracting is given by the following:

1. Align the significands by shifting the mantissa of the smaller number.
2. Add or subtract the significands.
3. Perform a carry on the exponent if necessary.
4. Normalize the result, and round the result if necessary.

[In addition, correct carrying and rounding, as well as overflow and underflow detection, require the use of additional *guard* bits during the calculation. These are not important for this particular example. In practice, IEEE implementation handles these and other subtle rounding issues correctly. Floating point arithmetic libraries that do not support the IEEE standard, such as those supplied on older mainframe computers, generally do not handle these issues correctly and should be avoided for statistical computation where possible. For more detail, see Overton (2001).]

We start with the first pair:

$$(1.11011100110101100101000)_2 \times 2^{29}$$
$$+ \quad (1.00000000000000000000000)_2 \times 2^1,$$

which is aligned to

$$(1.11011100110101100101000)_2 \times 2^{29}$$
$$+ \quad (0.00000000000000000000000)_2 \times 2^{29}$$

and then added and renormalized for the following result:

$$(1.11011100110101100101000)_2 \times 2^{29}.$$

In effect, the number 2 has simply been dropped from the calculation. The same thing occurs when we subtract the quantity 0.1. Finally, when we subtract

1000000000, we obtain

$$(1.11011100110101100101000)_2 \times 2^{29}$$
$$\underline{-(1.11011100110101100101000)_2 \times 2^{29}}$$
$$(0.00000000000000000000000)_2 \times 2^0$$

So in single-precision floating point

$$i = 1000000000 + 2 - 0.1 - 1000000000 = 0.$$

In contrast, when we perform these operations in a different order, we may obtain a completely different result. Consider the following order:

$$j = 1000000000 - 1000000000 + 2.0 - 0.1.$$

The first two quantities cancel out, leaving zero as a remainder. The next step is

$$(1.00000000000000000000000)_2 \times 2^1$$
$$-(1.10011001100110011001101)_2 \times 2^{-4}.$$

Aligned, this yields

$$(1.00000000000000000000000)_2 \times 2^1$$
$$\underline{-(0.00001100110011001100110)_2 \times 2^1}$$
$$(1.11100110011001100110011)_2 \times 2^0$$

Thus, we have an example of the violation of the associative law of addition:

$$1000000000 - 1000000000 + 2 + 0.1 \cong 1.9$$

and

$$1000000000 - 1000000000 + 2 + 0.1$$
$$\neq 10001000000000 + 2 + 0.1 - 1000000000.$$

There are three important lessons to take from this example. First, rounding errors occur in binary computer arithmetic that are not obvious when one considers only ordinary decimal arithmetic. Second, as discussed in general terms earlier in this chapter, round-off error tends to accumulate when adding large and small numbers—small numbers tend to "drop off the end" of the addition operator's precision, and what accumulates in the leftmost decimal positions is inaccurate. Third, subtracting a similar quantity from the result can then "cancel" the relatively accurate numbers in the rightmost decimal places, leaving only the least accurate portions.

2.4.2.2   Common Misconceptions about Floating Point Arithmetic

Higham (2002, p. 28) observes that researchers familiar with these basics of floating point arithmetic may still harbor some misconceptions that could lead to overconfidence and suggests guidelines (p. 27). We summarize these here:

- **Misconception:** Floating point errors can overwhelm a calculation only if many of them accumulate.

  Fact: Even one error, when propagated through the calculation, can cause wildly inaccurate results.

- **Misconception:** A short computation free from cancellation, underflow, and overflow must be accurate.

  Fact: Even a simple computation can be subject to rounding error.

- **Guideline:** Minimize the size of intermediate quantities relative to the final solution. (Attempt to keep calculations at the same scale throughout.)

- **Guideline:** Avoid subtracting quantities contaminated by error.

In addition, Higham (2002) notes researchers, misconceptions that could lead to *underconfidence* in their results. Sometimes, albeit rarely, rounding errors can help a computation, such that a computed answer can be *more* accurate than any of the intermediate quantities of the computation.

The moral is that a careful examination of the accuracy of each numerical *algorithm*, as a whole, is necessary for one to have the correct degree of confidence in the accuracy of one's results. Of course, a simple awareness of this is not enough to make one's results more accurate. Thus, in Chapters 3 and 4 we discuss various ways of testing and improving the accuracy of arithmetic.

2.4.2.3   Floating Point Arithmetic and Reproducibility

In addition to accuracy, one should be concerned with the reproducibility of floating point calculations. Identical software code can produce different results on different operating systems and hardware. In particular, although the IEEE 754 standard for floating point arithmetic (see Overton 2001) is widely honored by most modern computers, systems adhering to it can still produce different results for a variety of reasons.

Within the IEEE standard, there are a number of places where results are not determined exactly. First, the definition of extended-precision calculations specify only a minimum precision. Second, the standard (or at least, some interpretations of it) allows for the destination registers used to store intermediate calculations to be defined implicitly by the system and to differ from the precision of the final destination. Hence, on some systems (such as many Intel-based computers), some intermediate calculations in a sequence are performed at extended precision and then rounded down to double precision at the end. (There is some question whether this practice complies with the IEEE standard.) Third, IEEE does not completely specify the results of converting between integer and floating point data types. (This type of conversion is referred to as casting in some programming languages.) So variables defined as integers and later used in a floating point calculation can cause results to differ across platforms.

Compiler optimizations may also interfere with replication of results. In many programming languages, the compiler will rearrange calculations that appear on the same line, or are in the same region of code, to make them more accurate and/or to increase performance. For example, in the programming language C (using gcc 2.96), the result of

```
i =1000000000.0+2.0+0.1-1000000000.0;
```

is *not* identical to

```
i =1000000000.0;
i+=2.0;
i+=0.1;
i-=1000000000.0;
```

using the default settings, although the expressions are mathematically identical, and the ordering appears to be the same. Compilers may also invoke the use of extended intermediate precision as preceding, or fused, multiply–add operations (a hardware operation that allows efficient combination of adds and multiplies but has different rounding characteristics on hardware platforms that support these features). Although some compilers allow such optimizations to be disabled explicitly, many optimizations are applied by default. Differences among compilers and among the hardware platforms targeted by the compiler can cause the same code to produce different results when compiled under different operating systems or run on different hardware.

2.5   ALGORITHMIC LIMITATIONS

The problems discussed earlier were related to limits in implementation. For example, had all the steps in the summation been followed without mistake and with infinite precision, in the summation algorithm described earlier [the algorithm labeled SumSimple(S) in Section 2.2.2], the results would have been exactly equal to the theoretical quantity of interest. This is not the case with all algorithms. In this section we discuss the limitations of algorithms commonly used in statistical computations. These limited algorithms usually fall into one of the following four categories:

1. **Randomized algorithms** return a correct solution with some known probability, p, and an incorrect solution otherwise. They are used most commonly with decision problems.
2. **Approximation algorithms,** even if executed with infinite precision, are proved only to yield a solution that is known to approximate the quantity of interest with some known (relative or absolute) error.

3. **Heuristic algorithms** (or, simply, *heuristics*) are procedures that often work in practice but provide no guarantees on the optimality of their results. Nor do they provide bounds on the relative or absolute error of these results as compared to the true quantities of interest. Heuristics are most often used when the problem is too difficult to be solved with other approaches.

4. **Local search algorithms** comprise practically all general nonlinear optimization algorithms. Local search algorithms are guaranteed only to provide locally optimal solutions.

2.5.1 Randomized Algorithms

Randomized algorithms are rarely used in statistical computation but are a good example of how a perfect implementation may still yield incorrect results. The classic example of a randomized algorithm is Rabin's (1980) test for primality:

- **Problem:** Determine whether a particular number is prime.
- **Input:** An odd number $n > 4$.
- **Solution:** Return "true" if n is prime, "false" otherwise.

which is given in pseudocode by

```
Rabin(n)
    for i=1 to i {
        a = random(2..n-2)
        if n is not strongly pseudoprime to the base a {
            return false
        }
    }
return true;
```

This algorithm is based on the idea of a pseudoprime number. Fermat's (other) theorem states that a positive number n is prime if for every smaller prime number a greater than 1,

$$a^{n-1} = 1 \bmod n. \tag{2.13}$$

This theorem is readily provable, unlike its famous cousin, and it has been proven that there are infinitely many pseudoprimes (Sierpinski 1960).

For example, 11 is prime because for every prime less than 11 and greater than 1 there is some positive integer k such that $a^{n-1}/n = k$:

a	a^{n-1}/n	k
2	$(2^{11-1} - 1)/11$	93
3	$(3^{11-1} - 1)/11$	5,368
5	$(5^{11-1} - 1)/11$	887,784
7	$(7^{11-1} - 1)/11$	25,679,568

Of course, when $a^{n-1} \neq 1 \bmod n$, then n is not prime. The number n is *strongly pseudoprime* to a $(1 < a < n - 1)$ for arbitrary numbers s and t (t odd) if the condition

$$n - 1 = 2^s t \qquad (2.14)$$

holds, and

$$(a^t - 1) \bmod n = 0. \qquad (2.15)$$

So by this definition, although 341 is not prime ($31 \times 11 = 341$), it is strongly pseudoprime to the base $a = 2$ since for $s = 2$ and $t = 85$:

$$2^s = \frac{n-1}{t}, \qquad \frac{a^t}{n} = 1.13447584245361e + 23).$$

See McDaniel (1989) and Jaeschke (1993), as well as the references in Rotkiewicz (1972) for more details.

As important, Rabin's algorithm never returns a false negative, but returns a false positive with $p \cong 1 - 4^{-i}$. In practice, i is set such that the risk of a false negative is negligible. Therefore, although there is only a 0.75 probability that a positive result is really a prime on a single iteration, changing the base and rerunning provides arbitrarily high levels of reliability and is still faster than many competing (deterministic) algorithms for large numbers.

2.5.2 Approximation Algorithms for Statistical Functions

Some algorithms yield only approximations to the quantities of interest. Approximation algorithms are frequently used in the field of combinatorial optimization. For many combinatorial optimization problems, exact algorithms for finding the solution are computationally intractable, and approximations are used in their stead, if available. For example, linear programming, the minimization of m continuous variables subject to n linear inequality constraints, is computationally tractable. But *integer* linear programming, where each of the m variables is an integer, is computationally intractable.[10] Where available, approximation algorithms are sometimes used, even if the solutions produced by them are only guaranteed to be within a given percentage of the actual minimum (see, e.g., Hochbaum et al. 1993).

A milder form of approximation error is truncation error, the error introduced by using a finite series as an approximation for a converging infinite series. Use of infinite series expansion is very common in function evaluation. A similar form of truncation error can stem from using an asymptotic series expansion, which inherently limits the number of terms that can contribute to the accuracy

---

[10]Integer linear programming is *NP-complete*. Roughly, this means that no algorithm exists that is guaranteed to compute an exact answer to *any* legal instance of the problem in time less than k^n, For a precise definition of tractability, and comments on the tractability of integer programming, see Papadimitrious (1994).

of an approximation (Acton 1970). As Lozier and Olver (1994) point out in their extensive review of software for evaluating functions, before the construction (implementation) of software there are two stages. First, one must choose a suitable mathematical representation of the function of interest, such as asymptotic expansions, continued fractions, difference and differential equations, functional identities, integral representations, and Taylor series expansions. Second, one must choose a particular approach to evaluating the representations, such as through Chebyshev series, polynomial and rational approximations, Padé approximations, or numerical quadrature. Such methods are described in Kennedy and Gentle (1980), Thisted (1988), Lange (1999), and Press et al. (2002).

Consider a very simple method of approximating derivatives, the *forward difference method*, which is the common method used in statistical packages today:

$$f'(x) = \frac{f(x + \delta) - f(x)}{\delta}. \tag{2.16}$$

Even without rounding error the algorithm introduces truncation error of the form

$$\frac{1}{2}\delta^2 f''(x) + \frac{1}{6}\delta^3 f'''(x) + \cdots.$$

This error is sometimes sufficient to cause failures in difficult nonlinear optimization problems, as discussed in Chapter 8.

2.5.3 Heuristic Algorithms for Random Number Generation

Random numbers are fundamental to many types of mathematical simulation, including Monte Carlo simulation, which has become increasingly popular in the social sciences. Random numbers are also used in subsampling techniques, resampling techniques such as jackknife and the bootstrap (Efron 1982; Hall 1992; Shao and Tu 1995), and to pick starting parameters and search directions for some nonlinear optimization algorithms. A number of problems have been traced to inadequate random number generators, some occurring as early as 1968 (Knuth 1998), and poor generators continue to be used, recommended, and invented anew (L'Ecuyer 1994). Inadequate random number generators can cause problems beyond the fields of statistical and physical simulation. For example, computer-savvy gamblers have been known to exploit poor random number generators in gaming (Grochowski 1995), and an otherwise secure encryption implementation has been defeated for similar reasons (Goldberg and Wanger 1996). Peter Neumann (1995) reports on a variety of software system errors related to random number generators. We provide an overview of the topic here and a discussion of appropriate choice of generators for Monte Carlo and MCMC simulation in Chapter 5. For an extensive treatment of the design of PRNGs, see Gentle (1998) and Knuth (1998).

The numbers provided by computer algorithms are not genuinely random. Instead, they are *pseudo-random number generators* (PRNGs), deterministic processes that create a sequence of numbers. Pseudo-random number generators start with a single "seed" value (specified by the user or left at defaults) and generate a repeating sequence with a certain fixed length or period p. This sequence is statistically similar, in limited respects, to random draws from a uniform distribution. However, a pseudo-random sequence does not mimic a random sequence completely, and there is no complete theory to describe how similar PRNG sequences are to truly random sequences. In other words, no strict definition of approximation error exists with regard to PRNGs. This is a fundamental limitation of the algorithms used to generate these sequences, not a result of inaccuracy in implementation.

2.5.3.1 *Examples of Generators*
The earliest PRNG, and still in use, is the linear congruential generator (LCG), which is defined as

$$\text{LCG}(a, m, s, c) \equiv$$

$$x_0 = s,$$

$$x_n = (ax_{n-1} + c) \bmod m. \tag{2.17}$$

All parameters are integers, and exact integer arithmetic is used—correct implementation of this algorithm is completely accurate. The sequence generates a sequence of numbers between $[0, m-1]$ which appear uniformly distributed in that range. Note that in practice, x is usually divided by m.

Much of the early literature on PRNGs concerned finding good values of a, m, and c for the LCG. This is still an extremely popular generator, and modern versions of it very frequently use the choice of m and a attributed to Lewis et al. (1969), $x_n = (16807 x_{n-1}) \bmod 2^{31} - 1$. Even with these well-tested parameter values, the generator is now considered a comparatively poor one, because it has a short period, constrained by its modulus, and exhibits a lattice structure in higher dimensions (Marsaglia 1968).

For poor choices of a, m, and c, this lattice structure is extreme. The infamous RANDU generator, which was widely used in early computing and from which many other generators descended, is simply an LCG with values of 65,539, 2^{31}, and 0. Although the sequence produced appears somewhat random when subsequent pairs of points are two dimensions (Figure 2.2), the lattice structure is visually obvious when triples of the sequence are plotted in three dimensions, as in Figure 2.3. [See Park and Miller (1988) for a discussion.]

Other variations of congruential generators include multiple recursive generators, lagged Fibonnaci generators, and add with carry generators. Many variants of each of these types exist, but the simplest forms are the following:

- *Multiple recursive generators* (MRGs) take the form

$$x_n = (a_1 x_{n-1} + \cdots + a_k x_{n-k}) \bmod m.$$

Fig. 2.2 RANDU generator points plotted in two dimensions.

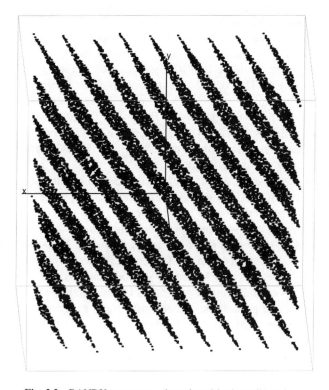

Fig. 2.3 RANDU generator points plotted in three dimensions.

- *Lagged Fibonnaci generators* take the form

$$x_n = (x_{n-j} + x_{n-k}) \bmod m.$$

- *Add with carry generators* take the form

$$x_n = (x_{n-j} + x_{n-k} + c_n) \bmod m,$$

where $c_1 = 0$ and $c_n = 0$ if $x_{n-j} + x_{n-k} + c_{n-1} < m$ and $c_n = 1$ otherwise.

Any of these generators, using appropriately chosen parameters for initialization, are likely to be better than the standard LCG in terms of both period length and distributional properties. [For a new class of recursive generators with long periods, see L'Ecuyer and Panneton (2000).] Still, each has limitations that may cause errors in simulation. Combined generators (see below) are good insurance against such defects, but even these may, in theory, result in periodicity effects for simulations that demand a large number of random numbers, as we show in Chapter 5 for Markov chain Monte Carlo methods.

A newer class of PRNGs, nonlinear generators, are promising because they appear to eliminate defects of previous PRNGs, although they are slower and less thoroughly tested and less well understood. One of the simplest examples is the *inversive congruential generator*, which is completely free of lattice structure:

$$x_n = (ax_{n-j}^{-1} + c) \quad \bmod m. \tag{2.18}$$

Some nonlinear generators, such as Blum et al. (1986), are intended for cryptographic applications and are extremely difficult to distinguish from true random sequences but are painfully slow for simulation purposes. Others, like the Mersenne twister (Matsumoto and Nishimura 1998), a variant of generalized feedback shift registers (see Lewis and Payne 1973) and variations of the inversive congruential generators (Eichenauer and Lehn 1986), are reasonably fast on today's computers but still much slower than congruential generators.

2.5.3.2   *Combinations and Transformations*

A judicious combination of generators can improve both the period and the distributional properties of the resulting sequence. An early technique (Maclaren and Marsaglia 1965), still in use, is to use one LCG to *shuffle* the output of another. The resulting stream has a longer period and a better lattice structure. Another combined generator, described by Wichmann and Hill (1982), combines three LCGs with carefully chosen moduli and coefficients to yield a generator with a period of 10^{12} and reasonable randomness properties:

$$WH_i = \left(\frac{\mathrm{LCG}(a_1, m_1, s_1)_i}{m_1} + \frac{\mathrm{LCG}(a_2, m_2, s_2)_i}{m_2} + \frac{\mathrm{LCG}(a_3, m_3, s_3)_i}{m_3} \right) \bmod 1$$

$$m_i = 30269, 30307, 3023, a_i = 171, 172, 170$$

Blind combination of multiple generators is not guaranteed to yield good results. Another approach to combining generators, originated by Collings (1987), is more straightforward. Collings *compounds* generators by maintaining a pool of k separate generators of different types and intermixing the results. A separate generator is used to generate a number i from $[1, k]$, and the ith generator is used to provide the next number in the sequence. This can greatly extend the period and can overcome serious distributional flaws in individual generators. If the periods of each generator used in the pool are p, the period of the combined generator is roughly p^2.

All of the PRNGs discussed are designed to generate uniformly distributed random numbers. Sampling of random numbers from other distributions is usually done by applying a transformation to a uniformly distributed PRNG. Two of the simpler techniques for such transformation are the *inverse cumulative distribution function* (CDF) *method* and the *rejection method*.

The inverse CDF method applies the transformation $X = P_X^{-1}(U)$, where U is uniformly distributed in $(0,1)$ and P_X is a univariate continuous CDF. When P_X^{-1} exists and is easy to compute, this is the most straightforward way of generating random numbers from nonuniform distributions.

The rejection method can be used where the inverse CDF is inapplicable and is straightforward to apply to multivariate distributions. Intuitively, it involves drawing a bounding box (or other bounding region) around the integral probability density function, uniformly sampling from the box, and throwing away any samples above the integral. More formally, the method is as follows:

1. Choose a distribution Y that resembles the distribution of x, from which it is easy to generate variates and that covers (or "majorizes") the probability density function of the desired distribution p_X such that $C \times p_y(x) \geq p_x(x), \forall x$ for some constant C.
2. Generate u from a uniform distribution on $(0,1)$ and y from Y.
3. If $u \leq p_x(y)/[c \times p_y(y)]$, return y as the random deviate; otherwise, repeat from step 2.

2.5.3.3 *Criteria for Building Pseudo-Random Number Generators*

For simulation or sampling results to be accurate, a PRNG should satisfy three criteria: long period, independence, and uniform distribution. In addition, all require a truly random "seed" to produce independent sequences (Ripley 1990; Gentle 1998; Knuth 1998).

First, a PRNG should have a long period. The recommended minimum length of the period depends on the number of random numbers (n) used by the simulation. Knuth (1998) recommends $p > 1000n$, while Ripley (1990) and Hellekalek (1998) recommend a more conservative $p \gg 200n^2$. Conservatively, PRNGs provided by most packages are inadequate for even the simplest simulations. Even using the less conservative recommendation, the typical period is wholly inadequate for computer-intensive techniques such as the double bootstrap, as McCullough and Vinod (1999) point out.

Second, a PRNG should produce numbers that are very close to independent in a moderate number of dimensions. Some PRNGs produce numbers that are apparently independent in one dimension but produce a latticelike structure in higher dimensions. Even statistically insignificant correlation can invalidate a Monte Carlo study (Gentle 1998). The identification of simple forms of serial correlation in the stream is an old and well-understood problem (Coveyou 1960), but see Chapter 5 for a discussion of problems that may arise in more complex simulation.

Third, the distribution of draws from the generator must be extremely close to uniform. In practice, we do not know if a PRNG produces a distribution that is close to uniform. However, as Hellekalek (1998) observes, any function of a finite number of uniformly distributed variables whose results follow a known distribution can suffice for a test of a PRNG. Good tests, however, constitute prototypes of simulation problems and examine both the sequence as a whole and the quality of subsequences (Knuth 1998).[11] Users should be cautioned that tests of random number generators are based on the null hypothesis that the generator is behaving adequately and may not detect all problems (Gentle 1998). Therefore, empirical tests should always be combined with theoretical analysis of the period and structure of the generator (Gentle 1998; Hellekalek 1998; Knuth 1998; L'Ecuyer and Hellekalek 1998).

Fourth, to ensure independence across sequences, the user must supply seeds that are truly random. In practice, statistical software selects the seed automatically using the current clock value, and users rarely change this. As encryption researchers have discovered, such techniques produce seeds that are not completely random (Eastlake et al. 1994; Viega and McGraw 2001), and much better solutions are available, such as hardware generators (see below).

Finally, for the purpose of later replication of the analysis, PRNG results must be reproducible. In most cases, reproducibility can be ensured by using the same generator and saving the seed used to initialize the random sequence. However, even generators that are based on the same PRNG algorithm can be implemented in subtly different ways that will interfere with exact reproduction. For example, Gentle (2002, pp. 251, 357) notes that the Super-Duper PRNG implementations differ sufficiently to produce slightly different simulation results in R and S-Plus.

In addition, more care must be used in parallel computing environments. Wherever multiple threads of execution sample from a single generator, interprocessor delays may vary during a run, affecting the sequence of random numbers received by each thread. It may be necessary to record the subsequences used in each thread of the simulation to ensure later reproducibility (Srinivasan et al. 1999).

If these conditions are met, there remains, inevitably, residual approximation error. This approximation error can also cause Monte Carlo algorithms to converge more slowly with PRNGs than would be expected using true random draws and may prevent convergence for some problems (Traub and Woznakowski

---

[11]Behavior of *short* subsequences is particularly important for simulations using multiple threads of execution. Entacher (1998) shows that many popular random number generators are inadequate for this purpose.

1992). Since the error of a PRNG does not dissipate entirely with sample size, traditional analysis of simulations based on asymptotic assumptions about sampling error overstates the accuracy of the simulation (Fishman 1996, Chap. 7). In other words, the accuracy of simulations *cannot* be increased indefinitely simply by increasing the number of simulations.

Developers of PRNG algorithms stress that there is no single generator that is appropriate for all tasks. One example of this is a set of "good" PRNGs that were discovered to be the source of errors in the simulation of some physical processes (Ferrenberg 1992; Selke et al. 1993; Vattulainen et al. 1994). PRNGs should be chosen with characteristics of the simulation in mind. Moreover, prudent developers of simulations should reproduce their results using several generators of different types [Gentle 1998; Hellekalek 1998; Knuth 1998; but see L'Ecuyer (1990) for a more optimistic view].

2.5.3.4 *Hardware Random Number Generation*

If true randomness must be ensured, random numbers can be generated through physical processes. A number of these hardware random generators or "true" random number generators (TRNGs) are inexpensive and suitable for use in personal computers.

Hardware generators are typically many orders of magnitude slower than PRNGs. (As of the time this book was written, the less expensive generators, produce roughly 10,000 random bytes per second.) Thus they are more often used in cryptographic applications, which require small amounts of extremely high quality randomness, than in Monte Carlo simulation. However, even the slowest generators can be used to provide high-quality seeds to PRNGs or to run many of the Monte Carlo simulations used by social scientists, if not for large MCMC computations. With forethought, large numbers of random bits from hardware random number generators can be stored over time for later use in simulation.

Moreover, the availability and speed of these generators have increased dramatically over the last few years, putting hardware random number generation within reach of the social scientist. We list a number of hardware random number generators and online sources for random bits in the Web site associated with this book. Several of the most widely available are:

- The Intel 800 series chip set (and some other series) contains a built-in hardware random number generator that samples thermal noise in resistors. Note that although many of these chip sets are in wide use in workstations and even in home computers, programming effort is needed to access the generator.
- The /dev/random pseudo-device, part of the Linux operating system, is a source of hardware entropy that can be used by any application software. This device gathers entropy from a combination of interkeystroke times and other system interrupts. In addition, the gkernel

<http://sourceforge.net/projects/gkernel/>

driver can be used to add entropy from Intel's TRNG to the device entropy pool.

- `Random.org` and `Hotbits` are two academic projects supplying large numbers of hardware-generated random bits online (Johnson 2001).

Some caution is still warranted with respect to hardware random number generators. Often, the exact amount of true entropy supplied by a hardware device is difficult to determine (see Viega and McGraw 2001). For example, the raw thermal noise collected by the Intel generator is biased, and firmware postprocessing is applied to make the results appear more random (Jun and Kocher 1999). Typically, some forms of TRNG hardware generation are subject to environmental conditions, physical breakage, or incorrect installation. Although most hardware generators check their output using the FIPS 140-1 test suite when the device starts up, these tests are not nearly as rigorous as those supplied by standard test suites for statistical software (see Chapter 3). Therefore, some testing of the output of a random number generator is warranted before using it for the first time.

2.5.4   Local Search Algorithms

We discuss search algorithms and how to choose them in more detail in Chapters 4 and 8. In Chapter 3 we show that modern statistical packages are still prone to the problems we describe, and in Chapter 10 we discuss some aspects of this problem with respect to nonlinear regression. The purpose of this section is to alert researchers to the limitations of these algorithms.

Standard techniques for programming an algorithm to find a local optimum of a function, which may or may not be the global optimum, typically involve examining the numerically calculated or analytic gradients of the likelihood function at the current guess for the solution, and then use these to determine a direction to head "uphill."

Like other algorithms, bugs and floating point inaccuracies may cause problems for nonlinear optimization algorithms, and poorly scaled data can exacerbate inaccuracies in implementation. Thus, numerical inaccuracies may prevent the location of local optima, even when the search algorithm itself is mathematically correct.

More important, nonlinear optimization algorithms suffer from a deeper limitation, that of finding the global optimum. The conditions for global optima for some classes of problems are known (e.g., quadratic functions). However, as one well-known set of practitioners in the field wrote: "Virtually nothing is known about finding global extrema in general" (Press et al. 1988, p. 290), and 14 years later wrote: "Finding a global extremum is, in general, a very difficult problem" (Press et al. 2002, p. 393). In addition, as Gentle (2002, p. 18) points out, the presence of multiple local optima may also raise conceptual problems concerning the estimation criterion itself. (See Chapter 4 for a discussion of inference in the presence of multiple optima.)

Techniques for finding global optima involve some degree of guesswork, or *heuristics*: Either the algorithm guesses at initial values for parameters and proceeds to find a local optimum from there, or it perturbs a local optimum in an attempt to dislodge the search from it. Alternatively, the problem itself is redefined in terms of local optima: Gentle (2002, Sec. 10.1) notes that in hierarchical clustering analysis, because of the computational difficulty, the definition of clusters is "merely what results from a specified algorithm."

For many nonlinear optimization problems, solving for the global optimum is provably computationally intractable (Garey and Johnson 1979). Furthermore, it has been proved that there is "no free lunch" for optimization—all optimizers must perform *no better than random search* (or better than any other heuristic) when averaged over all possible optimization problems (Wolpert and Macready 1997). These theorems apply to such popular black-box optimization techniques as neural networks, genetic algorithms, and simulated annealing. In addition, some of these methods raise other practical problems that render their theoretical properties invalid in all practical circumstances (see Chapter 4). In other words, all practical optimization algorithms are limited, and to choose or build an algorithm wisely, one needs to use specific knowledge about the structure of the particular problem to be solved by that algorithm.

In the absence of mathematical proofs of global optimality, prudent researchers may attempt to ascertain whether the solution given by the optimization algorithms are, in fact, global and whether the model is well specified. Although there are no guaranteed methods, a number have been developed:

- Finch et al. (1989) describe a test of global optimality, similar to grid search techniques, which is based on an evaluation of the local maximum likelihood function from randomly selected starting points. Veall (1990) describes another way of testing the hypothesis, using a similar set of local optima. Unlike a grid search, these tests provide a way to formally test the hypothesis that the optimum found is global.

- Derigs (1985) has developed a technique, which he uses for discrete optimization, of testing for global optima given an external, analytically derived upper bound for a function.

- Den Haan and Marcet (1994) describe a technique for testing the accuracy of complex econometric simulation results, through examination of the distribution of the residuals.

- White (1981,1982) has developed a test for misspecification of the maximum likelihood models based on the divergence between covariance matrices computed from the Hessian and from the cross-product of first derivatives (respectively).

We discuss these tests in more detail in Chapter 4. These tests are not in wide use and to our knowledge have not been incorporated into any statistical software package.

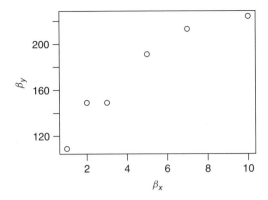

Fig. 2.4 BoxBOD data.

2.5.4.1 Optimization Example

The dual problems of numerical accuracy and algorithmic limitations are illustrated by a simple problem, first described in Box et al. (1978) and later incorporated into the NIST StRD test suite for optimization (see Chapter 3). The following data were collected (Figure 2.4): y represents biological oxygen demand (BOD) in mg/L, and x represents incubation time in days. The hypothesized explanatory model was

$$y = \beta_1(1 - \exp[-\beta_2 x]) + \varepsilon.$$

Figure 2.5 shows the contours for the sum-of-squared residuals for this model with respect to the data. (These contours are shown in the area very close to the true solution—a luxury available only after the solution is found.) Computing the parameters that best fit these data is not straightforward.

The contours illustrate some of the difficulty of discovering the solution. If one's initial starting values fall outside the top right quadrant, an iterative search algorithm is unlikely to be able to discover the direction of the solution. Furthermore, in the lower left quadrant there appears to be a small basin of attraction that does not include the real solution.

2.6 SUMMARY

We have discussed three types of error that affect statistical computing:

- *Bugs* are simple mistakes in implementing an otherwise correct algorithm. There are generally no tests that will prove the absence of bugs. However, comparing the results from multiple independent implementations of the same algorithm is likely to reveal any bugs that affect those particular results.

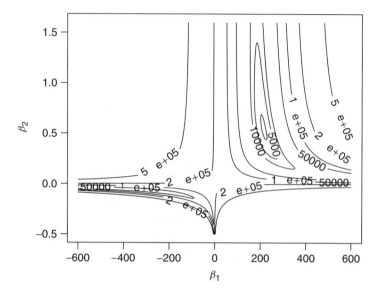

Fig. 2.5 BoxBOD SSR contours.

- *Inaccuracies* in implementation occur primarily because the arithmetic used by computers is not the perfect symbolic arithmetic used in algorithms. Particularly, when numerical algorithms are implemented using standard floating point arithmetic, *rounding* and *truncation errors* occur. These errors can accumulate during the course of executing the program, resulting in large differences between the results produced by an algorithm in theory, and its implementation in practice.
- *Algorithmic limitations* occur when the algorithm itself is not designed to compute the quantity of interest exactly. These limitations can be of different sorts. Algorithms for computing statistical distributions are often designed as *approximations*—even if executed with infinite precision, the algorithm will still yield an error that is only "close" to the desired solution. Other algorithms, such as those used in pseudo-random number generation, are *heuristic* algorithms for finding the solution to nonlinear optimization and are limited to finding local optima, although users of them often treat the results as if they were computed with respect to the global optimum.

Other pathological problems have also been reported, the sum of which tends to erode trust in statistical computing. For example, Yeo (1984) found that by simply reordering the input of data to SAS, he was able to get a noticeably different regression result.

After reading this chapter, one may be tempted to trust no computational algorithm. A more critical view of statistical computations is perhaps better than blind trust, but the lesson here is not as fatalistic as it may seem. The remainder of the book provides a variety of techniques for discovering inaccuracy and ameliorating it. The underlying theme of this chapter and book is that careful consideration of the problem, and an understanding of the limitations of computers, can guide researchers in selecting algorithms that will improve the reliability of the inference they wish to make.

CHAPTER 3

Evaluating Statistical Software

3.1 INTRODUCTION

Statistical software packages are collections of computer algorithms designed to facilitate statistical analysis and data exploration. Statistical software is a critical tool used in the vast majority of quantitative and statistical analysis in many disciplines. Without training in computer science, social science researchers must treat their package of choice as a black box and take it as a matter of faith that their software produces reliable and accurate results.

Many statistical software packages are currently available to researchers. Some are designed to analyze a particular problem; others support a wide range of features. Commercial giants such as SAS and SPSS, and large open-source packages such as R offer thousands of commonly used data manipulation, visualization, and statistical features. Although software may be distinguished by its features, less obviously, packages also differ significantly in terms of accuracy and reliability. These differences may make the difference between results that are robust and replicable and results that are not.

In this chapter we discuss strategy for evaluating the accuracy of computer algorithms and criteria for choosing accurate statistical software that produce reliable results. We then explore the methodology for testing statistical software in detail, provide comparisons among popular packages, and demonstrate how the choice of package can affect published results. Our hope is that this chapter leads social scientists to be more informed consumers of statistical software.

3.1.1 Strategies for Evaluating Accuracy

In an ideal world, we would be able to compute formal bounds on the accuracy and stability of every estimate generated by a computer program. For some distribution functions, and many individual computations in matrix algebra, it *is* possible to derive analytical bounds on the accuracy of those functions, given a particular implementation and algorithm. Methods such as interval arithmetic can be used to track the accumulated round-off errors across a set of calculations (Higham 2002). Unfortunately, the statistical models that social scientists are

Numerical Issues in Statistical Computing for the Social Scientist, by Micah Altman, Jeff Gill, and Michael P. McDonald
ISBN 0-471-23633-0   Copyright © 2004 John Wiley & Sons, Inc.

interested in estimating are typically too complex for bounds to be either derivable or informative. There are some specialized exceptions, such as the accuracy of linear regression in the presence of rounding error (Polasek 1987), and some bounds on Newton's method (Higham 2002, Sec. 25.2).

Furthermore, in an ideal world, all statistical computation would follow uniform standards for the best practices for treatment of data, choice of algorithm, and programming techniques. Best practices for many common analyses used by social scientists are discussed in Chapters 7 to 10. Unfortunately, these best practices are often ignored, even in large commercial statistical packages. While best practices can improve accuracy and stability, Chapter 2 explains why perfect accuracy and stability cannot be guaranteed for all algorithms, particularly random numbers generators and nonlinear optimization algorithms.

What can one do to assess or ensure the accuracy and stability of one's estimation procedure when formal bounds are not known? There are three general heuristics that can help identify potential computational problems:

1. **Test benchmark cases.** Correct estimates can sometimes be computed, exactly or to a known level of accuracy, for a particular model and set of test data. Artificial models and data, such as the standard deviation example in Chapter 2, may be constructed specifically to exploit known weaknesses in statistical algorithms. The estimates generated by a particular algorithm and implementation can then be compared to these known results. Discrepancies are an indication of potential computation problems. The NIST (National Institute of Standards and Technology) tests for accuracy of statistical software (Rogers et al. 2000) and other benchmark tests, such as those presented in this chapter, are examples of this approach.

   Benchmarks are useful and should be employed on publicly distributed software wherever feasible. However, benchmarks have three significant limitations. First, even for artificial data, benchmarks may be very expensive to create because of the amount of computation necessary to ensure accuracy. Second, realistic benchmarks, for which estimates can be calculated with known accuracy, are sometimes impossible to create. Third, benchmark testing can detect some inaccuracies but is valid only for the model and data tested. The performance of an algorithm for different models and data remains unknown. One can only hope that inaccuracy is unlikely where these models and data used in the tests are sufficiently similar to models and data being analyzed.

2. **Use separate information to confirm results or necessary/sufficient conditions for results.** In any statistical analysis, the researcher should always apply substantive knowledge of the model, data, and phenomena being analyzed to check that the results are plausible. Implausible results should be held up to extensive scrutiny.

   Besides this higher-level "gut check," there may be other techniques that can be used to confirm (or disconfirm) results. For example, estimates produced by maximum likelihood search algorithms may be checked by examining likelihood profiles and other diagnostics (see Chapter 8). Probabilistic tests may be applied to disconfirm the identification of a global optimum (see Chapter 4).

3. **Use sensitivity analysis.** One popular approach is to replicate the analysis keeping the data and model the same, but using many different algorithms, algorithmic parameters (such as starting values), and implementations (e.g., different PRNGs and/or different optimization software). If results disagree, one should investigate (applying the other techniques) until it is clear which set of results should be discarded. This is highly recommended where multiple implementations and algorithms are available. The effort required to create alternatives where none presently exist, however, can be prohibitively high.

A second popular and complementary approach is to replicate the analysis while perturbing the input data and to observe the sensitivity of the estimates to such perturbations. Sensitivity or *pseudoinstability* is not a measure of true computational stability, because values for the correct estimates are unknown. This has the advantage of drawing attention to results that cannot be supported confidently given the current data, model, and algorithm/implementation, and unlike the first method, is easy to implement. We discuss this in more detail in Chapter 4.

These two sensitivity tests can be combined fruitfully, as we show in Chapter 7. A potential drawback of the second method is that pseudoinstability detected by perturbing the data could be due to problems in the algorithm/implementation but may also be due to the interaction of model and data. For example, the results of a linear regression, running on data that is "almost" multicollinear, can be highly unstable with respect to very small amounts of noise in the data, even if the regression calculations are performed exactly (without numerical inaccuracy). The instability will not be reflected in the standard errors of the estimate. These can be arbitrarily small, even in the presence of multicollinearity (Beaton et al. 1976).

Combining the two methods can help to separate the portions of pseudoinstability due to model. By running multiple implementations/algorithms on the same sets of perturbed data, if one implementation is more stable than the other, the difference in pseudostability is a result of implementation and algorithm, not model and data, which are kept fixed by construction.

Note that the size and form of the noise is not what serves to differentiate numerical problems from model and data problems—even simple uniform noise at the level of machine round-off can affect analyses purely because of model and data problems. It is the combination of perturbations and varying implementations that allows one to gain some insight into sources of sensitivity. Nevertheless, regardless of the cause of sensitivity, one should be cautious if the conclusions are not pseudostable with respect to the amount of noise that is reasonably thought to be in the data.

These three approaches cannot be used to prove the accuracy of a particular method but are useful in drawing attention to potential problems. Further experimentation and analysis may be necessary to determine the specific cause of the problem. For example, if two software packages disagree on the estimates for the same model and data, the discrepancy could be a result of several factors:

- **Implementation issues.** One or both programs has a bug, one performs (some) calculations less accurately, or the results from each are conditioned on different implementation-level parameters (e.g., a difference in a convergence tolerance setting).
- **Algorithmic issues.** One or both programs may use an algorithm for which the required conditions are not met by the particular model and data. Algorithms may afford different levels of approximation error. Or the results are conditioned on different values for algorithm-specific parameters (e.g., starting values for local optimization algorithms).
- **Data and model issues.** The problem is ill-conditioned. (We discuss ill-conditioning in the next section.)

With the exception of standard software bugs from programming error, it is not obvious whether the programmer or end user is at fault for ignoring these issues. Users of statistical software should pay close attention to warning messages, diagnostics, and stated limitation of implementations and algorithms. Often, however, software developers fail to provide adequate diagnostics, informative warning messages, or to document the computational methods used and their limitations. Users should also examine data for outliers, coding errors, and other problems, as these may result in ill-conditioned data. However, users often have no a priori knowledge that a particular set of data is likely to cause computational problems given the algorithm and implementation chosen by the programmer.

3.1.2   Conditioning

Conditioning is worth discussing in more detail because while conditioning is often mentioned, its precise meaning is sometimes not generally well understood by social scientists.

Following Higham 2002 (Secs. 1.5-1.6): The most general definition of conditioning is "the sensitivity of the model to perturbations of the data" (p. 9). Condition numbers are used to represent the conditioning of a problem with respect to a particular set of inputs. For example, if a scalar function $f(\cdot)$ is twice differentiable, a useful way to define the relative condition number of $f(\cdot)$ is

$$c(x) = \left| \frac{xf'(x)}{f(x)} \right|. \tag{3.1}$$

When defined in this way, the accuracy of the estimate is [1]

$$c(x) \times \text{backward error}, \tag{3.2}$$

---

[1] These definitions and the example of $\log(x)$ below are both suggested by Higham (2002).

where backward error is defined as the minimum $|\Delta x|$ for which our computation of y, $\tilde{y}$ satisfies

$$\tilde{y} = f(x + \Delta x). \tag{3.3}$$

This is a formalization of the notion that the accuracy of the estimate is a function of the model, data, and the computational procedure. The condition number is a particularly useful formalization inasmuch as it is easier to derive the backward error of a computational method than its overall accuracy or stability.

Although conditioning is an important factor in the accuracy of any computation, social scientists should not assume that all computational inaccuracies problems are *simply* a matter of conditioning. In fact, a computation method with a large backward error will yield inaccurate results even where the problem itself is well conditioned.

Moreover, the conditioning of the problem depends on data, model, algorithm, and the form of perturbation. There is no such thing as data that is well conditioned with respect to every model. Although it might appear tempting to use condition number estimators produced by standard statistical software (such as MATLAB) to calculate condition numbers for a particular dataset, the results are bound to be misleading because the formulas used by these estimators are tailored to specific types of problems in linear algebra, such as matrix inversion. These formulas may be inappropriate when used to estimate the conditioning of another type of problem or computation procedure. Consider a simple example, using the condition number formula above: $x = 1$ is ill-conditioned for the function $\log(x)$ but is well conditioned for the function e^x.

In an example drawn from optimization, Higham (2002) shows that the accuracy of Newton's method depends on the condition of the Jacobian at the solution as well as the accuracy of the Jacobian and residual calculations, not (as one might naively assume) solely on the condition number of the data matrix.[2]

3.2 BENCHMARKS FOR STATISTICAL PACKAGES

Longley (1967) provided the most dramatic early demonstration that mainframe computers may not accurately reproduce estimates of a regression, whose solution was known. As a result, regression algorithms were resigned to avoid the flaws exposed by Longley. Following Longley, investigations of software inaccuracy and how to detect it have resurfaced regularly (e.g., Wilkinson and Dallal 1977; Wampler 1980; LeSage and Simon 1985; Simon and LeSage 1988, 1990), as our statistical methods have become more complex, and our software, more ambitious. Recently, software has been critically evaluated yet again (McCullough 1998, 1999a,b; McCullough and Vinod 1999; Altman and McDonald 2001).

---

[2]However, the conditioning of the data matrix will affect the conditioning of the residual calculation problem, as will the method of calculating the residual and the function being evaluated. In this case, a standard condition number can be informative, but more generally, such as shown in the previous example, does not necessarily shed light on conditioning for a specific problem.

As increasingly complex algorithms are implemented on statistical software, benchmarks serve as an important means to identify and correct potential flaws that may lead to inaccurate results in real applications.

In Chapter 2 we explain why the computer algorithms on which statistical software are built contain unavoidable numerical inaccuracies. Unfortunately, comprehensively testing statistical packages for all numerical inaccuracies is practically impossible given the amount of time that is required to investigate all algorithms.

In lieu of an exhaustive analysis of a statistical program, benchmarks serve as a basis for assessing their degree of numerical accuracy. Benchmarks are problems—models and data—with known answers that can be compared to the estimates produced by statistical packages. Here we describe three sets of benchmarks, each corresponding roughly to a source of inaccuracy as described in Chapter 2: floating point approximation error, inaccuracies in distribution functions, and inaccuracies in pseudo-random number generation. In addition, we benchmark an overlooked source of inaccuracy: basic processes of data input and export.

3.2.1 NIST Statistical Reference Datasets

The Statistical Engineering and Mathematical and Computational Sciences Divisions of the National Institute of Standards and Technology's (NIST) Information Technology Laboratory maintains a set of benchmark problems called the Statistical Reference Datasets (StRD), which are accompanied by certified values for a variety of statistical problems. These datasets provide benchmarks for assessing the accuracy of univariate descriptive statistics, linear regression, analysis of variance, and nonlinear regression (Rogers et al. 2000). The collection includes both generated and experimental data of varying levels of difficulty and incorporate the best examples of previous benchmarks, such as Simon and LeSage (1988) and Longley (1967). Since the release of the StRD, numerous reviewers have used them to assess the accuracy of software packages, and vendors of statistical software have begun to publish the test results themselves.

To gauge the reliability of a statistical package using the StRD, one loads the data into a program, runs the specified analysis, and compares the results to the certified values. Good performance on the StRD provides evidence that a software package is reliable for tested algorithms, but of course, provides no evidence for untested algorithms.

For each tested algorithm, the StRD contain multiple problems designed to challenge various aspects of the algorithm. The benchmark problems are named and assigned three levels of difficulty: low, average, and high. Although the degree of accuracy of the algorithm generally corresponds to the level of difficulty of the problem, the numerical difficulty of a problem is not independent of the particular algorithm used to solve it. In some cases software packages will fail spectacularly for low-difficulty problems but pass higher-level problems successfully.

Each StRD problem is composed of data generated by actual research or of data generated using a specific numeric function. For each problem, data are accompanied by values certified to be correct. For real-world data, the certified values are calculated on a supercomputer using the FORTRAN computer language. For data generated, the certified values are known quantities. The data, together with the algorithms and methodology to compute the certified values, can be found at the NIST Web site: <http://www.itl.nist.gov>.

As an example, consider the first and "easiest" univariate descriptives statistics problem, which are descriptive statistics for the digits of a familiar number, π. StRD provides π to 5000 digits of accuracy, and each digit serves as one of 5000 observations. The mean, standard deviation, and the one-observation lag autocorrelation coefficient are generated for the digits of π with a FORTRAN program at a very high precision level. Details of the formulas used to generate these descriptive statistics are available in the archive. The certified values of these descriptive statistics are then rounded to 15 significant digits, and are made available so that the performance of a statistical package's univariate statistics algorithms may be evaluated.

These 5000 observations are read into a statistical software package, and the same descriptive statistics are generated to verify the accuracy of the algorithm. For example, Stata correctly returned the certified values for the π-digits problem. We thus have confidence that Stata's univariate statistics algorithms are accurate in reproducing *this one benchmark problem*. We have nothing to say about untested algorithms, and indeed, most statistical packages are able to solve this problem correctly, even though some were unable to return correctly results close to the certified values for many other univariate statistics problems.

Although StRD tests are well documented and relatively easy to apply, some care is warranted in their application:

1. As we discuss in more detail below, some software packages may truncate the values in the data on first loading it—causing the results of subsequent tests to be inaccurate. In this case, the inaccuracy in data input may be attributed incorrectly to the statistical algorithm rather to than the data input process.

2. A software package may fail to display results with full internal precision or may truncate the values when they are exported into another program for analysis.

3. A software package may present errors in a graph or other visualization results even where the internal values are correct (see, e.g., Wilkinson 1994).

4. Often, a statistical package will provide multiple tools for computing the same quantity. Different tools, even within the same package, can provide different results. The Microsoft Excel standard deviation example in Chapter 2 shows how estimates from a built-in function may also be generated through other spreadsheet functions, which may yield different results. In other packages, even a substitution of x^2 for $x \times x$ may yield differences in performance on the benchmark tests.

5. For some benchmark tests, the StRD problem requires a function to be coded in the language of the statistical package. While NIST datasets include sample code to implement StRD problems, the syntax of statistical packages, like programming languages, have syntactic differences in precedence, associativity, and function naming conventions that can lead the same literal formula to yield different results. For example, "-2^2" will be interpreted as "$-(2^2)$" in many statistical and computer languages, such as R (R Development Core Team 2003) and Perl (Wall et al. 2000), but would be interpreted as $(-2)^2$ in a language such as smalltalk (Goldberg 1989), which uses uniform precedence and left associativity, and in FORTRAN (ANSI 1992), which among its more complex evaluation rules, assigns exponentiation higher precedence than unary minus (e.g., Goldberg 1989; ANSI 1992). Furthermore, an expression such as "$-2^{3^4} \times 5 \times \log(6)$" might be interpreted in many different ways, depending on the associativity of the exponentiation operator, the relative precedence of unary minus, multiplication, and exponentiation, and the meaning of *log* (which refers to the natural log in some languages and to the base 10 log in others). When in doubt, we recommend consulting the language reference manual and using parentheses extensively.

3.2.2 Benchmarking Nonlinear Problems with StRD

In addition to the issues just discussed, there are three important aspects of nonlinear algorithms that must be taken into consideration when performing benchmark tests: the vector of starting values passed to the search algorithm, the options that may affect the performance of the solver, and the type of solver used.

3.2.2.1 Starting Values

As described in Chapters 2, 4, and 8, nonlinear optimization is iterative. Every nonlinear optimization algorithm begins a set of *starting values* that are initially assigned, explicitly or implicitly, to the parameters of the function being optimized. The algorithm evaluates the function and then computes gradients (or other quantities of interest, such as trust regions) as appropriate so as to generate a new set of test values for evaluation until the convergence/stopping criteria are met.

Ideally, a nonlinear algorithm finds the global optimum of a function regardless of the starting values. In practice, when evaluating a function with multiple optima, solvers may stop at a local optimum, some may fail to find any solution, and some may succeed in finding a false (numerically induced) local optimum. An algorithm will generally find the global optimum if the starting values are "close" to the real solution, but may otherwise fail.

The StRD provides three sets of values to use as starting points for the search algorithm: start I is "far" from the certified solution, start II is "near" the certified solution, and start III is the certified solution itself. In addition, a nonlinear problem may be tested using the default starting values provided by the statistical package itself, which is often simply a vector of zeros or ones.

3.2.2.2   *Options*

Statistical packages usually offer the user more flexibility in specifying computational details for nonlinear optimization than for other types of analysis covered by the StRD. For example, a particular package may allow one to change the method by varying the following (see Chapter 4 for a more thorough treatment of these options):

1. The convergence/stopping criterion, such as the number of iterations
2. The convergence tolerance/stopping level
3. The algorithm for finding the optimum, such as Gauss–Newton
4. The method for taking derivatives

In Chapter 8 we provide an example of the accuracy of the `Excel 97` nonlinear solver in estimating the StRD nonlinear benchmark labeled "Misra1a." This benchmark problem is a real-world analysis encountered in dental research and may fool an algorithm because the function to be solved is relatively flat around the certified solution. McCullough demonstrates how changing the options described above dramatically affects the estimates produced by `Excel97`—in some cases `Excel97` does not even find the correct first digit—and further suggests useful diagnostic tests to identify when a solver may have stopped short of the true optimum.

The lesson is that nonlinear and maximum likelihood solvers may not find the true solution simply using the defaults of the program. Often, default options are chosen for speed over accuracy, because speedier algorithms typically perform no worse than more accurate, and slower, algorithms for easy problems and because the market for statistical software provides incentives to value speed and features over accuracy (see Renfro 1997; McCullough and Vinod 2000). For difficult problems, the defaults are not to be trusted blindly.

3.2.2.3   *Solver*

The type of solver used can affect the outcomes of the benchmarks. Statistical packages may offer nonlinear solvers tailored specifically to nonlinear least squares, maximum likelihood estimation (MLE), and constrained maximum likelihood estimation. These solvers will generally not yield identical results for the same StRD problems.

The nonlinear benchmark problems in the StRD are formulated as nonlinear least squares problems. Formally, nonlinear regression problems can simply be reformulated more generally as maximum likelihood problems, under the assumptions that data themselves are observed without error, the model is known, and the error term is normally distributed (see Seber and Wild 1989). In practice, however, nonlinear least squares solvers use algorithms that take advantage of the more restricted structure of the original problem that maximum likelihood solvers cannot.[3] Furthermore, reformulation as a MLE problem requires the additional

---

[3]For a detailed discussion of the special features of nonlinear regression problems, see Bates and Watts (1988). Most solvers dedicated to nonlinear least squares fitting use the Levenberg–Marquardt

estimation of a sigma parameter for the variance of the errors. For these reasons, although MLE solvers are not fundamentally less accurate than nonlinear least squares solvers, the former will tend to perform worse on the StRD test problems.

3.2.2.4 Discussion

Starting values, solver, and computational options should be reported along with the results of nonlinear benchmarks. The set of options and starting values that represent the most "realistic" portrayal of the accuracy of a statistical package, in practice, is the subject of some debate. McCullough (1998) recommends reporting two sets of test results for nonlinear regression problems. He reports the results for the default options using start I and also reports results using an alternative set of options, derived through ad hoc experimentation.[4] He reports results using start II values only if a program was unable to provide any solution using start I values. On the other hand, it is common for statistical software vendors to report results based on a tuned set of options using the easier start II values.

Clearly, a package's accuracy depends primarily on the circumstances in which it is used. Sophisticated users may be able to choose good starting values, attempt different combinations of options in search of the best set, and carefully scrutinize profile plots and other indicia of the solution quality. Users who are unfamiliar with statistical computation, even if they are experienced researchers, may simply use default algorithms, options and starting values supplied by the statistical software, and change these only if the package fails to give any answer. These erroneous results can even find their way into publication. (See Chapter 1 for examples of published research that were invalidated by naive use of defaults.) For this reason, when used with defaults, it is important that a software package not return a plausible but completely inaccurate solution. Warning messages, such as failure to converge, should be provided to alert users to potential problems.

3.2.3 Analyzing StRD Test Results

The accuracy of the benchmark results can be assessed by comparing the values generated by the statistical software to the certified values provided with the benchmark problems. The StRD reports certified values to 15 significant digits. A well-designed implementation will be able to replicate these digits to the same degree of accuracy, will document the accuracy of the output if it is less, or will warn the user when problems were encountered.

Following NIST recommendations, if a statistical software package produced no explicit warnings or errors during verification of a benchmark problem, the accuracy of the results is measured by computing the *log relative error* (LRE),

---

algorithm [see Moré (1978) for a description], which is used much less commonly in MLE estimation. See Chapters 4 and 8 for further discussion of these algorithms.

[4] See Chapter 7 for systematic methods for varying starting values and other optimization parameters.

given by

$$
\text{LRE} = \begin{cases} -\log_{10}(j(x-c)/(c)j) & \text{when } c \neq 0 \\ -\log_{10}(j \times j) & \text{when } c = 0. \end{cases} \tag{3.4}
$$

As mentioned in Chapter 2, the latter case is termed the *log absolute error* (LAE). When the test value $x = c$, the LRE (and LAE) is undefined, and we signify the agreement between the test and certified values to the full 15 significant digits by setting LRE equal to 15.

The LRE is a rough measure of the correct significant digits of a value x compared to the certified value c. LRE values of zero or less are usually considered to be completely inaccurate in practice, although technically, a statistical package that produced estimates with LREs of zero or less may provide some information about the solution. Often, where a single benchmark problem has many coefficients, reviewers of the software report the worst LRE for the set under the assumption that researchers are interested in minimizing the risk of reporting even one false value. However, for models used to make substantive predictions, inaccurate estimation of coefficients with minuscule predictive effect may be of little consequence.

3.2.4   Empirical Tests of Pseudo-Random Number Generation

A pseudo-random number generator (PRNG) is a deterministic function that mimics a sequence of random numbers but is not truly random and cannot be guaranteed as such. All PRNGs produce a finite sequence of numbers and have regularities of some sort. PRNGs cannot be made to fit every imaginable distribution of simulation well enough—for some transformation or simulation, the regularities in a PRNG are problematic.

Various statistical tests can be applied to the output of a PRNG or a generator to detect patterns that indicate violations of randomness. Strictly speaking, a PRNG cannot be proved innocent of all patterns but can only be proved guilty of non-randomness. When a generator passes a wide variety of tests, and more important, the tests approximately model the simulation in which the PRNG will be used, one can have greater confidence in the results of the simulation.

Most tests for PRNG follow a simple general three-part methodology:

1. Using the given PRNG and a truly random seed value, generate a sequence s of length N. A minimum length for N may be dictated by the test statistic used in step 2.
2. Compute a test statistic $t(s)$ on s.
3. Compare $t(s)$ to $E[t(s)]$ under the hypothesis that $t(s)$ is random. The comparison is usually made with a threshold value, an acceptance range, or a p-value.

In typical hypothesis testing, failure to accept the null is not proof of the truth of a rival hypothesis because there are an infinite number of alternative, untested hypotheses that may be true. In this case, the hypothesis being tested is that the PRNG is random, and the null hypothesis is that it is not. No test can prove that the generator is random, because a single test does not guarantee that a PRNG will be assessed similarly with regard to another test. A test for randomness may only prove the null hypothesis that the generator is not random. Failure to accept the null is evidence, not proof, of the hypothesis that the PRNG is indeed random.

It is important to note that tests such as these examine only one aspect of pseudo-random number generation: the distributional properties of the sequence created. As discussed in Chapter 2, there are two other requirements for correct use of PRNGs: adequate period length and truly random seed values. Furthermore, parallel applications and innovative simulation applications require special treatment.

In practice, a variety of tests are used to reveal regularities in known classes of PRNGs and capture some behavior of the sequence that is likely to be used by researchers relying on the PRNGS in their simulations. Some test statistics are based on simple properties of the uniform distribution, while other test statistics are designed explicitly to detect the types of structure that PRNG are known to be prone to produce. In addition, "physical" model tests, such as ISING model and random-walk tests, use PRNGs to duplicate well-known physical processes. Good physical tests have the structure of practical applications while providing a known solution. Two examples of common tests illustrate this general methodology.

3.2.4.1 Long-Runs Test
The long-runs test, a part of the Federal Information Processing Standard for testing cryptographic modules, is a relatively weak but simple test that sequence is distributed uniformly. It is defined on the bits produced by a PRNG, as follows:

1. Generate a random sequence of 20,000 bits.
2. Compute the distribution of all runs of lengths l in s. (A run is defined as a maximal sequence of consecutive bits of either all ones or all zeros.) The test statistic $t(s)$ is the number of runs of $l > 33$.
3. The test succeeds if $t = 0$.

3.2.4.2 Birthday Spacings Test
The birthday spacings test is performed on the set of integers produced by a PRNG. It examines the distribution of spacing among "birthdays" randomly drawn from a "year" that is n days long. Note that it tests only the distribution of the members of the set and does not test the ordering of the members. It is part of the DIEHARD test suite and proceeds roughly as follows:

1. Generate a sequence s of integers in $[1,n]$, for $n \geq 2^{18}$, of length m.
2. Compute a list of birthday spacings, $|S_i - S_{i+1}|$. Let $t(s)$ be the number of values that occur more than once in this sequence.
3. $t(s)$ is asymptotically distributed as Poisson, with mean $m^3/4n$, and a χ^2 test provides a p-value.

3.2.4.3 *Standard Test Suites*

Although there is no single industry-standard set of random number generation tests for statistical software, a number of test suites are available:

- The Federal Information Processing Standard (FIPS 140-1) for cryptographic software and hardware used with unclassified data (FIPS 1994) specifies four tests that an embedded PRNG must pass (runs, long runs, monobit, and poker). Although adequate for generating small cryptographic keys, the tests specified, are not extensive enough for statistical applications.

- Donald Knuth (1998) describes several empirical tests (including birthday spacings, coupon collector, collisions, frequency, gap, permutation, and serial correlations tests). No software is provided for these tests, but the tests described are reasonably straightforward to implement.

- George Marsaglia's (1996) DIEHARD software implements 15 tests, including birthday spacings, permutations, rank tests, monkey tests, and runs tests, among others. [Many of these tests are discussed more thoroughly in Marsaglia (1984, 1993).] In part because of its widespread availability, this test suite has been featured in numerous software reviews and has been adopted by a number of major software vendors.

- The NIST Statistical Test Suite for Random and Pseudo-random Number Generators for Cryptographic Applications (Rukhin et al. 2000) is one of the most recent and thorough test suites to emerge. It includes tests selected from the sources above, plus compression, complexity, spectral, and entropy tests.

- The Scalable Library for Pseudo-random Number Generation Library (i.e., SPRNG) (Mascagni and Srinivasan 2000), while designed primarily as a set of functions for generating pseudo-random numbers for parallel applications, contains tests specifically for parallel PRNGs that are not included in any of the other test collections (see below).

- The TESTU01 suite (L'Ecuyer and Simard 2003), proposed initially in 1992 by L'Ecuyer (1992), has recently been released to the public. This test suite encompasses almost all of the FIPS, Knuth, Marsaglia, and NIST tests noted above and adds many others. In practice, this appears to be the most complete and rigorous suite available (McCullough 2003).

The DIEHARD suite remains the easiest suite to run and has been used widely in software reviews in recent years. However, this venerable suite has been superseded by the TESTU01 set of benchmarks.[5]

---

[5]As of mid-2003, Marsaglia had announced that development had started on a new version of DIEHARD. This new version is being developed at the University of Hong Kong by W.W. Tsang and Lucas Hui with support from the Innovation and Technology Commission, HKSAR Government, under grants titled "Vulnerability Analysis Tools for Cryptographic Keys" (ITS/227/00) and "Secure Preservation of Electronic Documents" (ITS/170/01). The preliminary version examined provided two additional tests and allowed some of the parameters of existing tests to be varied. Based on this preliminary version, we still recommend TESTUO1 over DIEHARD.

In addition, although the SPRNG tests are designed for parallel generators, one does not need parallel hardware to run them. Many of the tests are simply designed to detect correlations among multiple streams extracted from a single generator. Moreover, the SPRNG suite includes a physical ISING model test that can be run to evaluate serial PRNG performance, which is not available from any other package.

Empirical tests, such as those just described, should be considered an essential complement for, but not a replacement of, theoretical analysis of the random number algorithm. In addition, new tests continue to be developed (see, e.g., Marsaglia and Tsang 2002).

3.2.4.4   Seed Generation

Recall that PRNGs are initialized with a set of random *seed* values, and from these seeds, generate a repeating pseudo-random sequence with fixed length p, the *period* of the generator. The seed used for initialization must be *truly* random. Non-random seeds may lead to correlations across multiple sequences, even if there is no intrasequence correlation.[6]

The PRNG supplied by a statistical package should always provide the user with a method to set the seed for the generator, although many packages also offer to set the seed automatically. Although it is possible to collect adequate randomness from hardware and from external events (such as the timing of keystrokes), many commonly used methods of resetting seeds, such as by using the system clock, are inadequate (Eastlake et al. 1994; Viega and McGraw 2001). The tests described previously are designed only to detect intrasequence correlations and will not reveal such problems with seed selection. Statistical packages that offer methods to select a new seed automatically should make use of hardware to generate true random values, and should thoroughly document the method by which the seed is generated.

3.2.4.5   Period Length

Even with respect to generators that pass the tests just described, it is generally recommended that the number of random draws used in a simulation not exceed a small fraction of the period (conservatively, $p \gg 200n^2$, where n is the number of random draws used in the simulation being run). The tests we just described do not measure the period of the generator, so it is essential that a statistical package documentation include the period of the PRNGs. According to documentation, many of the random number generators provided by statistical packages still have periods of $t < 2^{32}$, which is adequate only for small simulations. Modern PRNG algorithms have much longer periods. Even the simple multiply-with-carry generator has a period of 2^{60}. See Chapter 2 for more detail.

3.2.4.6   Multiple Generators

Developers of random number generation algorithms recommend the use of multiple generators when designing new simulations. A statistical package should

---

[6]Non-random seeds may make a sequence easier to predict—an important consideration in cryptography (see Viega and McGraw 2001).

provide the user with several generators of different types. Although some packages are now doing just this, it remains common for a package to provide a single generator.

3.2.4.7 *Parallel Applications*

Most PRNGs and tests of randomness are designed to operate within a single thread of program execution. Some statistical applications, however, such as parallelized MCMC, use multiple processors simultaneously to run different parts of the simulation or analysis. These parallel applications can pose problems for random number generation.

Use of standard PRNGs in a parallel application may lead to unanticipated correlations across random draws in different threads of the simulation. Typically, single-threaded PRNGs are split in one of three ways to support multiple streams of execution: (1) each thread may run its own PRNG independently using a different seed; (2) the PRNG can be executed in a single thread, with the other threads of the simulation drawing from it as needed; or (3) with the other threads being assigned systematically every ith random draw. However, this second scheme can lead to irreproducible results in another way, because variations in interprocessor timings across runs will result in different threads receiving different partitions of the sequence on each run (Srinivasan et al. 1998). This means that the simulation results may be different on each run, even using the same initial seeds. Recording all subsequences in each processor, and replaying these exactly during replication, will eliminate this problem.

These methods can lead to statistical problems, either because of intersequence correlation inherent in the PRNG or because a method inadvertently transforms long-run intrasequence correlations into short-run correlations. (These long-run correlations would not normally be a problem in single-threaded simulation if the number of draws were limited to a fraction of the period of the generator, as recommended previously.) For parallel applications, the SPRNG library (discussed previously) is a good source of tests and generators.

3.2.5 Tests of Distribution Functions

There are often multiple ways of evaluating a function for a particular statistical distribution that are often not solvable analytically. Different mathematical characterizations, evaluation algorithms, and implementations can yield results with wildly different accuracy. A method that produces results accurate to only three or four digits may be sufficient for significance tests but be entirely inadequate for simulation or when used as part of a likelihood function. A particular approach to approximation that works well in the center of a distribution may not be accurate in the tails, or vice versa.

Despite the importance of accurate statistical distributions for simulation and a number of common nonlinear models, there is no standard set of benchmarks for statistical distributions. Previous investigations of the statistical packages have noted frequent inaccuracies in common distribution functions (Knüsel 1995, 1998,

2002; McCullough 1999a,b), and McCullough (1998) has outlined a benchmarking methodology, but no standard set of benchmark exists.

Although no widely accepted set of benchmarks exists, there are a number of libraries and packages that may be used to calculate common statistical distribution with a high degree of accuracy:

- Knüsel's ELV package (1989) is a DOS program that computes the univariate normal, Poisson, binomial, both central and noncentral gamma, χ^2, beta, t, and F distributions. Distributions are computed to six significant digits of accuracy for probabilities of nearly 10^{-100}. The ELV package is designed as an interactive application and is quite useful as such. However, source code is not provided, so it cannot be used as a programming library, and support for the DOS platform in general is dwindling.
- Brown et al.'s (1998) DSTATTAB package computes a similar set of distributions to approximately eight digits of accuracy. DSTATTAB is available both as an interactive program and as a set of libraries that can be used by other programs.
- Moshier's (1989) Cephes library provides double-precision statistical distribution functions and quadruple-precision arithmetic and trigonometric functions, as well as others.
- A number of symbolic algebra packages, such as Mathematica (Wolfram 1999) and MATLAB, have the ability to compute a variety of functions, including statistical distribution functions, using "arbitrary precision" arithmetic (computation to a specified number of digits). Using this capability may produce highly accurate results for benchmark purposes, but may be relatively slow.

Constructing a set of tests for distributions of interest is relatively straightforward using these libraries. First, generate a set of p-values (and other distribution parameter values, as appropriate) across the region of interest for simulation, statistical analysis, or likelihood function. Second, use one of the aforementioned high-precision libraries to compute the correct value of the distribution for each p-value in the test set. Third, use the statistical package being tested to calculate the same distribution value using the same set of p-values. Finally, calculate the LREs (as described earlier) for the results produced by the statistical package.

A complementary strategy is to use the inverse functions to discover points where $f(x) \neq f^{-1}(f(x))$. Rather than searching at random, one might use a black box optimization heuristic, such as simulated annealing, to look for particularly large discrepancies, by finding local solutions to $\max_x |f(x) - f^{-1}(f(x))|$.

As discussed earlier, we believe that inaccurate and missing results should be distinguished in benchmark tests. Some statistical packages document that their distribution functions are valid for particular ranges of input and return missing values or error codes when asked to compute answers outside this range of input. We consider this acceptable and treat these results as missing, not as LREs of zero in our benchmarking analysis.

Using this general methodology, we create a set of benchmarks for five of the most commonly used statistical distributions: the normal distribution, the central t, F, χ^2, and gamma. These tests are archived at ICPSR[7] as Publication Related Archive Dataset 1243, and are also available from the Web site that accompanies this book. To our knowledge, this is the first and only set of benchmarks available for this purpose. In particular, we generate a set of p-values at regular and random intervals along { [1e-12,1-1e-12],0,1 }. We use Knüsel's (1989) package to generate reference values and checked these against identical computations using Brown et al.'s (1998) high-precision libraries. We calculate correct values for a number of central inverse distribution for those p-values (to five significant digits) and the corresponding p-values for those inverse values. Where distributions required ancillary parameters, we generate these at regular intervals in [1e-4,1e6], constrained by the legal range for the particular parameter.

Our benchmark is meant to probe only some key aspects of the accuracy of a statistical package. It is not comprehensive, and some applications may require accuracy in distributions that we do not test, or beyond the p (or other parameter range) that is covered by our procedure. For example, Knüsel (1995, 1998, 2002) and McCullough (1999a,b) clearly examine a number of areas we do not: the extreme tails $t < 10^{-12}$ (inaccuracies are reported for some distributions at very extreme tails, e.g., 10^{-70}), a number of discrete distributions, and a number of noncentral distributions. We believe our benchmarks are adequate for most normal use of a statistical package, but users with nonlinear models or simulations should use the libraries described previously to construct tests that better reflect their particular needs.

3.2.6 Testing the Accuracy of Data Input and Output

In the course of our research, we performed benchmark tests on over a dozen different packages and versions of packages. An unexpected lesson from our testing is that loading data into a statistical software package from text or binary files—a seemingly straightforward operation—may produce unexpected results. Most errors were obvious, but a few were subtle and caused statistical packages to fail the benchmarks. Once we diagnosed and corrected the problem, what initially appeared as inaccuracies of algorithms were revealed as inaccuracies of data input.

All statistical software packages are able to store a finite amount of data, either because of internal restrictions on the number of records and columns that a statistical packages is able to handle, or simply because of limited system resources, such as available hard disk space. Most statistical packages' storage limitations are documented, but in some cases they are not, and instead of providing a warning message when storage limits are reached, some of the packages silently truncated records, columns, and precision.

We create a simple set of benchmarks to detect inaccuracies in data input and output, the first benchmark of its kind, to our knowledge. To test the performance

[7]URL: <http://www.icpsr.umich.edu>.

Table 3.1 Results of Data Input Tests

Package	Input Precision	Exceeded Maximum Records	Exceeded Maximum Columns or Variables
1	Silent truncation	Memory error	Memory error
2	Silent truncation	Silent truncation	Silent truncation
3	Silent truncation	(No maximum record limit)	Hung the program
4	Silent rounding	(No maximum record limit)	Empty data file
5	Silent truncation	Memory error	Specific warning

of statistical programs in reading data, we create a small set of computer-generated data matrices, each with an unusually large number of rows, columns, or significant digits. We encode these matrices as tab-delimited flat ASCII text files and then use these data matrices to test error detection systematically in data loading. We consider a package to have "passed" the test as long as it reports when truncation or a related error occurred.

The results of our tests on recent versions of five major commercial packages are presented in Table 3.1. Since this book is not intended as a review of statistical packages, we have the software packages presented as "anonymous" in Table 3.1. Although the versions of the package were current at the time of testing, these results will undoubtedly be out of date by the time of publication. All packages reduced the precision of input data without warning, failing the data input tests. A number of packages failed to warn users when entire rows or columns of data were deleted.

Statistical packages should provide warnings when data are read improperly, even though we speculate that astute users will detect truncations of rows or columns by comparing descriptive statistics to known results for these data. In rare cases concerning original data, descriptive statistics may not be known *a priori*. In other cases, as we have found in reproducing published research (Altman and McDonald 2003), even astute users may "lose" cases from their replication data.

Although not as damaging as missing cases or variables, silent truncation of data precision may cause inaccuracies in later calculations and lead to results that are not replicable. McCullough suggests that "users of packages that offer single-precision storage with an option of double-precision storage should be sure to invoke the double-precision option, to ensure that the data are correctly read by the program" (McCullough 1999b, p. 151). This recommendation does not alleviate all precision issues; because some packages truncate whereas others round; reading the same data into packages using the same precision may still produce small differences.

Truncation or rounding data precision errors are unlikely to be detected by the user. Even if precomputed descriptive statistics are available, precision errors may be hidden by the limited numbers of digits displayed in the output of a statistical package. Moreover, descriptive statistics are not designed specifically to detect data alterations, and multiple errors can "compensate" to produce an identical statistic, even when the data have been altered.

3.2.6.1 Methods for Verifying Data Input

Essentially, what is needed is a function $f(\cdot)$ that maps each sequence of numbers (or more commonly, blocks of bits) to a single value:

$$f: \{i_0, i_1, \dots, i_n\} \to c. \tag{3.5}$$

To verify the data we would need to compute f once, when creating the matrix initially, then recompute it *after* the data have been read into our statistics package. For robust verification, we should choose $f(\cdot)$ such that small changes in the sequence are likely to yield different values in $f(\cdot)$. In addition, $f(\cdot)$ should be straightforward and efficient to compute.

The simplest candidate for our function is a *checksum*, which is a function of the form

$$f(\{i_0, \dots, i_n\}) = \sum i \mod K. \tag{3.6}$$

Although such checksums are simple to compute, they are, unfortunately, a weak form of verification. For example, checksums cannot detect when two blocks of the sequence have been transposed.

A more powerful test is provided by *cyclic redundancy checks* (CRCs). CRC functions treat the input as one large number, each bit of which represents a term in a long polynomial, and then take the remainder from a polynomial division of that number:

$$f(\{i_0, \dots, i_n\}) = (i_0 \times X^0 + \cdots + i_n \times X^n) \mod_p P. \tag{3.7}$$

Here P is often known as the *generator polynomial* and is often chosen to be

$$+X^{22} + X^{16} + X^{12} + X^{11} + X^{10} + X^8 + X^7 + X^5 + X^4 + X^2 + X^1 + 1. \tag{3.8}$$

Cyclic redundancy checks provide a powerful way to detect unintentional data corruption, and can be computed very quickly. These checks are used in most digital communication algorithms and are a more sophisticated version of the checksum. A CRC computes the remainder in the ratio of two polynomials, the numerator of which is a function of the data to be validated. The polynomials' design ensures that reasonable errors are detected. [See Ritter (1986) for a description of CRCs and Binstock and Rex (1995) for implementations.] In particular, a well-designed CRC will detect all one-, two-, and three-bit errors with certainty. It will also detect multiple-bit error patterns with probability $(1-1/2^n)$, where n is the number of bits in the checksum. We use a 32-bit checksum, so the detection probability is actually $>99.99999\%$.

Note that CRCs are designed to detect unintentional corruption of the data and can be fooled intentionally. To prevent intentional tampering, a strong cryptographic hash algorithm such as, MD5 (Rivest 1992) should be used for $f(\cdot)$.[8]

---

[8]TheMD5 algorithm pads the input stream and then divides it into blocks. Each block is processed three times using a variety of bitwise functions, and the results are combined with the preceding block.

CRCs and other verification functions are used routinely to detect changes in *files*, but they have not, to our knowledge, been used previously to verify the integrity of a *data matrix* within a statistics program. Computing verification functions on the input data files (not the data matrix) will not detect the truncation errors listed in Table 3.1, because the files themselves are intact.[9] On the other hand, computation of CRCs (or any other verification function) on the data matrix *in standardized form* and from *within* a program can be used to generate a distinctive signature for the data that can verify that the data matrix has been *read* accurately into another program. In other words, all previous applications of checksums in statistical software tested the integrity of *file transfer* but did not test the integrity of *information transfer*.

We have developed a small C library to compute CRCs on in-memory vectors of floating point numbers (see the dedicated Web page for this text). This library is usable from any other program that supports external libraries and may be used to verify that a data matrix has been read consistently across these programs. In addition, our library allows different checksums to be generated for representations of the vectors at various levels of precision. This allows the author of a data collection to generate checksums for different storage lengths (e.g., float, double-precision), or based on the accuracy of measurement instruments.

3.3 GENERAL FEATURES SUPPORTING ACCURATE AND REPRODUCIBLE RESULTS

Benchmarks can test only a small portion of the functionality of any statistical software package and reflect only on the precise version of the software under which the benchmarks are run. To have continuing faith in the reliability of a package, one must ask: Does the manufacturer of the package regularly test it for accuracy? Does the manufacturer document the algorithms used to compute statistical analyses and other quantities of interest?

In addition, benchmarks capture only a small portion of what aids a researcher in producing accurate and reproducible results. For example, ensuring accurate results may require augmenting the package with high-precision libraries. And reproducing an analysis may require running programs written in a previous version of the package. So when evaluating a statistical package, researchers should ask: Does the programming language support good programming practices, such as object-oriented design? Are the internal procedures of the package available for inspection and extension? Can the package make use of external libraries written in standard programming languages such as C, C++, and Java? How easy is it to use this package as part of a larger programming and data analysis environment? Does the package maintain backward compatibility across versions? Do programs written in previous versions of the package's syntax produce results identical to the original?

---

[9]Stata includes an internally callable function for creating checksums of data files, but it does not compute checksums on the data matrix per se and thus could not be used to detect truncation errors. It is functionally equivalent to an external checksum.

Furthermore, all software has bugs, and statistical software is no exception. Probably for commercial reasons, bug reports by commercial statistical software companies are typically not published, or even archived, making them difficult to cite in academic writing. To gather bug reports, we searched the Web sites of manufacturers of statistical software and searched the documentation and notes accompanying the software installation. Judging from available bug reports, most bugs in statistical software are unlikely to affect inference because they cause the program to fail outright, behave in an obviously incorrect manner, or affect a nonstatistical program function such as printing. Still, there are enough serious bugs reported in relatively recent versions of commonly used statistical software to warrant caution. Some examples culled from bug reports follow:

- In a recent version of a major commercial statistics package, t-statistics reported for maximum likelihood estimations were one-half of the correct values.
- A recent version of a second major commercial package produced incorrect results for regressions involving variables with long names, performed exponentiation incorrectly, and committed other statistical errors.
- A recent version of a third major commercial statistical package calculated t-tests incorrectly, and incorrectly dropped cases from cross-tabulations.

These are only samples of publicly reported and acknowledged problems, chosen to illustrate a point: Bugs in major statistical packages are serious enough to affect research. Users of statistical software are well advised to stay abreast of the latest updates to their programs and to seek out packages from publishers who fix bugs quickly, notify users of them actively, and maintain good records.

3.4 COMPARISON OF SOME POPULAR STATISTICAL PACKAGES

To illustrate the wide variation in accuracy across major commercial packages, we summarize accuracy benchmark results as published in software reviews (Knüsel 1995, 1998; McCullough 1999a,b; McCullough and Vinod 1999; Vinod 2000; Altman & McDonald 2001). All of these studies use the software benchmarking methodology described above.[10] Table 3.2 counts the number of packages out of those tested that produced "reliable" results overall. Eleven packages were reviewed, although some did not support or were not tested for every category. For purposes of this table, we define *reliable* to mean that the package passes *all* the benchmarks described above, or provided clear error messages whenever inaccurate results were presented in the benchmarks.

Table 3.2 hides a vast amount of detail but reveals a number of patterns. With the exception of pseudo-random number generation, most packages were

---

[10]DIEHARD results were not published in Altman and McDonald (2001) but were computed. Table 3.2 also incorporates these previously unpublished results.

Table 3.2   Number of Packages Failing Statistical Accuracy Benchmark Tests

	StRD Benchmark Tests			
	Univariate Statistics	ANOVA	Linear Regression	Nonlinear Regression
Low difficulty	0/11	0/6	0/11	4/11
Medium difficulty	0/11	3/6	2/11	4/11
High difficulty	4/11	1/6	2/11	5/11
Accuracy of Distributions		Random Number Generation		
Central	0/9	Failed $>$ 1 DIEHARD test		5/9
Tails	4/9	Period $\leq 2^3 1$		7/9
Extreme Tails	2/9	Single generator only		8/9

evaluated as accurate for most common tasks. However, a significant number of packages perform poorly on difficult nonlinear models and in the tails of statistical distributions. In addition, most of the packages tested did not provide robust random number generation capability: Periods were typically short, only a single generator was provided, and many generators failed distribution tests.

Care must be used to distinguish between differences among implementations of a particular algorithm and differences between algorithms used to compute the same quantity of interest. We expect that new versions of software will be made more accurate as implementations are improved and better algorithms found. Software writers have a responsibility not only to make improvements, but also to document the range of acceptable conditions for running their software and the accuracy that may be expected from it. Furthermore, as improvements are made, facilities should be provided to replicate the results from previous versions.

We note in closing this section that software companies update their programs continually and that any published investigation of their operation will necessarily be dated. Many software companies have become aware of the benchmarks since these reviews were published and have updated their software to pass the benchmark tests. We strongly suspect that if these benchmarks were tests with updated versions of these packages, many of the problems reported (and perhaps all of the problems reported with PRNGs, because these are easiest to remedy) will have been ameliorated. Still, we think that caution is warranted, especially as statistical methods continue to advance and continue to push packages into new areas of analysis.

3.5   REPRODUCTION OF RESEARCH

Benchmark tests are designed to exploit specific weaknesses of statistical software algorithms, and it may reasonably be argued that benchmarks do not reflect the situations that researchers will commonly encounter. Inaccuracies and implementation dependence is not simply an abstract issue. Here, we present an example drawn from a published article that illustrates some of the problems that can be caused by inaccurate statistical packages.

Our reproduction is of Nagler's (1994) skewed logit, or *scobit*, model, which offers a reexamination of Wolfinger and Rosenstone's (1980) seminal work on voter turnout, focusing on their conclusion drawn from a probit model that early closing dates of voter registration have a greater negative impact on people with less education.[11] Eschewing the assumption of typical binary choice models that the greatest sensitivity to change in the right-hand-side variables occurs at 0.5 probability of choice, Nagler develops a generalization of the logit that allows the steepest slope of the familiar logit probability density function to shift along the [0,1] interval, a model he calls *scobit*. Armed with the scobit model, Nagler reestimates the Wolfinger and Rosenstone probit model using the SHAZAM statistical program, and contrary to the earlier findings, estimates no statistically significant interactive relationship between registration closing dates and education.

Nagler follows Burr (1942) in adding an additional parameter to the familiar logit distribution. The log likelihood of the Scobit estimator is given by

$$\log L = \sum_{i=1}^{N} (1 - y_i) \log[F(-\mathbf{X}_i \boldsymbol{\beta})] + \sum_{i=1}^{N} y_i \log[1 - F(-\mathbf{X}_i \boldsymbol{\beta})], \qquad (3.9)$$

where $F(-\mathbf{X}_i \boldsymbol{\beta})$ represents what Nagler refers to as the *Burr-10 distribution*:

$$F(-\mathbf{X}_i \boldsymbol{\beta}) = (1 + e^{(\mathbf{X}_i \boldsymbol{\beta})})^{-\alpha}. \qquad (3.10)$$

In this setup, logit is a special case of the scobit model, when $\alpha = 1$. Scobit is an attractive example here because unlike logit, its functional form is not proved to be single peaked. There may be multiple local maxima to the log-likelihood function of the Scobit estimator, particularly if nonlinear terms, such as interaction terms, appear in the linear $\mathbf{X}_i \boldsymbol{\beta}$.

We encountered difficulties in our replication of scobit that do not necessarily relate to numerical accuracy issues but might be misconstrued as such if not accounted for. In our replication we found discrepancies in the selection of cases for the analysis. Evidence suggested that the original data provider had corrected errors in the original data subsequent to Nagler's. Fortunately, Nagler achieved a verification data set for his personal use and made these data available to us for the analysis presented here. We also discovered errors in the published derivatives for the scobit model which if undetected could lead to further numerical computation problems. For a full discussion, see Altman and McDonald (2003).[12]

We reproduce Nagler's analysis using SHAZAM, Stata, and Gauss statistical packages. The program and results file using SHAZAM v6.2 were generously supplied to us by Nagler. We ran the original program on the most recent version

---

[11]Nagler generously supplied the original SHAZAM code and data used in his analysis as well as the original output.

[12]Also see Achen (2003) for a discussion of the drawbacks of model generalization in Scobit.

of SHAZAM then available (v9.0), on a Linux platform. In addition, we reproduced the analysis on Stata v6.0 running on Windows and Gauss v3.2.43 running on a HP-Unix platform. We chose Stata because scobit is implemented as a built-in function *scobit* within that program and Gauss because of its early popularity among political methodologists.

In describing the NIST benchmarks earlier, we note that starting values for the MLE may affect the performance of optimization routines. We found evidence here, too. Stata's implementation of scobit uses logit estimates as starting values for the scobit model. Using this methodology, Stata finds a credible maximum to the likelihood function. By default, Gauss assigns a vector of zeros as starting values if no other starting values are provided by the user, and scobit would not converge using these defaults. When we followed the lead of Stata to provide a vector of logit results, Gauss converged, albeit slowly. Gauss provides a number of different maximum-likelihood algorithms: the steepest descent method, BFGS (Broyden, Fletcher, Goldfarb, Shannon), DFP (Davidson, Fletcher, Powell), Newton–Raphson, BHHH (Berndt, Hall, Hall, and Hausman), and the Polak–Ribiere conjugate gradient method (Aptech 1999). In our analysis we used the default Gauss algorithm, BFGS. (See Chapters 4 and 8 for descriptions of these algorithms.) SHAZAM provides BFGS and DFP, but Nagler did not specify which was used for his analysis.

Different optimization routines may find different maxima. One important option, described previously, for many optimization routines is the choice between analytic gradients, often requiring painstaking coding by the user, or internally generated numerical approximations. Nagler's (1994) original implementation of scobit in SHAZAM uses a numerical approximation for the gradient and Hessian. Stata's implementation employs an analytic gradient and Hessian. For Gauss's maxlik package, we code an analytic gradient and use a numeric approximation for the Hessian.

The results of our replication analysis are presented in Table 3.3. When we simply plug the same program and data used in the original analysis into a later version of SHAZAM, we find a surprisingly statistically strong interactive relationship between registration closing date and education, stronger than reported in the original output. (No errors or warnings were issued by any of the packages used.) When we estimate the model using Stata and Gauss we find a much weaker interactive relationship than originally published. Most of the reported coefficients are the same, and the log likelihood at the reported solution is consistent across packages. The exception is "Closing Date." The estimated coefficient reported by Gauss is much different from the coefficients reported by the other programs, although the variable itself is not statistically significant.

Although the three programs generally appear to agree with the original Nagler analysis in locating the same optimum, we find disagreement among the statistical software packages in the estimate of the standard errors of the coefficients. The original SHAZAM standard errors are roughly a tenth smaller, and the newer version reports standard errors roughly a third smaller than those reported by Stata and Gauss. These discrepancies in the standard errors are not simply a choice of

Table 3.3   Comparison of Scobit Coefficient Estimates Using Four Packages

	Published	SHAZAM Verification	Stata Verification	Gauss Verification
Constant	−5.3290	−5.3290	−5.3289	−5.3486
	(0.3068)	(0.2753)	(0.3326)	(0.3340)
Education	0.3516	0.3516	0.3516	0.3591
	(0.1071)	(0.0788)	(0.1175)	(0.1175)
Education^2	0.0654	0.0654	0.0654	0.0647
	(0.0135)	(0.0097)	(0.0146)	(0.0145)
Age	0.1822	0.1822	0.1822	0.1824
	(0.0098)	(0.0102)	(0.0107)	(0.0108)
Age^2	−0.0013	−0.0013	−0.0013	−0.0013
	(0.0001)	(0.0001)	(0.0001)	(0.0001)
South	−0.2817	−0.2817	−0.2817	−0.2817
	(0.0239)	(0.0209)	(0.0313)	(0.0314)
Gov. election	0.0003	0.0003	0.0003	0.0004
	(0.0219)	(0.0178)	(0.0313)	(0.0314)
Closing date	−0.0012	−0.0012	−0.0012	−0.0006
	(0.0088)	(0.0065)	(0.0095)	(0.0095)
Closing date × education	−0.0055	−0.0055	−0.0055	−0.0058
	(0.0038)	(0.0028)	(0.0042)	(0.0042)
Closing date × education^2	0.0002	0.0002	0.0002	0.0003
	(0.0004)	(0.0003)	(0.0005)	(0.0005)
α	0.4150	0.4150	0.4151	0.4147
	(0.0324)	(0.0330)	(0.0341)	(0.0342)
Log likelihood	−55,283.16	−55,283.16	−55,283.16	−55,283.17
No. of observations	98,857	98,857	98,857	98,857

analytic or numeric to calculate the Hessian, which in turn is used to calculate standard errors, as our reproduction on SHAZAM and Gauss both use a numerically calculated Hessian. We note further that the numerical calculation of Gauss is close to the analytic calculation of Stata, leading us to speculate that numerical inaccuracies are present in SHAZAM's numerical approximation of the Hessian. However, lacking access to the internals of each optimization package, we cannot state this finding with certainty. These discrepancies in standard error estimates have a pattern here, but in replication analysis of other research, we discovered that in some cases SHAZAM's estimates of standard errors were larger, and sometimes they were smaller than Stata and Gauss, while those two software packages consistently produced the same estimates (Altman and McDonald 2003).

Although we experimented with a variety of computation options within SHAZAM, the results could not be made to match the originals or those of the other package. However, given that Stata and Gauss confirm Nagler's original findings with his original data, and that all three programs confirm the substantive results when using a corrected version of the source data, we have some confidence that the original results are valid.

This case study shows that substantive conclusions may be dependent on the software used to do an analysis. Ironically, in this case, if the original author had performed his analysis on the most recent version of the computing platform he chose, he would have failed to discover an interesting result. We cannot say how extensive this problem is in the social sciences because we cannot replicate all research. If benchmarking serves as a guide, we are reasonably confident that most simple analysis, using procedures such as descriptive statistics, regression, ANOVA, and noncomplicated nonlinear problems, is reliable, while more complex analysis may be require the use of computational algorithms that are unreliable.

We have chosen this particular case study because the complexity of its functional form may tax statistical software. As one researcher confided, a lesson may be to keep implementing a model on different software until one gets the answer one wants. Of course, the correct answer should be the one desired. Evidence of implementation dependence is an indicator of numeric inaccuracies and should be investigated fully by researchers who uncover them.

3.6 CHOOSING A STATISTICAL PACKAGE

Social scientists often overlook numerical inaccuracy, and as we have demonstrated, this may have serious consequences for inference. Fortunately, there are steps that researchers can follow to avoid problems with numerical accuracy. In much research, especially that involving simple statistical analysis, descriptive statistics, or linear regression, it may only be necessary to choose a package previously tested for accuracy and then to apply common sense and the standard diagnostics. In other research, especially that involving complex maximum likelihood estimation or simulation, one is more likely to encounter serious numerical inaccuracies.

A number of packages now publish the results of numerical benchmarks in their documentation, and benchmarks of accuracy are becoming a regular component of statistical software reviews. Be wary of software packages that have not been tested adequately for accuracy or that do not perform well. Choose regularly updated software that buttresses itself with standard tests of accuracy and reliability, but be aware that standard benchmarks are limited and that performance may change across versions.

Choose statistical software that provides well-documented algorithms, including any available options. The documentation should include details on where algorithms may be inaccurate, such as in the tails of distribution functions. Preference should be given to software that generates warning messages when problems are encountered, or fails to produce an estimate altogether, over software that does not provide warning messages. These messages, too, should be carefully documented.

When using a new model or a new statistical package, researchers can use methods discussed in this chapter to apply standard benchmarks to new software, or create their own. For those that wish to run their own tests or benchmarks, we suggest:

- Use the benchmark presented in this chapter to verify that data are properly input into a program.
- Use the NIST and SPRNG tests for pseudo-random number generators.
- Use the resources listed in this chapter for tests of the accuracy of distribution functions.
- Use the NIST StRD benchmarks for univariate statistics, regression, ANOVA, and nonlinear regression; keeping in mind that the nonlinear benchmarks may be used for maximum likelihood search algorithms but are not designed specifically for them.

We have found that statistical packages that allow calls to external libraries are friendlier to do-it-yourself benchmarks. Those that provide an open source are even more agreeable, as users can code and substitute their own algorithms, such as a pseudo-random number generator, and compare performance. [Of course, one must be alert to the limitations of borrowed code. As Wilkinson (1994) cautions, such code may not be designed for accuracy, and even published examples of high-precision algorithms may not contain complete error-handling logic.] In addition, alert users of open-source software are able to identify and diagnose bugs and other inaccuracies, too, and thus serve to increase the reliability of the software package.

Unfortunately, most statistical software packages do not make their code available for inspection, for the obvious reason that creating and maintaining a statistical software package is a costly enterprise and companies need to recover costs. Notable exceptions are Gauss, which exposes some, but not all, of its source code, and R, which is "open source." Other statistical software packages must be treated like a black box, and we can imagine that some consumers of statistical software are happy with this arrangement, because they are social, not computer, scientists. Still, the free market dictates that supply will meet demand. Paying attention to the concepts discussed in this chapter and book increases demand for numerically accurate statistical software, which will result in more reliable social science research.

CHAPTER 4

Robust Inference

4.1 INTRODUCTION

In the preceding chapters we have discussed the manner in which computational problems can pose threats to statistical inference, and how to choose software that avoids those threats to the greatest extent possible. For linear modeling and descriptive statistics, choosing a reliable statistic package (see Chapter 3) and paying attention to warning messages usually suffices. In this chapter we discuss some advanced techniques that can be used in hard cases, where one suspects that software is not providing accurate results. In chapters subsequent to this, we examine the best practices for accurate computing of specific types of problems, such as stochastic simulation (Chapter 5), common nonlinear model specifications (Chapter 8), misbehaved intermediate calculations (Chapter 6), spatial econometric applications (Chapter 9), and logistic regression (Chapter 10).

This chapter can also be thought of as addressing a variety of computational threats to inference, both algorithmic and numerical. Some of these threats are illustrated in Figure 4.1, which shows an artificial likelihood function and some of its characteristics, such as flat regions, multiple optima, discontinuities, and non-quadratic behavior that can cause standard computational strategies to yield incorrect inference. In the following sections we discuss how to deal with these problems.

4.2 SOME CLARIFICATION OF TERMINOLOGY

Robustness evaluation is the systematic process of determining the degree to which final model inferences are affected by both potential model misspecification by the researcher through assumptions and the effect of influential data points. *Global robustness* evaluation performs this systematic analysis in a formal manner to determine the range of the inferences that can be observed given the observations. Conversely, *local robustness* uses differential calculus to determine the volatility of specific reported results. It is important to remember that undesirable instability can be the result of data characteristics (Guttman et al. 1978) or model specification (Raftery 1995).

Numerical Issues in Statistical Computing for the Social Scientist, by Micah Altman, Jeff Gill, and Michael P. McDonald
ISBN 0-471-23633-0   Copyright © 2004 John Wiley & Sons, Inc.

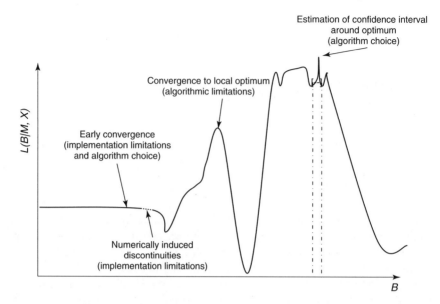

Fig. 4.1 Some computational threats to inference in finding maximum likelihood estimates.

Classical robustness analysis centers on the linear model's sensitivity to potentially influential outliers, with the idea that one should mitigate this effect.
Researchers sometimes actually incorporate robustness into model specifications
as part of the initial setup (Lange et al. 1989; Marín 2000). The literature is vast,
including Andrews et al. (1972), Huber (1972, 1973, 1981), Andrews (1974),
Hampel (1974), Hampel et al. (1986), Barnett and Lewis (1978), Belsley et al.
(1980), Cook and Weisberg (1982), Emerson and Hoaglin (1983), Rousseeuw
and Leroy (1987), and Hamilton (1992).

Interestingly, there is a common misuse of the word '*robust*' that is generally
tolerated. The appropriate term for analyzing outlier effects is actually "resis-
tance," in the sense of a model's ability to *resist* (or not) the influence that these
values exert. In contrast, robustness is concerned directly with the model's ability to absorb violations of the underlying assumptions. [See the classic text by
Hoaglin et al. (1983) for additional contrasts.] For instance, it is well known that
the linear regression model is not very *resistant* to outliers in the x-axis direction
but is *robust* to mild levels of heteroscedasticity. We are, however, sometimes
not terribly concerned with the distinction inasmuch as robustness occurs increasingly with rising sample size. However, in other instances it is important to find
and scrutinize outliers, and many authors provide detailed guides, including Kit-
igawa and Akaike (1982), Polasek (1984, 1987), Pettit and Smith (1985), and
Liseo, et al. (1996).

The most thorough treatment of robustness is done in the Bayesian literature because these authors often have to go to great lengths to justify their
prior assumptions (Berger 1984, 1990; Kass et al. 1989; Kadane and Srinivasan

1996). As just mentioned, a common distinction for Bayesians is between *global robustness*, which systematically changes the prior specification to see the effect on the posterior, and *local robustness*, which looks at the rate of change of posterior inferences for infinitesimal perturbations of the prior (Sivaganesan 2000). The global approach is discussed in general terms by Berger (1985), Wasserman (1992), O'Hagan (1994), and Gill (2002), for specific classes of priors by West (1984) and Moreno (2000), and in detail with the principal method of using *contamination priors* by a small number of authors: Berger and Berliner (1986), Moreno and González (1990), Moreno and Cano (1991), Moreno and Pericchi (1991, 1993), Moreno et al. (1996). Details about local robustness are given in Cuevas and Sanz (1988), Gelfand and Dey (1991), Ruggeri and Wasserman (1993), Sivaganesan (1993), Bose (1994a,b), Delampady and Dey (1994), Gustafson and Wasserman (1995), Berger and O'Hagan (1988), Dey and Micheas (2000), and Gustafson (2000).

Sensitivity analysis is the process of changing various model assumptions and parameters in some ad hoc but intuitive way to observe subsequent implications on inferences. If reasonably large changes in the model assumptions have a negligible effect on calculated quantities, this is evidence of low sensitivity. On the other hand, mild changes in model assumptions and parameters that produce dramatic changes in inferences are certainly a cause for alarm. This process is differentiated from robustness analysis because it is necessarily smaller in scope and usually limited to a few areas of concern. Sensitivity is also a central concern among Bayesians, for similar reasons (Leamer 1978, 1984, 1985; Lavine 1991a,b; Zellner and Moulton 1985; Weiss 1996).

In Section 4.3 we discuss how to test a given model analysis for possible sensitivity to computational details. This is followed in Section 4.4 by an overview of techniques for performing more accurate analyses. In Section 4.5 we discuss methods of inference that can be used in the face of obvious, and perhaps unavoidable, computational problems.

4.3 SENSITIVITY TESTS

Recall from Chapter 2 that sensitivity analyses are valuable because of their wide applicability and ease of use. Sensitivity analysis can be performed on the data and model being analyzed without extensive formal analysis or custom programming. Sensitivity analysis can draw attention to potential problems in algorithm, implementation, or model. There are several common approaches.

4.3.1 Sensitivity to Alternative Implementations and Algorithms

From the theoretical standpoint typically presented in textbooks, statistical estimates are determined entirely by model and data and are not dependent on algorithmic or implementation programming. This assumption is often true in practice, but is not guaranteed to hold. Gill et al. (1981; p. 319) recommend a simple heuristic test to determine if model and data are sensitive to programming issues by varying the conditions under which an estimate is generated: by varying the program used, the operating system and hardware environment, algorithm options, and even substituting new programming code, where possible. The production of different estimates serve as an indication of algorithm or implementation dependence.

The variety of existing statistical software packages indicates the demand for statistical computing solutions. General-purpose programs such as `Excel`, `Gauss`, `S-Plus`, `R`, `SAS`, `SHAZAM`, `SPSS`, and `Stata` provide users with a common set of basic statistical tools. Some offer more tools than others, and specialized programs such as `BUGS`, `E-Views`, `Limdep`, and `Rats` exist for specific applications. Moreover, multiple implementations of the same statistical program are often available on different computing platforms, including MacOS, Windows, Unix, and Linux. Different programs will often have options that allow users control over some aspects of the implementation of an algorithm, and often these programs do not commonly share the same scope of options or default options. Some programs, such as `R`, are open source and allow users the greatest option control by allowing users to substitute their own programming code where they see fit. Others, such as `Gauss`, `S-Plus`, and `Stata`, expose the source code used for some methods but provide the core of the program in opaque binary format only.

4.3.1.1 Varying Software Packages

Most general-purpose statistical software packages include heavily used models such as regression, and nonlinear models, including probit and logit. As the demand for more sophisticated models has increased, statistical software companies have keep pace with the state of the discipline by implementing new models as they are published. For open-source programs, new models may also be available in software archives produced by their authors. Most statistical packages that implement maximum likelihood solvers also allow users to program custom nonlinear models. With such a wide range of models, it is often not difficult for users with proficiency in more than one statistical software package to estimate a model with their data on multiple statistical packages.

The variety of statistical software packages provides the opportunity to conduct the crudest sensitivity test: inputting data and model into different statistical programs and observing the results. Differing results are an indication of some form of programming issues that affect the estimation.

The software sensitivity test is "crude" in that it treats statistical software as a black box and does not identify the cause of the discrepancies. As we discuss in Chapters 2 and 3, differences between the results produced by different packages may be attributable to differences in the underlying algorithms, differences in the implementation of algorithms, formula parsing, and/or data input, ill-conditioning of the data with respect to the model, or a combination of any of these factors.

Sophisticated programs allow users more control over the algorithm used in estimation and features of its implementation. When the implementation and algorithm are sufficiently well documented and under the user's control, the user can contrast the results of estimation using different algorithms, or using the same algorithm on different platforms. On the other hand, when such details are hidden, the user may still witness that the estimation produces different results when run in different packages but may not be able to determine why this is so.

Sensitivity to changing software packages may be caused by software bugs, but they may also reveal data sensitivity to the numeric computation issues discussed in Chapter 2. The computation issues involved may depend on the type of model being estimated and the algorithms used to estimate it.

For PRNGs, algorithm sensitivity tests may reveal inadequate randomness or period of a PRNG. When random numbers are generated, a small amount of disagreement in results is natural, especially in simulations where simulation variance alone will inevitably cause differences. If results disagree substantively across alternatives, however, one should investigate until one clearly understands which set of results should be discarded.

For nonlinear optimizers, changing default options may identify multiple local optima in the function to be solved and is generally a good idea when such multiple optima are suspected to exist.

Sometimes the same program may be implemented on two or more computing platforms, such as Linux, Windows, and Unix. This provides the opportunity to examine sensitivity to underlying operating system architecture. Assuming that the statistical package, in fact, uses the same code on each platform, operating system sensitivity may be a consequence of the way in which different computing platforms subtly handle underlying computations. In rare circumstances it may even be a consequence of a bug, although more often such differences are due to subtle differences between implementations of floating point operations or numerical library functions.

What does a sensitivity test reveal? When results are confirmed, a researcher may have greater confidence that the results are not sensitive to the issues discussed here. However, there is always the remote possibility that two results generated through one of these tests are both incorrect. When results differ, the researcher is faced with a conundrum: Which answer is the "truth"? In this case, the researcher should ask: Is one implementation known to be more accurate than another? Is one algorithm better suited for the type of problem being analyzed? What diagnostics are available?

If the researcher is not sure of the answer to these questions, we suggest reading on, as in the next section we discuss another sort of sensitivity test that may shed light on the numerical accuracy of an algorithm. In subsequent sections we discuss methods for directly improving the accuracy of your results and how to deal with threats to inference caused by numerical problems.

4.3.2 Perturbation Tests

Perturbation tests replicate the analysis of interest while changing the input data to a small extent. The purpose is to observe the subsequent sensitivity of the estimates to such perturbations. Noticeably large changes in the resulting inferences are a sign that the original model depends greatly on the unique structure of this particular dataset. Note that sensitivity, as defined here, is not a measure of true computational stability, because values for the correct estimates are unknown.

In addition, these tests can be used in combination with alternative computing implementations such as different packages or different hardware platforms. This approach has the potential to separate out the portions of pseudoinstability that are due to both model and computation. By running multiple implementations and algorithms on the same sets of perturbed data, the differences in software and hardware implementation plus algorithms are highlighted since the model and underlying data are fixed.

The magnitude and distributional form of the perturbations is not always important: Even the basic approach of using uniform disturbances, "noise," can identify model and data problems. However, it is important to gauge the perturbations relative to known or suspected levels of measurement error that exist in the data at hand.

4.3.2.1 Perturbation Strategy

An exploratory test for sensitivity of a given problem to measurement and estimation error is to introduce small random perturbations to these data, on the order of the measurement error of the instruments used to collect it, and recalculate the estimate. This technique is analogous to bootstrapping. However, in bootstrapping the sample selection is randomly perturbed, but individual observations are not, whereas in our strategy the sample selection is not perturbed but the individual observations are.

Perturbation tests were first used for assessing numerical accuracy by Beaton et al. (1976), who develop a stability index based on it. Gill et al. (1981, particularly Sec. 8.3.3) recommend a similar method, although it is informally described and suggested as a pragmatic method for gauging whether a program was stable. Also, whereas Beaton et al. perturb only the explanatory variables in the model, Gill et al. do not distinguish among the inputs to the computation.

To see how these perturbations affect the estimation process, consider two likelihood functions: a standard form based on the observed data $\ell(\theta, \mathbf{x})$, and an identical specification but with perturbed data $\ell_{\mathbf{p}}(\theta, \mathbf{x_p})$. Here $\mathbf{p}$ denotes an individual perturbation scheme: $\mathbf{p} = [p_1, p_2, \ldots, p_n] \in \mathfrak{R}^n$ applied to the data: $\mathbf{x} = [x_1, x_2, \ldots, x_n] \in \mathfrak{R}^n$. Thus we can show that comparing the two likelihood functions is analogous to comparing an unweighted likelihood function $\ell(\theta, \mathbf{x}) = \sum_i \ell_i(\theta, \mathbf{x}_i)$ to a weighted version $\ell_{\mathbf{p}}(\theta, \mathbf{x_p}) = \sum_i p_i \ell_i(\theta, \mathbf{x}_i)$. Or we could define the unperturbed likelihood function to be one in which there are null perturbations or weights: $\ell_{\mathbf{p}_0}(\theta, \mathbf{x_{p_0}}) = \sum_i p_{0i} \ell_i(\theta, \mathbf{x}_i)$, where $\mathbf{p}_0$ is simply a vector of 1's. This setup gives us two maximum likelihood vectors to compare: $\hat{\theta}$ and $\hat{\theta}_{\mathbf{p}}$.

In this context, our approach is to evaluate the range of $\hat{\theta}$ produced by multiple samples of $\mathbf{x_p}$ generated by random production of $\mathbf{p}$ disturbances across different datasets, **x**. The idea builds on the mechanical approach of Cook (1986), who looks for maximizing and minimizing perturbances, and roughly follows a simpler test of logistic regression given by Pregibon (1981). Lawrence (1988) applies Cook's method to develop diagnostics for linear regression.

In addition, although this evaluation methodology does not require that the likelihood function be statistically well-behaved, it does have a natural interpretation for well-behaved maximum likelihood estimations. If the likelihood function for an MLE is well behaved, there is a simple mapping between perturbations of data and perturbations of the model. For example, small normally distributed noise added to the data should induce a corresponding small mean shift in the likelihood curve (St. Laurent and Cook 1993).

Cook (1986) continues on to define the *likelihood displacement*:

$$\text{LD}_\mathbf{p} = -2 \left[\ell_\mathbf{p}(\theta, \mathbf{x}_\mathbf{p}) - \ell(\theta, \mathbf{x}) \right], \tag{4.1}$$

which measures the statistical influence that different perturbation schemes have on the estimation process. Not surprisingly, it can be shown that $\text{LD}_\mathbf{p}$ defines a large-sample confidence region distributed χ_k^2, where k is the number of parameters specified (Cox and Hinkley 1974; Cook and Weisberg 1982).

If the likelihood surface is steeply curved (in the multidimensional sense) at a MLE solution, then clearly we will see large values of $\text{LD}_\mathbf{p}$ for even small perturbation schemes. Contrary to intuition, a sharply spiked likelihood function in this way across *every* dimension is not a sign of a reliable result, it indicates a fragile finding that is heavily dependent on the exact form of the data observed. This is because the change in the likelihood function is not due to different values for the estimate (where sharp curvature is desired), it is due to changes in the data (perturbations) where sharp changes indicate serious sensitivity of the likelihood function to slightly different data: a model that is "nonresistant."

Cook also ties this curvature definition back to the idea of statistical reliability by a geometric interpretation. Define $\mathbf{p}_a = \mathbf{p}_0 + a\mathbf{v}$, where $a \in \Re$ and $\mathbf{v}$ is a unit-length vector. The interpretation of $\mathbf{p}_a$ is as a line that passes through $\mathbf{p}_0$ in the direction of the vector $\mathbf{v}$, where a gives the n-dimensional placement relative to $\mathbf{p}_0$. The geometric curvature of $\text{LD}_{\mathbf{p}_a}$ in the direction of the vector $\mathbf{v}$ starting at $\mathbf{p}_0$ is given by

$$C_\mathbf{p} = 2|(\Delta\mathbf{v})' H^{-1} (\Delta\mathbf{v})|, \tag{4.2}$$

where Δ is the $k \times n$ matrix given by $\Delta_{ij} = \partial^2 \ell_{\mathbf{p}_0}(\hat{\theta}, \mathbf{x}_\mathbf{p}) / \partial \hat{\theta}_i \, p_{0j}$ (i.e., evaluated at the MLE and $\mathbf{p}_0$) and H is the standard Hessian matrix. Cook (1986, p. 139) suggests calculating the maximum possible curvature by obtaining the $\mathbf{v}$ vector that maximizes (4.2): $C_{\max} = \text{Max}_\mathbf{v} C_\mathbf{v}$, which gives the greatest possible change in the likelihood displacement. Other strategies include perturbing in one direction at a time as a means of understanding which dimensions (i.e., parameters) are more sensitive than others.

Recent work by Parker et al. (2000) formalizes a variant of this idea, which they call *Monte Carlo arithmetic*. Essentially, they replicate an analysis while randomly perturbing (in the form of random rounding) all calculations. This approach is more widely applicable than formal analysis (which can be practically impossible to apply to complex problems). Even where formal analysis is possible, Parker et al. show that MCA can also yield tighter practical bounds. They argue that this is a very successful "idiot light" for numerical inaccuracy.

4.3.2.2 Perturbations and Measurement Error

Perturbations can be considered in a manner similar to measurement error. The effects of measurement error on statistical models have been known for quite some time in a number of contexts (Wald 1940; Nair and Banerjee 1942; Reiersol 1950; Durbin 1954; Madansky 1959) and are well-studied today (Fedorov 1974; Schneeweiss 1976; Fuller 1987, 1990; Stefanski 1989; Wong 1989; Gleser 1992; Cragg 1994; Carrol et al. 1995, 1999; Jaccard and Wan 1995; Buonaccorsi 1996; Schmidt and Hunter 1996; Sutradhar and Rao 1996; Zhao and Lee 1996; Gelfand et al. 1997; Lewbel 1997; Mukhopadhyay 1997; Wang and Wang 1997; DeShon 1998; Li and Vuong 1998; Marais and Wecker 1998; Skinner 1998; Cheng and Van Ness 1999; Iturria et al. 1999). Essentially there are two problems: zero and nonzero mean measurement error. The nonzero case leads obviously and immediately to biased coefficients in the opposite direction of the data bias. That is, in a linear model, multiplying some nontrivial $\delta > 1$ to every case of an explanatory variable, $\mathbf{X}$, implies that larger increases in this variable are required to provide the same effect on the outcome variable, thus reducing the magnitude of the coefficient estimate. Another way of thinking about this is that a 1-unit change in $\mathbf{X}$ now has a smaller expected change in $\mathbf{Y}$. This effect is also true for GLMs where there is the additional complexity of factoring in the implications of the link function (Carroll et al. 1995).

Zero mean measurement error is the more important and more common situation. Furthermore, the effects found for zero mean measurement error also apply to nonzero mean measurement error in addition to the effects discussed above. Here we shall make our points primarily with the linear model, even though many solutions are certainly not in this group, purely to aid in the exposition.

Suppose that the true underlying (bivariate for now, $k = 2$) linear model meeting the standard Gauss–Markov assumptions is given by

$$\underset{(n \times 1)}{\mathbf{Y}} = \underset{(n \times k)(k \times 1)}{\mathbf{X} \ \boldsymbol{\beta}}, \tag{4.3}$$

but $\mathbf{X}$ and $\mathbf{Y}$ are not directly observable where we instead get $\mathbf{X}^*$ and $\mathbf{Y}^*$ according to

$$\mathbf{Y}^* = \mathbf{Y} + \xi \qquad \text{and} \qquad \mathbf{X}^* = \mathbf{X} + v, \tag{4.4}$$

where $\xi \sim \mathcal{N}(0, \sigma_\xi^2)$ and $v \sim \mathcal{N}(0, \sigma_v^2)$, and σ_ξ^2 and σ_v^2 are considered "small." It is typically assumed that ξ and v are independent both of each other and of $\mathbf{X}$ and $\mathbf{Y}$. These are standard assumptions that can also be generalized if necessary.

Let's look first at the ramifications of substituting in the error model for **Y**:

$$(\mathbf{Y}^* - \xi) = \mathbf{X}\boldsymbol{\beta} + \boldsymbol{\epsilon}$$

$$\mathbf{Y}^* = \mathbf{X}\boldsymbol{\beta} + (\boldsymbol{\epsilon} + \xi). \tag{4.5}$$

Since ξ has zero mean and is normally distributed by assumption, the effect of measurement error is to attenuate the overall model errors but not to violate any assumptions. That is, measurement error in **Y** falls simply to the regression residuals. Therefore, there is now simply a composite, zero mean error term: $\boldsymbol{\epsilon}^* = \boldsymbol{\epsilon} + \xi$. Unfortunately, the story is not quite so pleasant for measurement error in **X**. Doing the same substitution now gives

$$\mathbf{Y} = (\mathbf{X}^* - \nu)\boldsymbol{\beta} + \boldsymbol{\epsilon}$$

$$= \mathbf{X}^*\boldsymbol{\beta} + (\boldsymbol{\epsilon} - \boldsymbol{\beta}\nu). \tag{4.6}$$

Although this seems benign since ν is zero mean, it does in fact lead to a correlation between regressor and disturbance, violating one of the Gauss–Markov assumptions. This is shown by

$$\text{Cov}[\mathbf{X}^*, (\boldsymbol{\epsilon} - \boldsymbol{\beta}\nu)] = \text{Cov}[(\mathbf{X} + \nu), (\boldsymbol{\epsilon} - \boldsymbol{\beta}\nu)]$$

$$= E[(\mathbf{X} + \nu)(\boldsymbol{\epsilon} - \boldsymbol{\beta}\nu)]$$

$$= E[\mathbf{X}\boldsymbol{\epsilon} - \mathbf{X}\boldsymbol{\beta}\nu + \nu\boldsymbol{\epsilon} - \boldsymbol{\beta}\nu^2]$$

$$= -\boldsymbol{\beta}\sigma_\nu^2. \tag{4.7}$$

To directly see that this covariance leads to biased regression coefficients, let us now insert the measurement error model into the least squares calculation:

$$\hat{\beta}_1 = \frac{\text{Cov}[(\mathbf{X} + \nu), \mathbf{Y}]}{\text{Var}[\mathbf{X} + \nu]} = \frac{E[\mathbf{X}\mathbf{Y} + \nu\mathbf{Y}] - E[\mathbf{X} + \nu]E[\mathbf{Y}]}{E[(\mathbf{X} + \nu)^2] - (E[\mathbf{X} + \nu])^2}$$

$$= \frac{E[\mathbf{X}\mathbf{Y}] - E[\mathbf{X}]E[\mathbf{Y}] + E[\nu\mathbf{Y}] - E[\nu]E[\mathbf{Y}]}{E[\mathbf{X}^2] + 2E[\mathbf{X}\nu] + \sigma_\nu^2 - (E[\mathbf{X}])^2 - 2E[\mathbf{X}]E[\nu] - (E[\nu])^2}$$

$$= \frac{\text{Cov}(\mathbf{X}, \mathbf{Y}) + E[\nu\mathbf{Y}]}{\text{Var}(\mathbf{X}) + 2\,\text{Cov}(\mathbf{X}, \nu) + \sigma_\nu^2}, \tag{4.8}$$

where we drop out terms multiplying $E[\nu]$ since it is equal to zero. Recall that ν is assumed to be independent of both **X** and **Y**, and note that things would be much worse otherwise. So the $2\,\text{Cov}(\mathbf{X}, \nu)$ term in the denominator and the $E[\nu\mathbf{Y}]$ term in the numerator are both zero. Thus σ_ν^2 is the only effect of the measurement error on the slope estimate. Furthermore, since this is a variance and therefore nonnegative, its place in the denominator means that coefficient bias will always be downward toward zero, thus tending to diminish the relationship observed. In extreme cases this effect pulls the slope to zero, and one naturally implies no relationship.

Of course, the discussion so far has considered only the rather unrealistic case of a single explanatory variable (plus the constant). Regretfully, the effects of measurement error worsen with multiple potential explainers. Consider now a regression of $\mathbf{Y}$ on $\mathbf{X}_1$ and $\mathbf{X}_2$. The true regression model expressed as in (4.3) is

$$\mathbf{Y} = \beta_0 + \mathbf{X}_1\beta_1 + \mathbf{X}_2\beta_2 + \epsilon, \tag{4.9}$$

and assume that only $\mathbf{X}_1$ is not directly observable and we get instead $\mathbf{X}^* = \mathbf{X} + v$ with the following important assumptions:

$$
\begin{array}{cc}
v \sim \mathcal{N}(0, \sigma_v^2) & \mathrm{Cov}(\mathbf{X}_1, v) = 0 \\
\mathrm{Cov}(v, \epsilon) = 0 & \mathrm{Cov}(v, \mathbf{X}_2) = 0.
\end{array} \tag{4.10}
$$

Now as before, substitute in the measurement error component for the true component in (4.9):

$$
\begin{aligned}
\mathbf{Y} &= \beta_0 + (\mathbf{X}_1 - v)\beta_1 + \mathbf{X}_2\beta_2 + \epsilon \\
&= \beta_0 + \mathbf{X}_1\beta_1 - \beta_1 v + \mathbf{X}_2\beta_2 + \epsilon \\
&= \beta_0 + \mathbf{X}_1\beta_1 + \mathbf{X}_2\beta_2 + (\epsilon - \beta_1 v).
\end{aligned} \tag{4.11}
$$

It should be intuitively obvious that problems will emerge here since β_1 affects the composite error term, leading to correlation between regressor and disturbance as well as heteroscedasticity.

To show more rigorously that this simple measurement error violates the Gauss–Markov assumptions, we look at the derivation of the least squares coefficient estimates. For the running example with two explanatory variables, the normal equations are

$$
\begin{bmatrix}
n & \sum X_{i1} & \sum X_{i2} \\
\sum X_{i1} & \sum X_{i1}^2 & \sum X_{i1}X_{i2} \\
\sum X_{i2} & \sum X_{i1}X_{i2} & \sum X_{i2}^2
\end{bmatrix}
\begin{bmatrix}
\hat{\beta}_0 \\
\hat{\beta}_1 \\
\hat{\beta}_2
\end{bmatrix}
=
\begin{bmatrix}
\sum Y_i \\
\sum X_{i1}Y_i \\
\sum X_{i2}Y_i
\end{bmatrix}. \tag{4.12}
$$

So the 3×3 inverse matrix of $\mathbf{X}'\mathbf{X}$ is

$$
\det(\mathbf{X}'\mathbf{X})^{-1}
\begin{bmatrix}
\sum X_{i1}^2 \sum X_{i2}^2 - \sum X_{i1}X_{i2} \sum X_{i1}X_{i2} \\
\sum X_{i1}X_{i2} \sum X_{i2} - \sum X_{i1} \sum X_{i2}^2 \\
\sum X_{i1} \sum X_{i1}X_{i2} - \sum X_{i1}^2 \sum X_{i2},
\end{bmatrix}
$$

$$\sum X_{i2} \sum X_{i1} X_{i2} - \sum X_{i1} \sum X_{i2}^2$$

$$n \sum X_{i2}^2 - \sum X_{i2} \sum X_{i2}$$

$$\sum X_{i1} \sum X_{i2} - n \sum X_{i1} X_{i2}$$

$$\sum X_{i1} \sum X_{i1} X_{i2} - \sum X_{i2} \sum X_{i1}^2$$

$$\left. \sum X_{i2} \sum X_{i1} - n \sum X_{i1} X_{i2} \right]$$

$$n \sum X_{i1}^2 - \sum X_{i1} \sum X_{i1} \right],$$

where

$$\det(\mathbf{X}'\mathbf{X}) = n \left(\sum X_{i1}^2 \sum X_{i2}^2 - \sum X_{i1} X_{i2} \sum X_{i1} X_{i2} \right)$$

$$+ \sum X_{i1} \left(\sum X_{i1} X_{i2} \sum X_{i2} - \sum X_{i1} \sum X_{i2}^2 \right)$$

$$+ \sum X_{i2} \left(\sum X_{i1} \sum X_{i1} X_{i2} - \sum X_{i1}^2 \sum X_{i2} \right).$$

Using $\hat{\beta} = (\mathbf{X}'\mathbf{X})^{-1}\mathbf{X}'\mathbf{Y}$, and paying attention only to the last coefficient estimate, we replace X_{i1} with $X_{i1} + v$:

$$\hat{\beta}_2 = \det(\mathbf{X}'\mathbf{X})^{-1} \left[\left(\sum (X_{i1} + v) \sum (X_{i1} + v) X_{i2} - \sum (X_{i1} + v)^2 \sum X_{i2} \right) \right.$$

$$\times \left(\sum (X_{i1} + v) Y_i \right) + \left(\sum (X_{i1} + v) \sum X_{i2} \right.$$

$$- n \sum (X_{i1} + v) X_{i2} \right) \left(\sum X_{i2} Y_i \right) + \left(n \sum (X_{i1} + v)^2 \right.$$

$$\left. - \sum (X_{i1} + v) \sum (X_{i1} + v) \right) \left(\sum X_{i2}^2 \right) \right], \tag{4.13}$$

where the same adjustment with v would have to be made to every X_{i1} term in the determinant above as well. The point here is to demonstrate that the measurement error in $\mathbf{X}_1$ can have a profound effect on the other explanatory variables, the extent of which is now obvious only when looking at the full scalar calculation of the coefficient estimate $\hat{\beta}_2$, even in just a linear model example.

An easier but perhaps less illustrative way to show this dependency is with the handy formula

$$\hat{\beta}_2 = \frac{\beta_{-1} - \beta_{-2}\beta_{21}}{1 - r_{21}^2}, \tag{4.14}$$

where: β_{-1} is this same regression leaving out (jackknifing) $\mathbf{X}_1$, β_{-2} is the regression leaving out $\mathbf{X}_2$, β_{21} is a coefficient, and r_{21}^2 is the R^2 measure obtained by regressing $\mathbf{X}_2$ on $\mathbf{X}_1$. What this shows is that the extent to which the variable with measurement error "pollutes" the coefficient estimate of interest is governed by the linear association between these explanatory variables, and the only time that measurement error in $\mathbf{X}_1$ will not have an effect is when there is no bivariate linear relationship between $\mathbf{X}_2$ and $\mathbf{Y}$ or no bivariate linear relationship between $\mathbf{X}_1$ and $\mathbf{X}_2$.

Although we have made these points in a linear model context, corresponding effects of measurement are also present in the generalized linear model and are more complicated to analyze, due to the link function between the systematic component and the outcome variable. In generalized linear models, the additive, systematic component is related to the mean of the outcome variable by a smooth, invertible function, $g(\cdot)$, according to

$$\boldsymbol{\mu} = \mathbf{X}\boldsymbol{\beta} \quad \text{where} \quad E(\mathbf{Y}) = g^{-1}(\boldsymbol{\mu}). \tag{4.15}$$

This is a very flexible arrangement that allows the modeling of nonnormal, bounded, and noncontinuous outcome variables in a manner that generalizes the standard linear model (McCullagh and Nelder 1989; Gill 2000; Fahrmeir and Tutz 2001). Using the link function we can change (4.11) to the more general form

$$E(Y_i) = g^{-1}\left(\beta_0 + \beta_1(\mathbf{X}_{i1} - v) + \beta_2\mathbf{X}_{i2} + \epsilon_i\right)$$
$$= g^{-1}\left(\beta_0 + \mathbf{X}_1\beta_1 + \mathbf{X}_2\beta_2 + (\epsilon - \beta_1 v)\right), \tag{4.16}$$

which reduces to the linear model if $g(\cdot)$ is the identity function. Common forms of the link function for different assumed distributions of the outcome variable are $g(\mu) = \log(\mu)$ for Poisson treatment of counts, $g(\mu) = -1/\mu$ for modeling truncation at zero with the gamma, and logit $(\log[\mu/1 - \mu])$, probit $[\Phi^{-1}(\mu)]$, or cloglog $[\log(-\log(1 - \mu))]$ for dichotomous forms.

By including a link function, we automatically specify interactions between terms on the right-hand side, including the term with measurement error. To see that this is true, calculate the marginal effect of a single coefficient by taking the derivative of (4.16) with regard to some variable of interest. If the form of the model implied no interactions, we would obtain the marginal effect free of other variables, but this is not so:

$$\frac{\partial Y_i}{\partial \mathbf{X}_{i2}} = \frac{\partial}{\partial \mathbf{X}_{i2}} g^{-1}\left(\beta_0 + \mathbf{X}_1\beta_1 + \mathbf{X}_2\beta_2 + (\epsilon - \beta_1 v)\right)$$
$$= (g^{-1})'\left(\beta_0 + \beta_1\mathbf{X}_{i1} + \beta_2\mathbf{X}_{i2} + (\epsilon - \beta_1 v)\right)\beta_2. \tag{4.17}$$

The last line demonstrates that the presence of a link function requires the use of the chain rule and therefore retains other terms on the right-hand side in addition to β_2: With the generalized linear model we always get partial effects for a given variable that are dependent on the levels of the other explanatory variables. Furthermore, here we also get partial effects for $\mathbf{X}_2$ that also depend on the measurement error in $\mathbf{X}_1$.

In this section we have analyzed measurement error effects in detail and demonstrated that outcome variable measurement error is benign, and explanatory variable measurement error is dangerous. Perturbations are essentially researcher-imposed measurement error. Having discussed the deleterious modeling problems with measurement error, we advise against intentionally including data perturbations in published estimates, although some researchers have explicitly modeled known data measurement error using an approach similar to perturbation analysis (Beaton et al. 1976). Instead, the perturbations act like unintended, but completely known measurement error that provide a means of testing the behavior of estimators and algorithms. That is, models that react dramatically to modest levels of measurement error are models that one should certainly be cautious about (particularly in the social sciences).

In summary, perturbation may introduce bias, but if the problem is well-conditioned and the algorithm and implementation accurate, the bias should be small. Moreover, any bias introduced by perturbations should be the same when the same model and perturbed data are used in different implementations. So if two implementations of the same model show marked differences in pseudostability with respect to similar perturbation analyses, the root cause is asserted to be computational and not statistical.[1]

4.3.2.3 *Some Ramifications of Perturbed Models*

Using the core idea of random perturbation, we can assess whether results are reliable, whether they are consistent with respect to small perturbations of the data. This methodology complements standard diagnostic plots in two ways. First, one can use the strictly numerical results as an unambiguous check: Simply evaluate whether the range of results across input perturbations still fits the original substantive conclusions about the results. Second, this methodology may sometimes reveal numerical problems that may be missed in standard diagnostic plots.

With regard to input perturbation, what is considered "small" for any particular case is a matter of subjective judgment. There is an obvious lower limit: Perturbations of the data that are at the level below the precision of the machine should not be expected to cause meaningful changes in output. The upper limit on perturbations is less clear, but should be bounded by the accuracy of data measurement.

In many of the social sciences, measurement error certainly dominates machine precision as a source of input inaccuracy. Many of the significant digits of macro data are reported as rounded, for example, to thousands. Introducing perturbations to the rounding error of these data is a tractable problem to solve.

Sometimes data are bounded, which introduces complications to perturbations. The simplest way of avoiding the bounding problem is to truncate any illegal value generated by perturbations to the constraint, but this introduces mass at the boundary points.

---

[1]This approach is complementary to the one proposed by Judge et al. (2002). Their approach uses instrumental variables, where available, to reduce the effects of measurement error. Our approach provides a diagnostic of the results, sensitivity to it.

Choosing the *number* of perturbed datasets to generate is also something of an art. The extant literature does not specify a particular number of samples that is guaranteed to be sufficient for all cases. Parker (1997) and Parker et al. (2000) use as many as 100 and as few as four samples in their Monte Carlo arithmetic analysis. Parker (1997) also shows that in all but pathological cases, the distribution of the means of coefficients calculated under random rounding are normal, which suggests that 30 to 50 samples should be adequate. Moreover, since the perturbation technique can be replicated indefinitely, one can simply rerun the analysis, increasing the number of samples, until the variance across replications is acceptable for the substantive problem at hand.

4.3.3 Tests of Global Optimality

Another sensitivity "test" involves noting the responsiveness of a search algorithm to starting values. Researchers who analyze complex functions are probably already aware that a poor choice of starting values may lead the search algorithm horribly astray: for example, leading the search algorithm onto a discontinuity with no maximum. Here, we briefly explain the problem that multiple optima poses to a search algorithm, discuss some strategies for selecting starting values that might reveal a global optimum, and explore some tests that have been proposed to identify if a global optima has been found.

4.3.3.1 Problem—Identifying the Global Optimum

Once an algorithm reaches convergence, how certain can a researcher be that the optimum reached is truly the global and not merely a local optimum? If the maximization function is globally concave, the local optimum and the global optimum are the same (Goldfeld and Quandt 1972; Gill et al. 1981; Nocedal and Wright 1999). Unfortunately, if the function is known to have several local optima, there is no known way of guaranteeing that a global optimum will be found in a finite amount of time (Veall 1990, p. 1460). Commonly used search algorithms can get stuck at a local optimum, thereby missing the true global maximum. Searching the entire parameter space for global optimum is often too computationally intensive, especially when there are a large number of right-hand-side variables in the equation to be estimated.

Multiple optima may exist more frequently than most political scientists realize. Consider this example from Goldfeld and Quandt (1972, pp. 21–22):

$$Y_i = a + a^2 x_i + u_i \quad \text{where} \quad u_i \text{ is iid } N\left(0, \sigma^2\right). \tag{4.18}$$

This simple nonlinear equation has three local optima. Fortunately, such heavily used models as the binomial and ordered probit and logit (but not multinomial probit) have been shown to have globally concave log-likelihood functions (Pratt 1981). In general, however, log-likelihood functions are not guaranteed to be concave and can have many unbounded local optima (e.g., Barnett 1973).[2]

---

[2]Even those functions identified by Pratt are not guaranteed to be concave when nonlinear terms such as interaction terms are included in the linear form of the equation.

In addition, if the log-likelihood function is not concave, the log likelihood may not be a consistent estimator, even at the global maximum (Haberman 1989).

Even if a function has a unique optimum, an optimization algorithm may not be able to find it. If the curvature of the function around the global maximum is "flat," less than the precision of the computer on which the estimation is performed, the algorithm can fail to find the global optimum (Gill et al. 1981, p. 300). Since small errors may be propagated and magnified, it is not inconceivable that an observed local optima may be purely a consequence of numerical inaccuracies. Ultimately, the solution provided by search algorithms will depend on the shape of the machine representation of the function, not the theoretical shape.

Researchers employing increasingly sophisticated maximum likelihood functions with multiple local optima should find these statements troubling. Simply programming an equation, plugging in data, and letting the software find a solution will not guarantee that the true global optimum is found, even if the "best" algorithm is used. Knowing when a function has reached its true maximum is something of an art in statistics. Although feasibility can be a reality check for a solution, relying solely on the expected answer as a diagnostic might bias researchers into unwittingly committing a type I error, falsely rejecting the null hypothesis in favor of the hypothesis advocated by the researcher. Diagnostic tests are therefore needed to provide the confidence that the solution computed is the true solution.

4.3.3.2 *Proposed Diagnostic Tests for Global Optimality*

Search algorithms depend on the starting point of their search. A poor choice of starting values may cause a search algorithm to climb up a discontinuity of the function and either break down completely or never find convergence. In other cases the search algorithm may settle at a local optimum far from the global optimum. The choice of good starting values can be critical to finding the global optimum.

One popular method used to choose good starting values is first to estimate a simple model that is both well behaved and comparable (sometimes nested within) to the model to be estimated and then use the solution from the simple model as the starting values for the parameters of the more complex model.

Locating a set of starting values that produce a solution may lull the unwitting researcher into a false sense of security. Although a search algorithm that fails to find a solution sends a clear warning message that something has gone wrong, one that produces a seemingly reasonable solution does not alert the user to potential problems. Researchers who might search for alternative starting values, and know that their function breaks down in certain regions, may in the course of avoiding trouble fail to perform a thorough search of the parameter space for the starting values that lead to the true global optimum. "There may be a temptation, one suspects, to choose alternative sets of starting values fairly close to either the initial set or the solution, easing computational difficulties but making it less likely that another maximum would be found if it existed" (Veall 1990, p. 1459).

One diagnostic commonly used to determine if the global maximum of a function has been found is to conduct an exhaustive search of the parameter space around the suspected global maximum of the function. The parameter space may be either deterministically or randomly searched. If the evaluation of the function at an alternative parameter vector yields a higher value for the function, the researcher knows immediately that the global optimum has not been found. The vector of parameters generated may also be used as starting values for the search algorithm, to see if it settles back at the suspected global optimum.

The reliability of this method of searching the parameter space depends on the proportion of the parameter space that can be searched, which may be quite small for unconstrained functions with a large number of variables. So an effective search across the parameter space should select alternative starting values in such a manner as to attempt to climb the function from different "sides" of the basin of attraction (the region of the parameter space that leads to an optimum). For example, if all the starting values are smaller than the solution found by the search algorithm, the researcher should try starting values that are all greater than the solution. If alternatively, the starting values consistently locate either the candidate optimum or another local optimum, the researcher can be more confident that the solution is the global maximum.

Strategies for finding starting values that lead to a global optimum have been formalized in tests for global optimality. Finch et al. (1989) propose selecting random starting values and using the distribution of the value of the likelihood function at their associated convergence points as a way to gauge the probability that an optimum, which may be local or global, has not been observed. Drawing on a result presented by de Haan (1981), Veall (1990) suggests that by using a random search and applying extreme asymptotic theory, a confidence interval for the candidate solution can be formulated.

The intuition behind Finch et al.'s statistic is to estimate the number of unobserved basins of attraction through the number of basins of attraction observed. The greater the number of basins of attraction observed, the lower the probability that a global optimum has been located. The statistic is first attributed to Turing (1948), and Starr (1979) provides a later refinement:

$$V_2 = \frac{S}{r} + \frac{2D}{r\,(r-1)}. \tag{4.19}$$

Here V_2 is the probability that a convergence point has not been observed, and r is the number of randomly chosen starting points. S is the number of convergence points that were produced from one (a single) starting value, and D is the number of convergence points that were produced from two (double) different starting values. Starr's result is further generalizable for triples and higher-order observed clumping of starting values into their basins of attraction, but Finch et al. assert that counting the number of singles and doubles is usually sufficient.

Finch et al. (1989) demonstrate the value of the statistic by analyzing a one-parameter equation on a [0, 1] interval for $r = 100$. Although the statistic proposed by (4.19) is compelling, their example is similar to an exhaustive grid search on the [0, 1] interval. The practicality of computing the statistic for an unbounded parameter space with a high degree of dimensions for a computationally intensive equation is not demonstrated. However, the intuition behind the statistic is still sound. If multiple local optima are identified over the course of a search for good starting values, a researcher should not simply stop once an apparent best fit has been found, especially if there are a number of local optima that have basins of attraction that were identified only once or twice.

The second method of parameter searching begins from the location of the candidate solution. A search for the global optimum is conducted in the parameter space surrounding the candidate solution, either deterministically (e.g., grid search) or randomly. As with the search for starting values, there is no good finite number of parameter vectors or bounds on the parameter space that will guarantee that the global optimum will be identified.

This is particularly true for the grid search approach. The grid search is best applied when the researcher suspects that the search algorithm stopped prematurely at a local optimum atop a flat region around the global maximum, a situation that Gill and King (2000) refer to as locating the highest piece of "broken glass on top of a hill." In this case the grid search neighborhood has a clear bound, the flat region around the candidate solution, that the researcher can employ in searching for the global maximum.

Drawing on a result presented by de Haan (1981), Veall (1990), suggests that by using a random search and applying extreme asymptotic theory, a confidence interval for the candidate solution can be formulated. According to Veall (1990, p. 1460) the method is to randomly choose a large number, n, of values for the parameter vector using a uniform density over the entire parameter space. Call the largest value of the evaluated likelihood function L_1 and the second largest value L_2. The $1 - p$ confidence interval for the candidate solution, L', is $[L_1, L^p]$, where

$$L^p = L_1 + \frac{L_1 - L_2}{p^{-1/\alpha} - 1} \qquad (4.20)$$

and $\alpha = k/2$, where k is a function that depends on n such that $k(n) \to 0$, as $k(n)$, $n \to \infty$ (a likely candidate is $k = \sqrt{n}$).

As Veall (1990, p. 1461) notes, the bounds on the search of the parameter space must be large enough to capture the global maximum, and n must be large enough to apply asymptotic theory. In Monte Carlo simulations, Veall suggests that 500 trials are sufficient for rejecting that the local optimum identified is not the a priori global optimum. This procedure is extremely fast because only the likelihood function, not the model, is calculated for each trial. As with starting value searches, researchers are advised to increase the bounds of the search area and the number of trials if the function to be evaluated has a high degree of dimensionality, or a high number of local optimum have been identified.

4.3.3.3 Examples: BoxBOD and Scobit

In practice, can these proposed diagnostic tests work? The example cases in the aforementioned articles are usually quite simplistic, with one or two right-hand-side variables. The practicality of these tests is not shown when a high number of right-hand-side variables are present. To put these tests through their paces, we program these tests for two examples. One by Box et al. (1978, pp. 483–87), which we treat in detail in Chapter 2, we refer to here as BoxBOD. This equation is selected because it is known to have multiple optima, and some maximizer algorithms may become stuck at a local rather than a global optimum. The second example is Nagler's (1994) scobit model, discussed in Chapter 3.

In our evaluation of the scobit, we ran the model hundreds of times. We found that the estimation procedure was slowed by the nearly 100,000 observations in the 1984 Current Population Survey dataset, which the scobit model is run in conjunction with. To speed our analysis, we subset these data. What we found, however, was that the scobit model had significant convergence problems with the small dataset. Sometimes the maximum likelihood optimization program would fail to converge, and when it did converge, it failed to converge at the same point. Here, we investigate the properties of the scobit model using a small dataset of 2000 observations.

To evaluate random starting values, we select a lower and upper bound of starting values for each coefficient. We then randomly select starting values in the given range. In practice, any distribution may be used to select starting values, but in our examples we choose starting values using a uniform distribution, because we want to choose starting values near the extreme ends of the ranges with a relatively high probability. The practical limit of the number of random starting vectors of coefficients is determined by the complexity of the equation to be estimated. For a small number of observations and a relatively simple likelihood function, such as BoxBOD, MLE analysis is quick. Thus, a large number of random starting values can be tested. For a function like scobit, a smaller number of random start values is more practical, as the more complex likelihood function and higher number of observations slows the speed of the search algorithm.

We implement the two tests for global optima discussed above, which we call the *Starr test*, as described by Finch et al. (1989), and the *Veall test* (Veall 1990). For the Starr test, we calculate the probability that an optimum, local or global, remains unobserved. For the Veall test, we simply provide an indicator if the observed highest optimum is outside the 95% confidence interval for the global optimum, which we report as a "1" if the observed highest optimum is outside the 95% confidence interval for the global optimum.

The implementation of the Starr test reveals a computational complexity that is unrecognized in the literature but must have been addressed in programming. In our many replications, the MLE often settles on a mean log-likelihood value of -2.1519198649, which thanks to high-precision estimation on a supercomputer by NIST, we know to be the global optimum. However, at the eleventh digit to the right of the decimal and beyond the reported log likelihood, the log likelihood varies across runs. Finch et al. (1989) are silent on how these distinct log likelihoods should be treated. We suspect that they counted these optima as one optimum. However, it may be possible that the optimizing algorithm has settled on two distinct local optima that are coincidentally near the same height of the likelihood function.

So we should distinguish among distinct multiple optima, even when the difference among their log likelihoods is small, if they imply substantially different estimates. To do this, we calculate the Euclidean distance between pairs of estimated coefficients. If the distance is small, we determine that the log likelihoods are drawn from the same optima. We identify distances less than 0.1 to be distinct solutions, but in practice this tolerance level will depend on the number of coefficients and the values of estimates that the researcher is concerned about. We consider the case of three (or generally, more) optima in n-dimensional space, where two may be less than the critical value from the third, but more than the critical value from each other, as two distinct optima.

The results of our two tests of global optimum for the two example cases are presented in Table 4.1. For each model, we present the number of starting coefficient vectors that we used, the number of successful convergences, the Starr probability that an optimum has not been observed, and the Veall test. We run these tests for 10, 50, and 100 starting coefficient vectors.

For the BoxBOD model, the search algorithm has mild convergence difficulties. While the search algorithm finds a convergence point for all 10 random starting values, it finds convergence for 42 of 50 and 95 of 100 random starting values. The Veall test of global optimality indicates that at least one of these convergence points is outside the 95% confidence interval for the global optimum. Among these convergence points, two unique optima are identified in 10 and 50 runs of random starting values, three in 100 runs. Accordingly, the Starr test for global optimality expresses a positive probability that another optimum may exist.

Table 4.1   Results of Tests for Global Optima

	Number of Starting Coefficient Vectors		
	10	50	100
BoxBOD			
Successful convergences	10	42	95
Distinct optima	2	2	3
Veall	1.000	1.000	1.000
Starr	0.010	0.024	0.210
Scobit			
Successful convergences	4	32	71
Distinct optima	4	32	71
Veall	1.000	1.000	1.000
Starr	1.000	1.000	1.000

For the scobit model, the search algorithm has greater difficulty in reaching convergence. In only four of 10, 32 of 50, and 71 of 100 times does the algorithm find convergence. The Veall test for global optimum indicates that at least one of these convergence points is outside the 95% confidence interval for the global optimum. Every instance of convergence is distinct from another—for 10, 50, and 100 runs—and the Starr test accordingly indicates that another optimum is possible.

The scobit example merits further discussion. Even though every convergence point is unique, if only one run was performed successfully, and a researcher did nothing further, there would have been indications that the search algorithm encountered difficulty in converging. For example, at the end of each run, the Hessian is non-invertible, and no standard errors are produced (see Chapter 6). Among the runs that exceed the maximum number of iterations, the log likelihood is often given as an imaginary number, or infinity.

Among the multiple convergence points of the scobit example, many report the same log likelihood (to eight or fewer digits), but the coefficients are still "far" from one another, sometimes varying over 1000 percent from one another. Thus, the likelihood surface is flat in a broad region around the optimum. What should we do in this circumstance? One possible solution is to gather more data, and for this instance, we can do exactly that because we have subset these original data. Our discussion in Chapter 3 indicates that different programs have difficulty in producing consistent estimates for the full scobit dataset, which includes nearly 100,000 observations. We witness similar results reported here for subsets of these data as large as 50,000 observations. Scobit appears to be a fragile model, nonresistant to data issues in "small" datasets such as the National Election Survey. In other circumstances where scobit estimates are generated, we recommend that careful attention be is paid to all aspects of the numerical algorithms underpinning the resulting estimates.

Although the Veall test indicated for all of our runs that an optimum outside the 95% confidence interval for the global optimum has been found, the usefulness of this test should not be discounted. The Veall and Starr tests are complementarity. If the range of starting values are tight around the attraction of basin for an optimum, the Veall test will probably capture the optimum within the 95% confidence interval, and fail to reject that the optimum has not been found. In calibrating our analysis we encountered this result and increased the range of our starting values as a consequence.

For these complex likelihood functions, there is a trade-off between increasing the range of the starting values and the performance of the search algorithm. Often, there is a "core" region, which may contain multiple basins of attraction, where the search algorithm will find a convergence point. Outside this region, the likelihood surface is convex, and the search algorithm will merrily iterate away toward infinity. This behavior requires careful monitoring, which can be problematic for automated procedures such as ours. Thus, some experimentation is required to find the right range of starting values such that the global optimum may be identified without causing the search algorithm to break down.

4.3.3.4 *Examining Solution Properties*

The preceding tests for global optima rely on evaluation of the likelihood function. Another class of diagnostic tests examine the properties of an optimum to determine if it is the global optimum.

As we have seen, many methods for finding optima declare convergence when the gradients are close to zero and the Hessian is positive definite. Computational methods may also look at local properties near the solution, such as decrease in the change of step size, little relative change in parameters, sum of squares, or likelihood (Bates and Watts 1988, p. 49). None of these properties distinguish a local from a global solution. But there are other properties that under certain conditions are necessary conditions for a global optimum but are not necessary conditions for a local solution. These properties provide the basis for two additional tests.

Den Haan and Marcet (1994) describe a technique for testing the accuracy of complex econometric simulation results, through examination of the distribution of the residuals. They note that in many economic models, the solution will have the property that the expectation of the cross product of the residuals from the estimation and any arbitrary function will be zero: $E[u_{t-1} \times h(x_t)] = 0$. They use this observation, along with a choices of h and a distribution for $E[\dots]$, to form a test of global optimality.

Gan and Jiang (1999) note that White's misspecification test can alternatively be interpreted as a test of global optimality. Any global root to the likelihood function must satisfy $(\partial \ell / \partial \theta)^2 + (\partial^2 \ell / \partial \theta^2) \approx 0$. If this does not hold for a given solution, either the model is misspecified or the solution given for the log-likelihood problem is not global.

4.4 OBTAINING MORE ACCURATE RESULTS

Inasmuch as there is a trade-off between speed and accuracy, we may choose to perform many analyses with less accuracy than it is possible to obtain. However, when statistical software fails (e.g., reporting that a Hessian was not invertible, as in Chapter 6), or sensitivity tests indicate potential problems, one may want to replicate the analysis with special attention to numerical accuracy. Furthermore, many practitioners will want to replicate preliminary results using high levels of numerical analysis prior to submitting them for final publication.

We assume that most social scientists will have neither the time nor the inclination to write their own libraries of high-precision statistical functions, matrix algebra subroutines, or optimizers. For those who wish to pursue this route, we recommend such texts as Kennedy and Gentle (1980), Thistead (1988), Nocedal and Wright (1999), and Higham (2002). [See also Press et al. (2002) for an introduction to a broad spectrum of numerical computing, although it lacks the attention to accuracy of the books just mentioned.]

Instead, in this section we discuss three approaches to increasing the accuracy of a statistical analysis that can be applied easily to a wide variety of models: utilizing specialized libraries for computing statistical functions and for matrix algebra, increasing the precision of intermediate calculations, and choosing appropriate optimization algorithms. In addition to these general techniques, there are a variety of specialized techniques for dealing with individual functions and models. We discuss the techniques that are most applicable to popular social science models in Chapters 5, 6, 8, 9, and 10.

4.4.1 High-Precision Mathematical Libraries

If the mathematical functions, distributions, or linear algebra operations supplied by a statistical package are not sufficiently accurate, an option may exist to replace them with more accurate substitutes. Most statistical packages allow users to supply user-defined functions if they are written within the statistical programming environment provided by the package. In addition, many packages permit users to call external libraries, written in other languages, such as C or FORTRAN, or even to "overload" (dynamically replace) functions supplied by the package.

There are a variety of popular libraries available for high-performance computing. NIST's "Guide to Available Mathematical Software (GAMS)" [first described in Boisvert et al. (1985)] is a good place to start. GAMS is a large online guide to and repository of mathematical and statistical libraries that points to many different libraries (available online at <http://math.nist.gov/>). In addition, the Web archive accompanying Lozier and Olver's (1994) catalog of methods for evaluating special functions also has an online extended bibliography referring to dozens of high-performance libraries and algorithms. (This is available from <http://math.nist.gov/mcsd/Reports/2001/nesf>.)

Some individual code libraries of particular interest include:

- Brown et al.'s (1998) DSSTAB package computes a similar set of distributions to approximately eight digits of accuracy. DSSTAB is available both as an interactive program and as a set of libraries that can be used by other programs.
- Moshier's (1989) Cephes library provides double-precision statistical distribution functions and quadruple-precision arithmetic and trigonometric functions, as well as other functions.
- The journal *Transactions on Mathematical Software*, which is published by the Association for Computing Machinery (ACM), has a large archive of software implementing each algorithm described in print. This includes many high-precision distribution functions and other functions of mathematical interest. The archive is available from <http://www.acm.org/calgo/>.
- LAPACK is a very popular, high-quality FORTRAN library for solving systems of simultaneous linear equations, least squares solutions, eigenvalue problems, and singular-value problems. It is built on BLAS, which provides vector and matrix operations (see, Lawson, et al. 1979; Dongarra et al. 1990; Dongarra and Walker 1995).

- The Gnu Scientific Library (see Galassi, et al. 2003) is written in C++ and provides BLAS support for matrix and vector operations, as well as accurate implementations of FFTs, pseudo- and quasirandom number generation, and special mathematical functions (among other features). It is available from <http://www.gnu.org/software/gsl/gsl.html>.

- The previously mentioned Scalable Library for Pseudo-Random Number Generation Library (SPRNG) and TESTU01 suite (discussed in Chapter 3) provides a set of high-quality functions for generating pseudo-random numbers.

Even when using a high-precision library, care should be exercised when examining the results, and users should consult the documentation for limitations on the intended use (such as a range of input parameters) for individual functions supplied in it.

4.4.2 Increasing the Precision of Intermediate Calculations

Accumulated rounding errors in intermediate calculations, although not the sole source of inaccuracy in statistical computation (see Chapter 2), contribute to it heavily. Numerical accuracies can interact with optimization algorithms, causing false convergence (or lack of convergence) when inaccuracy causes the function surface to falsely appear to be discontinuous or completely flat. A straightforward, although computationally expensive way of ameliorating this major source of inaccuracy is to use multiple-precision arithmetic—in essence calculating fundamental arithmetic operations to an arbitrary number of digits. Furthermore, where intermediate chunks of the problem being solved can be described and solved in closed form, rounding error can be eliminated entirely.

4.4.2.1 Multiple-Precision Software

In almost all computer hardware, and in almost all standard programming languages, the precision of arithmetic of built-in operations is limited. There are, however, *multiple-precision software libraries* that allow calculations to be performed above the precision of the built-in operators. [The first such library was due to Brent (1978).] These libraries can perform calculations beyond 1000 decimal digits of precision, given sufficient computing resources, such as available memory and storage space.

A common approach to implementing multiple-precision numbers is to use arrays to represent numbers in the standard floating point form (as described above), with the base and mantissa of each number spread across elements of the array. Modern FORTRAN libraries that support multiple precision are described in Smith (1988, 2001) and Bailey (1993). The GMP (Gnu Multiple Precision) library is also available in C++ and is open source (http://www.gnv.org/). A number of commercial software packages also support multiple-precision arithmetic. These include most of the computer algebra systems: Mathematica, Maple, MuPad, and Yacas. The popular MATLAB computing environment also supports multiple-precision computations.

Some sophisticated systems such as Mathematica support *arbitrary precision*. Rather than specifying a high but fixed level of multiple precision for a set of calculations, the system tracks the precision and accuracy of all input parameters and subsequent operations. The user specifies the desired precision of the function being evaluated, and the system then adjusts the precision of each intermediate calculation dynamically to ensure that the desired precision (or maximum precision given the precision of the inputs) is obtained. If, for some reason, the user's precision goal cannot be met, the system warns the user of the actual level of accuracy associated with the results.

There is a considerable trade-off between accuracy and efficiency using multiple-precision techniques. Computing answers to several times the normal precision of built-in operators may multiply execution time by a factor of 100 or more (see, e.g., Bailey et al. 1993). Still, multiple-precision arithmetic packages can be a practical way of improving the accuracy of standard statistical analyses.

4.4.2.2 Computer Algebra Systems

A way of avoiding rounding and truncation errors is to use computer algebra systems. Computer algebra is a set of mathematical tools for computing the exact solution of equations, implemented in such programs as Maple (Kofler 1997) and Mathematica (Wolfram 2001).

Computer algebra uses symbolic arithmetic—it treats all variables and quantities as symbols during intermediate calculations. Symbolic algebra is limited both by the tractability of the problem to closed-form solution and by the efficiency of algorithms to perform it. If a problem cannot be solved in closed form, it cannot be solved within a computer algebra system. Other problems, such as the solution of large systems of linear equations, although theoretically analytically tractable, are hundreds or thousands of times slower to solve using current modern algebra algorithms than by using numerical approaches. [For a thorough introduction to computer algebra systems, see von zur Gathen and Gerhard (1999).] Although symbolic algebra may rarely be a tractable approach for an entire estimation, it can be used judiciously to simplify numerical computation and reduce error. For example:

- Many algorithms for solving nonlinear optimization problems require derivatives. By default, they approximate these derivatives numerically (see Chapter 8). A symbolic algebra program can be used to calculate the exact formula for the derivative.
- Symbolic manipulation may be used to rationalize input data, or functional constants, enabling further calculations to be performed without additional rounding error.
- Symbolic manipulation may be used to simplify expressions, including the likelihood function.
- Numeric integration is susceptible to both approximation error and rounding error. If possible, symbolic integration, yields a function that can be evaluated directly, with no approximation error and reduced rounding errors.

Table 4.2 **Accuracy of BoxBOD Estimates Using Mathematica**

Digits of Precision	B1 Estimate	B2 Estimate
NIST certified results	213.80940889	0.54723748542
10 (reduced precision)	213.775	0.547634
16 (default)	213.808	0.547255
30	213.809	0.547237
30 (plus input rationalization)	213.80940889	0.54723748542

4.4.2.3 Combining CAS and MP

Many computer algebra systems, such as Mathematica, Maple, and Yacas, combine symbolic algebra and multiple-precision arithmetic capabilities. This can be a powerful combination for solving statistical problems accurately. For example, McCullough (2000) reports that when operating at high levels of precision, there was no statistical package that was as accurate and reliable as Mathematica.

Table 4.2 demonstrates the effect of multiple-precision arithmetic and symbolic computing. To create the table we used Mathematica's *NonLinear-Regression* procedure to solve the NIST Strd BoxBOD benchmark (see Table 4.2) while varying the levels of arithmetic precision used by the operators.[3] At a reduced precision, results are accurate to the fourth digit. As working precision is increased, the results become increasingly accurate. Rationalizing the input data yields more accuracy still—in fact, it produces the correct, certified result.

4.4.3 Selecting Optimization Methods

Nonlinear optimization is essential to estimating models based on nonlinear regression or maximum likelihood. Most statistical packages implement some form of unconstrained optimization algorithm for solving general nonlinear problems. (Constrained optimization often requires extensions of fundamental unconstrained algorithms, whereas nonlinear least squares is often solved with specializations.)

All practical general nonlinear algorithms are iterative. Given a set of starting values for variables in the model, the algorithm generates a sequence of iterates. This sequence terminates either when the algorithm converges to a solution (a local optimum or stationary point of the function) or when no further progress can be made. The determination that an algorithm has failed to progress involves comparisons of differences in gradients and objective functions at various points. So numerical precision is an important limiting factor on the ability of an algorithm to confirm progress. In addition, ad hoc practical considerations, such as a

---

[3]Note that Mathematica's definition of precision is compatible with (but not identical to) precision as it was used in Chapter 3. For the sake of replication, in the table we list the levels of precision as identified in Mathematica.

limit on the maximum number of iterations allowed, are also used in practice to determine if the algorithm has failed to progress.

The types of nonlinear optimization problems that social scientists most commonly encounter involve the maximization of a single function of continuous variables. Most commonly this function is unconstrained, twice differentiable, not computationally expensive, and has associated gradients that can be computed directly or approximated through differences. In addition, the datasets that social scientists tend to use are small enough that they can be kept in main memory (RAM) during computation. Optimization of such problems is discussed in detail in Chapter 8, and we present only a few recommendations and general precautions in this section and review alternatives for difficult problems.

The class of algorithms designed specifically to solve this problem are based on computing gradients and using the information in the gradient to inform the search. Most of these algorithms involve one of two fundamental search strategies: line search and trust region. *Line search algorithms* use gradient information to choose a search direction for the next iteration, and then choose a step length designed to optimize the search in that direction. *Trust region algorithms* essentially reverse these steps by choosing a length first and then a direction. In particular, trust region methods choose a region radius, construct an approximate model of the behavior of the function within the multidimensional region defined by that radius, and then choose an appropriate direction along which to search within that region.

Steepest descent and conjugate gradient methods use only the gradient information and do not compute a Hessian, so they are sometimes used to save memory on large problems. *Steepest descent* is the simplest of the line search methods, moving along the negative of the gradient direction at every step. It is however, very slow on difficult problems and can be less accurate than quasi-Newton methods. *Conjugate gradient methods* use search directions that combine gradient directions with another direction, chosen so that the search will follow previously unexplored directions. They are typically faster than steepest descent but susceptible to poor scaling and less accurate than quasi-Newton methods.

Newton's method converges much more rapidly than steepest descent when in the neighborhood of the solution. The basic *Newton's method* forms a quadratic model (using a Taylor series expansion) of the objective function around the current iterate. The next iterate is then the minimizer of the quadratic model approximation. This approach requires the Hessian of the approximation function to be computable and to be positive definite (see Chapter 6 for details), which in turn requires that starting values be "close enough" to the optimum. In addition, the computation of the Hessian at each iteration is often expensive.

Various hybrids and fixes have been attached to Newton's method to deal with this problem. The most popular, quasi-Newton methods address both these problems by building an approximation to the Hessian based on the history of gradients along each step of the algorithm. This approximation of the Hessian is less expensive to compute and can be adjusted to be positive definite, even when the exact Hessian is not.

4.4.3.1 Recommended Algorithms for Common Problems

For the unconstrained, continuous, differentiable, moderate-sized problem described above, the quasi-Newton method proposed by Broyden, Fletcher, Goldfarb, and Shannon (BFGS) is generally regarded as the most popular method (Nocedal and Wright 1999, p. 194) and as the method most likely to yield best results (Mittelhammer et al. 2000, p. 199). The algorithm is, in theory, at least as accurate as other competing line search algorithms, both as to scaling of input and to deviations from quadratic shape in the likelihood function. In practice, it has very often proved to be more accurate than popular alternatives, such as conjugate gradient, steepest descent, and DFP (Fletcher 1987, pp. 68–71, 85–86; Nocedal and Wright 1999, pp. 200, 211). (See Chapters 7 and 8 for examples of real-world models where the choice of optimization method strongly affects the estimates generated.)

A promising extension of BFGS has emerged from the work of Ford and Moghrabi (1994). They use the BFGS formula but replace the vectors used in it. The new vectors are determined by interpolating curves over the gradients of several previous iterations. In limiting testing, incorporation of this additional information improves the performance and accuracy of the original algorithm. However, this approach has not yet been tested extensively, and does not, to our knowledge, appear as yet in any commercial statistical packages.

A good statistics package should provide more than one type of optimization algorithm, even for the unconstrained case. A package should offer a trust region algorithm in addition to BFGS and may offer another algorithm specifically for nonlinear regression:

- Trust region algorithms may be able to find solutions where BFGS fails because they do not necessarily rely on the Hessian to find a search direction (Fletcher 1987, p. 95; Kelley 1999, p. 50). BFGS can fail to converge when the Hessian is excessively nonpositive definite in the region of the starting values. Unlike BFGS, however, trust region algorithms are sensitive to scaling.

- Nonlinear least squares regression is a special case of unconstrained optimization. Because of the structure of the nonlinear least squares problem, the Jacobian matrix is a good approximation of the Hessian matrix, especially near the solution. The Gauss–Newton and Levenberg–Marquardt variants of the line search and trust region algorithms discussed above take advantage of this structure to reduce computing expense and speed convergence. One should be aware, however, that these algorithms do not work particularly well for cases where the solution has large residuals. So a hybrid approach using one of these algorithms with BFGS (or other Newton or quasi-Newton algorithm) as a fallback is usually recommended (Dennis and Schnabel 1982, pp. 225, 233; Fletcher 1987, pp. 114–17; Nocedal and Wright 1999, p. 267)

Unfortunately, however, some algorithms persist in statistical packages for what seem to be historical reasons. For example, the BHHH optimization algorithm (Berndt et al. 1974) is offered in some econometrics packages, although it is generally ignored by modern optimization researchers. Another example, the downhill simplex method, invented by Nelder and Mead (1965), is most appropriate where derivatives are absent, but is sometimes seen in statistical routines being applied to differentiable functions, possibly because of its brevity and appearance in the popular *Numerical Recipes* library.[4]

4.4.3.2 Cautions: Scaling and Convergence

One important source of difficulty in common optimization practice is *scaling*. In unconstrained optimization, a problem is said to be poorly scaled if changes in one independent variable yield much larger variations in the function value than changes in another variable. (Note that scaling resembles, but is not identical to, conditioning, which refers to the change in function value as all inputs change slightly. See Chapter 2 for a discussion of conditioning.) Scaling is often a result of differences in measurement scale across variables, although such differences are neither necessary nor sufficient for a problem to be poorly scaled. A problem's scaling is affected by changing the measurement units of a variable.

Poor scaling results in two types of problems. First, some algorithms, such as gradient-based descent methods, conjugate gradient methods, and trust region methods, are affected directly by scaling level. Other algorithms, such as Newton and quasi-Newton algorithms (of which BFGS is a member) are not directly affected and are said to be *scale-invariant*, Second, even scale-invariant algorithms, however, are susceptible to rounding problems in their implementations, and the likelihood of rounding problems often becomes greater when problems are poorly scaled, since poor scaling can lead indirectly to the addition or subtraction of quantities of widely different magnitudes. So, in practice, it is best to rescale poorly scaled problems even when using scale-invariant algorithms. (Note that some software packages offer automatic rescaling as an optimization option. Although potentially convenient, we suggest using the techniques described in Chapter 3 to gauge the reliability of the package when this option is enabled. In the course of our testing of a variety of statistical packages, for example, we found cases, particularly in Microsoft Excel, in which the automatic scaling routines used in the program actually led to less accurate optimization results.)

Convergence criteria are another common source of difficulty in the use of optimization software. We note some precautions here and discuss convergence criteria selection and convergence diagnostics in detail in Chapter 8 (See also Chapter 10 for examples of how a poor choice of stopping rule can cause false convergence, and thus incorrect estimates, in logistic regressions.)

Three precautions are warranted with respect to algorithm convergence. First, software packages often offer convergence criteria that do not distinguish between lack of algorithmic progress (such as lack of change in the objective function) and true convergence (in which the gradients of the function are tested for

---

[4]The Nelder–Mead algorithm is intended for optimization of noisy functions. In such problems, gradients cannot be used. Nelder–Mead performs well in practice for such problems, although it possesses no particularly good theoretical convergence properties (Kelley 1999, p. 168).

sufficiency). Second, some algorithms, such as Nelder–Mead, are not globally convergent. Even using the proper convergence criterion, these algorithms must be started within the neighborhood of the solution to be effective. Choosing a steepest descent, trust region, or quasi-Newton algorithm can lead to convergence from starting conditions under which Newton, inexact Newton, Nelder–Mead, or conjugate gradient methods would fail to find a solution. Finally, even with the correct convergence criterion, globally convergent algorithms such as modified Newton and quasi-Newton methods[5] convergence criteria will stop at a *local* optimum (or, technically, a stationary point).

4.4.3.3 Implementation of Optimization Algorithms

When choosing an optimization algorithm, one should remember that both algorithm and the implementation thereof are vitally important to finding an accurate solution. We will not review the distinction between implementation and algorithm in this chapter, as we discussed it thoroughly in Chapter 2. Note, however, that numerical issues such as round-off error can cause a theoretically scale invariant and globally convergent algorithm to fail.

Note also that because of the sophistication of optimization algorithms, and the many variations available for the standard algorithms, implementation is especially likely to encompass critical design choices. Often, the developers of a particular software package will expose some of these choices or even parameterize them for the user to control. Thus, it is not uncommon for software packages to offer choices over algorithm family, line search method, derivative evaluation, and convergence/stopping criteria. Even with such choices, however, many important implementation choices remain subterranean.

A comparison by Maros and Khaliq (2002) is especially revealing: They note that a commercial implementation of the simplex algorithm for solving linear systems (distributed by Wang in 1969) was a logically correct implementation of the basic algorithm which comprised only 30 lines of code, whereas modern implementations are typically 1000 times as long. Striking differences remain among modern implementations of optimization algorithms. Nonlinear optimization algorithms show similar variations in implementation. For example, the simulated annealing algorithm in Press et al. (2002) consists of approximately 50 lines of code, whereas the popular public domain library for simulated annealing (and algorithmic variants) due to Ingber (1989) is approximately 12,000 lines.

4.4.3.4 Optimization of More Difficult Nonlinear Problems

In optimization, provably "there is no free lunch"; all algorithms perform equally well (or poorly) across the set possible objective function. Thus, explicitly or implicitly, successful optimization, exploits some of the structure of the function that one wants to optimize (Wolpert and Macready 1997) The algorithms described above were designed to exploit the structure of well-behaved,

---

[5]Technically, BFGS has thus far been proved to be globally convergent only under limited assumptions (see Nocedal and Wright 1999).

continuous, nonlinear functions. When the objective function does not have these characteristics, these algorithms will not be effective.

When a problem does not fit into the intended domain of the algorithms available in a statistical package, we strongly recommend searching for a more appropriate algorithm rather than attempting to shoehorn a problem to fit a particular package. For example, some constrained optimization problems can sometimes be transformed into unconstrained problems through the use of barrier functions (such as logarithms). This converts the constraints into continuous unconstrained variables with values that approach infinity as the original constraint is approached. This new function can then be optimized using an unconstrained optimization algorithm. *This is not recommended.* Converting the constrained problem into an unconstrained problem in this way can make the problem ill-conditioned, which can lead to inaccurate solutions. Native treatment of constrained problems (e.g., through Lagrangian methods) are more efficient, robust, and accurate (see Nocedal and Wright 1999).

Good catalogs of specialized optimization algorithms for different categories of problems include Moré and Wright (1993) and its online successor, and the online guides in Mittelmann and Spellucci (2003). These resources can readily be accessed at <http://www.ece.northwestern.edu/OTC/> and <http://plato.asu.edu/guide.html> (respectively).

Whereas some specialized algorithms are designed to exploit the structure of a particular type of function, black box heuristics take the opposite approach. Black box algorithms (some of which are technically heuristics) gain their sobriquet by avoiding, as much as possible, explicit assumptions about the function being optimized. [For example, black box algorithms make use only of function evaluations (rather than derivatives, etc.) and can be applied to any function with at least some hope of success.] Their popularity comes from their wide applicability, and their practical success in a broad range of very difficult problems, especially discrete (combinatoric) optimization problems, for which specialized algorithms are unavailable. The most popular of the black box optimization algorithms— simulated annealing, genetic algorithms, and artificial neural networks—all draw upon analogies to physical processes in their design. See Aarts and Lenstra (1997) and Michalewicz and Fogel (1999) for good introductions, and Kuan and White (1994) or Gupta and Lam (1996) for specific application to common problems such as missing data estimation and financial prediction.

Simulated annealing (SA) exploits an analogy between the way in which molten metal freezes into a minimum-energy crystalline structure (the annealing process) and the search for a function optimum. SA was first modeled by Metropolis et al. (1953) and popularized in Kirkpatrick et al. (1983). At each iteration, simulated annealing randomly generates a candidate point (or set of points) within a local neighborhood of the current solution. The probability of moving from the current solution to one of the candidate points is a function of both the difference in the value of the objective function at each point and a temperature parameter. At high temperatures, candidate points that are "worse" than the current solution can be selected as the solution in the next iterate. This helps the algorithm to avoid local optima. At each iteration, the temperature is reduced gradually so that the probability of heading downhill becomes vanishingly small.

SA first made its appearance in social science in the field of finance and has since appeared in economics, sociology, and recently, political science, showing considerable success on problems too difficult for standard optimization methods in these fields (Ingber 1990; Ingber et al. 1991; Goffe et al. 1992, 1994; Carley and Svoboda 1996; Cameron and Johansson 1997; Altman 1998). [For an elementary introduction, see Gill (2002 Chap. 11).] Recently, a number of the statistical packages commonly used by social scientists have incorporated simulated annealing, notably R and GaussX.

Genetic algorithms (GAs) are a form of algorithm inspired by analogies between optimization (and adaptation) and the evolution of competing genes. The classic GA is attributed to Holland (1975), but GAs have become part of a wider field of "evolutionary computation" combining Holland's work with contemporaneous independent work by Rechenberg (1973) and Schwefel (1977), and Fogel et al. (1967).

In a genetic algorithm, one starts with a population comprising a set of candidate solutions to the optimization problem. Each solution is encoded as a string of values. At each iteration each member of the population is subject, at random, to mutation (an alteration of the solution vector) and hybridization (a reshuffling of subsequences between two solutions). In addition, each round undergoes selection, where some solutions are discarded and some are duplicated within the population, depending on the fitness (function evaluation) of that member. Research in social science that makes use of GA techniques for optimization is just now beginning to appear (Dorsey and Mayer 1995; Altman 1998; Sekhon and Mebane 1998, Everett and Borgatti 1999; Wu and Chang 2002), although similar techniques are popular in agent-based modeling.

Artificial neural networks (ANNs) are inspired by a simplified model of the human brain. The model was first proposed by McCulloch and Pitts (1943) but was not applied to optimization until much later (Hopfield 1982; Hopfield and Tank 1985; Durbin and Wilshaw 1987; Peterson and Soderberg 1989). An ANN is a network of elementary information processors called *neurons* or *nodes*. Each node has a set of inputs and outputs, and each node processes its inputs with a simple but usually nonlinear activation function to determine whether to "fire," or activate its outputs. Nodes are connected together via *synapses*, which are weighted so that the output of one neuron may be attenuated or amplified before becoming the input of another neuron. Connections among nodes may be arbitrarily complex, but they are often simplified through the use of layering. Finally, ANNs are "trained" to solve a particular problem by feeding instances, known as the *training set*, of the problem in as initial inputs to an ANN, observing the resulting final output, and then iteratively adjusting the synapses and activation functions until the ANN produces correct outputs for all (or most) of the inputs in the training set.

Although ANNs can be used for function optimization (see Aarts and Lenstra 1997), they are primarily used directly to generate predictions, as an alternative to fitting an explicit function. In the latter context they have become relatively popular in social science (Tam and Kiang 1992; Kuan and Liu 1995; Hill et al. 1996; Semmler and Gong 1996; Nordbotten 1996, 1999; West et al. 1997; Beck et al. 2000; King and Zeng 2001b.) Some critics warn, however, of the prevalence of overfitting in neural network applications and, the difficulty of explanation-oriented interpretation of neural network results [but see Intrator and Intrator (2001)] and note that far simpler statistical models can perform comparably (or better) on some problems to which neural nets have been applied (Schumacher et al. 1996; Nicole 2000; Paik 2000; Xianga et al. 2000).

While black box optimization methods such as simulated annealing and genetic optimization can be applied across an astounding domain of optimization problems, this flexibility comes with a number of important drawbacks. The black box approach is both a strength and a weakness—by assuming next to nothing about the problem structure, black box algorithms can often make progress where algorithms that require derivatives (or other types of structure) would simply fail. On the other hand, where a problem does have an exploitable structure, these algorithms take little advantage of it. In particular, although obscured by some advocates, the following limitations will come as no surprise to those who take the "free lunch" theorem to heart:

- SAs, GAs, and ANNs are extremely sensitive to algorithmic parameters and implementation details (Michalewicz and Fogel 1999; Mongeau et al. 2000; Paik 2000).
- Although the proofs of convergence associated with SAs, GAs, are vaguely reassuring, they do not offer much practical guidance. With appropriate cooling schedules and selection rules (respectively), SAs and GAs both converge to the *global* function optimum. Convergence proofs, however, are either completely nonconstructive or yield only asymptotic convergence results (with respect to the most general class of continuous nonlinear functions) (Lundy and Mees 1986; Eiben et al. 1991; Locatelli 2000). For example, with proper cooling, the probability that SA reaches the optimum at each iteration approaches 1 asymptotically as iterations approach infinity. Unfortunately, asymptotic convergence results such as these generally guarantee only that the algorithm achieves the right answer in infinite time and does not imply that the intermediate results improve over iterations monotonically. (Given that most of the objective functions are implemented using finite-precision arithmetic, simply enumerating all possible combinations of function value input would take only finite time and guarantee the optimum with certainty.)
- Similarly, while ANNs are often touted as being able to approximate any function to any degree desired, the same can be said of spline regression, polynomial regression, and other fitting techniques. In these other techniques, susceptibility to overfitting is a cost of flexibility (Schumacher et al. 1996; Paik, 2000). Furthermore, the ability to approximate a particular type of function also depends critically on the structure of the network. In fact, a

common type of neural network is statistically equivalent to logistic regression (Schumacher et al. 1996). Finally, "optimal" training of networks is computationally intractable in general (Judd 1990), and practical training methods run the risk of getting trapped in local optima (Michalewicz and Fogel 2000).

- ANNs, SAs and GAs use convergence criteria that are neither sufficient nor necessary for convergence to either global or local optima (Michalewicz 1999; Locatelli 2000; Mongeau et al. 2000; Paik 2000). In contrast, quasi-Newton algorithms can make use of information in gradients to determine whether necessary/sufficient conditions for a local optimum have been met.

For example, if a problem satisfies the requirements for BFGS, BFGS will be more efficient than simulated annealing. More important, for most such problems, convergence properties and the convergence criteria will yield either (efficient) arrival at a local optimum or an unambiguous declaration of failure. SA is only asymptotically convergent, and the convergence criteria used in practice fail to distinguish true convergence from simple lack of progress.

These limitations notwithstanding, the popularity of black box optimization algorithms is, in large part, a direct result of their astounding success across a remarkable variety of difficult problems. Although they should be used with skill, and the results should be examined carefully, they often work well where standard optimization algorithms fail miserably.

4.5   INFERENCE FOR COMPUTATIONALLY DIFFICULT PROBLEMS

Nonlinear statistical models may be ill behaved in at least three basic ways that affect the resulting estimates. First, the neighborhood around the function optimum may be highly nonquadratic. Second, the objective function may have multiple optima. Third, poor scaling, ill conditioning, and/or inaccuracy in algorithmic calculations can lead to general instability.

These are serious problems when they occur, in that the resulting conclusions are dubious, and worse yet, the user may not be aware of any problem. When the neighborhood around the optimum is nonquadratic, confidence intervals calculated in the standard way will yield wildly incorrect results because standard errors are calculated from the square root of the diagonals of the negative inverse of the Hessian (second derivative matrix) at the MLE solution. The Hessian (discussed extensively in Chapter 6) measures the curvature around the numerically determined likelihood optimum. Suppose that rather than having the classical quadratic form, typically displayed in econometric texts, there are modal features like those displayed in Figure 4.1. Then the calculation of t-statistics and confidence intervals that use standard errors from the Hessian calculation will be wrong.

Multiple optima present a special challenge. If there is one dominant optimum of interest and other obvious suboptima, the problem is merely to make sure that root-finding algorithms do not satisfice on one of the local solutions. If there are multiple optima that compete as a final solution, special caution is warranted and it is important to fully describe the posterior surface with HPD intervals (explicated below) or other thorough procedures. In such cases, relying simplistically on point estimates and standard errors gives a deceptive view of the posterior results because it assumes unimodality and symmetry.

General numerical accuracy counts in this context as well as in those described in Chapter 1.4. Consider an algorithm that exaggerates or understates the magnitude of standard errors around the found mode, as well as one that is sufficiently unreliable to produce different values on different runs, under the same circumstances or with slightly varied conditions. Not only will the inferred statistical reliability be misguided, but the user may have no indication that this circumstance has occurred.

4.5.1   Obtaining Confidence Intervals with Ill-Behaved Functions

Posterior/likelihood results may be nonnormal and nonquadratic around the mode(s). Most of the time this occurs because of small sample size and model complexity, or perhaps both. In this section we review some of the tools for describing and possibly correcting for these sorts of problems. The primary problem is that most standard calculations for coefficient standard errors and confidence intervals make assumptions about the shape of the posterior, such as symmetry and unimodality. In the more traditional sense, significance levels can be seriously misleading if calculated under incorrect t or normal distribution assumptions.

4.5.1.1   Wald versus LR Intervals

One of the most common methods of generating confidence regions is simply to use the t-statistic or, more generally, the Wald interval (Engle 1984). As discussed in detail in Chapter 8, for the standard constraints of the form $h(\theta) = \mathbf{r}$, the 95% Wald interval from the MLE values $\hat{\theta}$ from the log-likelihood function $\ell(\mathbf{X}, \theta)$ is given by values of $h(\hat{\theta})$ such that

$$(h(\hat{\theta}) - \mathbf{r})' \left[\frac{\partial h(\theta)}{\partial \theta'} \bigg|_{\hat{\theta}} \left(\frac{\partial^2 \ell(\mathbf{X}, \theta)}{\partial \theta \partial \theta'} \bigg|_{\hat{\theta}} \right)^{-1} \frac{\partial h(\theta)'}{\partial \theta'} \bigg|_{\hat{\theta}} \right]^{-1} (h(\hat{\theta}) - \mathbf{r}) \leq \chi^2_{k, 1-\alpha} s^2$$

(4.21)

for selected $(1 - \alpha)$- and k-dimensional θ (Amemiya 1985, Sec. 4.5.1; McCulloch and Searle 2001, Sec. 5.5). The distributional assumption is an asymptotic property under the null assumption that $h(\theta) - \mathbf{r} = \mathbf{0}$.[6] Once $\hat{\theta}$ is determined, this calculation proceeds directly for the parametric form, $\ell(\mathbf{X}, \theta)$. Consider what

---

[6]It is sometimes also useful for GLM results to report the Wald confidence intervals on the mean response point vector: $\overline{\mathbf{X}}$, which is straightforward (Myers and Montgomery 1997; Myers et al. 2002, Chap. 7).

happens, though, if $\hat{\theta}$ is reported with inaccuracy or if the derivative and matrix algebra calculations are stored with inadequate numerical precision. The final calculation is the product of a $1 \times k$ vector with a $k \times k$ matrix and a $k \times 1$ vector, so there are many intermediate values to store. As discussed in Chapter 2 and illustrated elsewhere, such rounding and truncating decisions "ripple" through subsequent calculations. As a specific case, Long (1997, p. 98) reminds us that if an estimation quantity such as this is hand-calculated from typically rounded reported values (either in published form or from a regression printout), the results can be very misleading. Furthermore, if the curvature at the mode is nonquadratic, the underlying justification for this procedure is flawed since the second derivative calculation in (4.21) will not meet the conditions required. [See Serfling (1980) for details]. Harmful deviations include flat areas, "broken glass," asymmetry, and other pathologies. The Wald test also has the drawback that it is not invariant under reparameterization, unlike the two other classical tests, likelihood ratio and Rao—but this is not of direct concern here.

When such problems arise, the likelihood ratio test (LRT) can be used as a replacement for the Wald test (one may want to use it for other reasons as well, such as its simplicity). Likelihood ratio intervals are more robust than Wald intervals because they are less reliant on the shape of the likelihood surface. If the constraints are simple, such as $\theta = 0$ or some other constant for a subset of the vector, the restricted likelihood is calculated simply by using these constant values for the restricted coefficients and letting the others be determined by estimation. In the case of more complex, nonlinear constraints, it is necessary to take the derivative of $\ell(\mathbf{X}, \boldsymbol{\theta}) - \boldsymbol{\Upsilon}'h(\boldsymbol{\theta})$ and set it equal to $\mathbf{r}$, where $\boldsymbol{\Upsilon}$ is the Lagrange multipliers to produce the restricted log-likelihood value: $\ell(\mathbf{X}, \boldsymbol{\theta}_r)$. This is often avoided in practice with the specification of simple restrictions. The likelihood ratio interval is given by values of $h(\hat{\boldsymbol{\theta}})$ such that

$$-2\left(\ell(\mathbf{X}, \boldsymbol{\theta}_r) - \ell(\mathbf{X}, \hat{\boldsymbol{\theta}})\right) \le \chi^2_{k, 1-\alpha}. \tag{4.22}$$

This is more flexible in one sense since the gradient and Hessian calculations are not required, but it is also asymptotically justified by the second (quadratic) term in a Taylor series expansion (Cox and Hinkley 1974, Sec. 9.3; Lehmann 1999, Sec. 7.7). Therefore, ill-behaved likelihood surfaces that result from small-sample problems are also suspect, although the direct implications are not as extreme as with the Wald statistic.

4.5.1.2 *Bootstrapped Corrections*

Alternative methods for obtaining confidence intervals exist for the worried (and persistent) researcher. One method is to bootstrap the sample in such a way as to obtain a reasonable standard error and impose a normal or t-distribution. The general idea is to produce a $1 - \alpha$ confidence interval of the form $\hat{\theta} \pm z^{1-\alpha/2}\hat{se}$, where $\hat{se}$ is the bootstrap calculation of the standard error. More generally, from Efron and Tibshirani (1993), generate B bootstrap samples of the data: $\mathbf{x}^{*1}, \mathbf{x}^{*2}, \dots, \mathbf{x}^{*B}$, and for each sample, $b = 1, 2, \dots, B$, calculate

$Z_b^* = (\hat{\theta}_b^* - \hat{\theta})/\hat{se}_b^*$. Here $\hat{\theta}_b^*$ is the bth bootstrap calculation of $\hat{\theta}$ with associated bootstrap standard error $\hat{se}_b^*$. These Z_b^* are then ordered and the desired quantiles for a specific α-level are picked out: $[Z_{\alpha/2}^*, Z_{1-\alpha/2}^*]$. That is, for $B = 1000$, $Z_{\alpha/2}^*$ is the 25th ordered replicate and $Z_{1-\alpha/2}^*$ is the 975th ordered replicate. So the bootstrap-t confidence interval is $[\hat{\theta} - Z_{1-\alpha/2}^*\hat{se}, \hat{\theta} - Z_{\alpha/2}^*\hat{se}]$, using $\hat{se}$ from the original calculation.

This is elegant and well grounded in standard theory, (Efron 1979; Bickel and Freedman 1981; Freedman 1981; Hinkley 1988; Hall 1992; Shao and Tu 1995), but in practice some problems appear. The empirical construction means that outliers can have a tremendous and unwanted effect. This basic procedure is also not transformation invariant and is only *first-order* accurate: errors in probability matching for the interval go to zero on the order of $1/\sqrt{n}$. Efron and Tibshirani (1993, Chap. 14) develop the *bias-corrected and accelerated* (BC$_a$) version, which is slightly more complicated to produce but is now invariant to transformations and *second-order* accurate: Errors in probability matching for the interval got to zero on the order of $1/n$. Unfortunately, BC$_a$ is not uniformly better than the straight bootstrap t-test or other alternatives (Loh and Wu 1991; DiCiccio and Tibshirani 1987).

Other relevant work includes the *bootstrap calibration confidence set* of Loh (1987, 1988, 1991); the Bayesian bootstrap (Rubin 1981; Stewart 1986; Lo 1987; Weng 1989); the weighted likelihood bootstrap, which extends the Bayesian bootstrap (Newton and Raftery 1994); various performance aspects with economic data (Horowitz 1997); and special considerations for longitudinal settings (Runkle 1987; Datta and McCormick 1995; Ferretti and Romo 1996). Recently, Davison et al. (1986a, b), and Hinkley (1988), and Davison and Hinkley (1988, 1997) have added extensions to enhance the reliability bootstrap estimates for likelihoods.

4.5.2 Interpreting Results in the Presence of Multiple Modes

Standard econometric theory states that in general we should worry primarily about the distinction between posterior[7] inferences based on a t-distribution due to limited sample size and others that can comfortably be calculated on normal tail values. Greene (2003), for instance, distinguishes between "Finite Sample Properties of the Least Squares Estimator" and "Large Sample Properties of the Least Squares and Instrumental Variables Estimators." The extension to generalized linear models uses this same distinction in measuring coefficient reliability by tail values under the null hypothesis. Suppose, however, that this limitation to two forms is not a reasonable assumption.

There are essentially two types of problems. Some empirically observed likelihood functions are multimodal because of the small-sample effects or poorly implemented simulation studies. In these cases the theoretical likelihood distribution is asymptotically normal and unimodal (usually, t) as sample sizes

---

[7]We continue to term coefficient sampling distributions and Bayesian posteriors as just posteriors since the former is based on uniform prior assumptions whether the researcher knows it or not.

increase [see Berger and O'Hagan (1988) and O'Hagan and Berger (1988) for a Bayesian view] but does not display these characteristics with the problem at hand. Conversely, some likelihood/posterior forms are inherently multimodal, such as the multinomial probit specifications (see McFadden and Ruud 1994), Bayesian mixture models (Gelman et al. 1995, Chap. 16; Cao and West 1996), mixtures of bivariate normals (Johnson 1996), and Cauchy log-likelihood functions with small samples (Brooks and Morgan 1995), just to name a few. In the second example it is obvious that a point estimate badly misses the point of the intended analysis.

As pointed out in Section 4.3.3.1, it is easy to envision substantively important general functions that are multimodal. The primary reason that this can be a "problem" is that most empirical researchers are accustomed to unimodal likelihood functions or posteriors. Because this is almost always a safe assumption with commonly specified models such as linear forms, many GLMs, log-linear specifications, and others, it is easy to become lulled into the belief that this is not something to worry about when specifying more complex or more particularistic models.

One type of multimodal posterior that can result from certain economic models is the $m - n$ poly-t parametric forms, which result from the ratio of m to n multivariate t-distributed densities (Dreze and Richard 1983; Bauens and Richard 1985; Oh and Berger 1993).

4.5.2.1 Example of Induced Multimodality

Brown and Oman (1991) give a simple example where the researcher controls a parameter ξ in an experimental setting and observes its effect on an outcome variable of interest Y. The distribution of Y conditional on ξ is given by

$$Y = h(\xi) + \epsilon \quad \text{where} \quad E[\epsilon] = 0, \text{Var}[\epsilon] = \gamma, \tag{4.23}$$

where predictions of future values of ξ are given by inverting this distribution through estimates $\hat{h}$ and $\hat{\gamma}$. The prediction distribution of ξ is then

$$f(\xi) = (Y - \hat{h}(\xi))'\hat{\gamma}^{-1}(Y - \hat{h}(\xi)). \tag{4.24}$$

If $\hat{h}$ is nonlinear in ξ, this distribution can be multimodal and Brown and Oman give a specific case from Brown (1982) where the posterior has three modes in the allowable range resulting from a quadratic form of $\hat{h}$, shown here in Figure 4.2. Also shown is a 95% asymptotic critical level, with the dotted line indicating the difficulty in making inferential claims in the standard manner of finding the mode and measuring the curvature around it. This is really a marking of the highest posterior density interval described below and in Chapter 6.

The inferential problem that arises here is what mode we consider to be most important. In the standard econometric setup, unimodality stems directly from the model assumptions, and this problem never arises (by design). The resulting confidence region in this particular case is the discontinuous interval determined

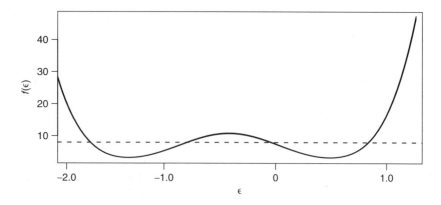

Fig. 4.2   Posterior for $\hat{h}$ Quadratic in ϵ.

by the points where $f(\epsilon)$ falls below the horizontal line in Figure 4.2, which is Brown's estimate of the region that contains the lowest 95% of the density (the concern is with calibration and with the goal of minimizing ϵ). How should we interpret such interval estimations? Should one area be preferred over the other? It is entirely possible that multimodal states are consistent with the data and the model and should therefore not by ignored simply by picking one mode, which might be only slightly more prominent than others.

Recall that the *likelihood principle* (Birnbaum 1962) shows that all necessary sample information that can be used for estimating some unknown parameter is contained in the likelihood function: a statement about the most likely value of the parameter given the observed data.[8] Once the data are observed, and therefore treated as given, all the available evidence for estimating $\hat{\theta}$ is contained in the (log) likelihood function, $\ell(\theta|\mathbf{X})$, allowing one to ignore an infinite number of alternates (Poirer 1988, p. 127). So the shape of this likelihood function contains the relevant information, whether or not it is unimodal. What does this say about multimodal forms? Clearly, summarizing likelihood information with a point estimate given by the highest mode is consistent with the likelihood principle, but providing no other information about the likelihood surface conceals the level of certainty provided by this point estimate alone. In such cases we recommend describing the likelihood surface or posterior distribution to the greatest extent possible as a means of telling readers as much as possible in a succinct format.

Furthermore, a multimodal posterior often causes problems for standard mode-finding algorithms that become satisfied with suboptimal mode solutions. This means that the choice of starting points becomes very important. It is possible to change the inferential conclusions dramatically by picking very good or very

---

[8]The Bayesian approach uses this same likelihood function (albeit with the addition of a prior distribution on the unknown parameters (Gill 2002). Thus every point in this section also pertains to posterior summary.

bad starting points. One solution is to "grid up" the posterior space and map out the approximate region of the largest mode (if that is the ultimate goal), but this becomes very difficult and computationally expensive with large dimensions. Oh and Berger (1993) suggest using importance sampling (described in detail in Chapter 6) from mixture distributions to describe fully these sorts of posterior forms, with the advantage that simulation work helps describe features that are difficult to visualize. One helpful strategy is the use of *product grids* (Naylor and Smith 1982, 1988a,b; Smith et al. 1985, 1987), which involves reparameterizing the posterior distribution into a form that is recognizable as a polynomial times a normal form. The advantage of this approach is that the new form can be "orthogonalized" (to some extent) and numerical integration becomes easier.

4.5.2.2 Highest Posterior Density Intervals

The last few sections have identified problems that can occur so as to seriously affect the quality of inference. In this section we argue that a posterior (or likelihood) summary procedure which describes features that would be missed by standard reporting mechanisms is justified as a *standard* procedure in social science inference.

The standard Bayesian method for describing multimodal posteriors (or others, for that matter) is the *highest posterior density* (HPD) *interval*. For marginal distribution summaries this is a noncontiguous interval of the sample space for the parameter of interest, and for joint distribution summaries this is a region in the multidimensional sample space. Marginal HPD intervals are easier to display and discuss, and therefore this is the norm. It is important to note that while HPDs are a Bayesian creation, the core idea does not depend on any Bayesian assumption and non-Bayesian treatments, *highest density regions* (HDRs), are just as easily constructed and interpreted.

More specifically, the $100(1 - \alpha)\%$ HPD interval is the subset of the support of the posterior distribution for some parameter, ξ, that meets the criteria

$$C = \{\xi : \pi(\xi|\mathbf{x}) \geq k\}, \tag{4.25}$$

where k is the largest number that guarantees

$$1 - \alpha = \int_{\xi : \pi(\xi|\mathbf{x}) > k} \pi(\xi|\mathbf{x}) \, d\xi \tag{4.26}$$

(Bernardo and Smith 1994, p. 395). This is the smallest possible subset of the sample space where the probability that ξ is in the region is maximized at the chosen $1 - \alpha$ level. Unlike the more common confidence interval, the Bayesian approach is to treat ξ like the random quantity as opposed to the interval. Figure 4.3 shows a 95% HPD interval for a bimodal density where k and C from (4.25) are labeled. The figure illustrates the utility of an HPD interval as opposed to a confidence interval where one starts at the mean and marches out a

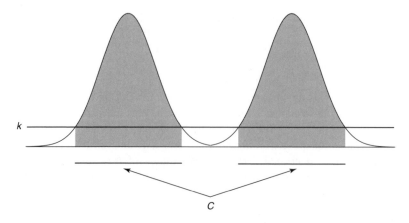

Fig. 4.3 HPD region and corresponding density.

certain number of standard errors in both directions.[9] In the example here, with two poles, one mode, and two antimodes, the standard confidence interval would include the low-density region in the middle and exclude regions on either end that had higher density than this "valley." Hyndman (1996) presents a number of alternative ways to display HPD intervals.

HPD intervals can also be used in more integrative ways. Joseph et al. (1995) show how sample size determination is really a Bayesian process since it requires the use of prior assumptions before collecting the prescribed data. They develop a method that ties desired HPD coverage to alternative sample sizes with the following alternative criteria:

- *Average coverage criterion.* Fix the desired total width of the HPD interval (w) and find the minimum sample size (n) that gives the expected coverage probability (averaging over the sample) of $1 - \alpha$:

$$\int \left(\int_{a(\mathbf{x},n)}^{a(\mathbf{x},n)+w} \pi(\xi | \mathbf{x}, n)\, d\xi \right) f(\mathbf{x})\, d\mathbf{x} \geq 1 - \alpha, \qquad (4.27)$$

where $a(\mathbf{x}, n)$ is the lower limit of the HPD interval.

- *Average length criterion.* Now suppose that we want to fix the coverage probability and find the minimum sample size for an *average* total HPD width (w):

$$\int_{a(\mathbf{x},n)}^{a(\mathbf{x},n)+w(\mathbf{x},n)} \pi(\xi | \mathbf{x}, n)\, d\xi = 1 - \alpha \quad \text{such that} \quad \int w(\mathbf{x}, n) f(\mathbf{x})\, d\mathbf{x} \leq w,$$

$$(4.28)$$

where $w(\mathbf{x}, n)$ is the average function.

[9]The Bayesian version of a confidence interval is the credible interval, where the same approach is taken but the interpretation is different since parameters are assumed to possess distributional qualities rather than being fixed.

- *Worst outcome criterion.* A more conservative approach is to require simultaneous limits for both coverage probability and width over possible sample characteristics, minimizing n:

$$\inf_{\mathbf{x}} \left(\int_{a(\mathbf{x},n)}^{a(\mathbf{x},n)+w} \pi(\xi|\mathbf{x}, n)\, d\xi \right) \geq 1 - \alpha, \tag{4.29}$$

where the infimum is taken over the support of $\mathbf{x}$.

These are straightforward for HPD intervals that are contiguous but obviously require more care in the case of noncontiguous intervals since the multiple integrals need to be done separately.

As an example of routine reporting of HPD intervals, Wagner and Gill (2004) extend Meier's standard model of the education production function by making it an explicitly Bayesian design, creating priors from Meier's previous work (Meier and Smith 1995; Smith and Meier 1995; Meier et al. 2000) as well as from his detractors. This is a linear model with modestly informed priors, a random effects component, and an interaction term, according to the specification:

$$
\begin{aligned}
Y[i] &\sim \mathcal{N}(\lambda[i], \sigma^2), \\
\lambda[i] &= \beta_0 + \beta_1 x_1[i] + \cdots + \beta_k x_k[i] + \epsilon[i] \\
\epsilon[i] &\sim \mathcal{N}(0.0, \tau) \\
\beta[i] &\sim \mathcal{N}(0.0, 10)
\end{aligned}
\tag{4.30}
$$

for $i = 1 \cdots n$, from a pooled panel dataset of over 1000 school districts in Texas for a seven-year period (the last term includes interacted data). The outcome variable is the standardized-test pass rate for the district, and the covariates include demographics, class size, state funding, teacher salary, and experience. The central theoretical interest is whether the education bureaucracy is the product or cause of poor student performance, and Meier measures bureaucracy as the total number of full-time administrators per 100 students and lag the variable so as to create a more likely causal relationship.

The following semi-informed normal priors are derived by Wagner and Gill directly from the conclusions in Smith and Meier (1995):

$$\beta_{\text{constant}} \sim \mathcal{N}(0.0, 10) \qquad \beta_{\text{lag of student pass rate}} \sim \mathcal{N}(-0.025, 10)$$

$$\beta_{\text{lag of bureaucrats}} \sim \mathcal{N}(0.0, 10) \qquad \beta_{\text{low-income students}} \sim \mathcal{N}(0.23, 10)$$

$$\beta_{\text{teacher salaries}} \sim \mathcal{N}(0.615, 10) \qquad \beta_{\text{teacher experience}} \sim \mathcal{N}(-0.068, 10)$$

$$\beta_{\text{gifted classes}} \sim \mathcal{N}(0.0, 10) \qquad \beta_{\text{class size}} \sim \mathcal{N}(-0.033, 10)$$

$$\beta_{\text{state aid percentage}} \sim \mathcal{N}(0.299, 10) \qquad \beta_{\text{funding per student}} \sim \mathcal{N}(0.0, 10)$$

$$\beta_{\text{class size} \times \text{teacher salaries}} \sim \mathcal{N}(0.0, 10).$$

Table 4.3 Posterior Summary, Interaction Model

Explanatory Variable	Posterior Mean	Posterior SE	95% HPD Interval
Constant term	4.799	2.373	[0.165: 9.516]
Lag of student pass rate	0.684	0.008	[0.667: 0.699]
Lag of bureaucrats	−0.042	0.261	[−0.557: 0.469]
Low-income students	−0.105	0.006	[−0.117: −0.094]
Teacher salaries	0.382	0.099	[0.189: 0.575]
Teacher experience	−0.066	0.046	[−0.156: 0.025]
Gifted classes	0.096	0.021	[0.054: 0.138]
Class size	0.196	0.191	[−0.180: 0.569]
State aid percentage	0.002	0.004	[−0.006: 0.010]
Funding per student (×1000)	0.049	0.175	[−0.294: 0.392]
Class size × teacher salaries	−0.015	0.007	[−0.029: −0.002]

Posterior standard error of $\tau = 0.00071$

The model is estimated (for analytical convenience) with `WinBUGS` and the results are given in Table 4.3. Here all of the 95% HPD intervals are contiguous, although it would be easy simply to list out the interval set if this were not so. Most important, notice that this presentation of coefficient estimate (posterior mean here; see Chapter 6), standard error, and interval summary has a very comfortable and familiar feeling even though this is a fully Bayesian model reporting density regions instead of confidence intervals or p-values. Furthermore, it is straightforward to scan down the column of HPD intervals and judge statistical reliability by the conventional interpretation of zero being included in the interval (although Bayesians typically consider other posterior characteristics as well).

4.5.2.3 Simulation Methods for Producing HPD Intervals

Sometimes the integral in (4.26) can be difficult to calculate analytically. In these cases it is often possible to obtain the HPD interval via standard Monte Carlo simulation. The general procedure is to generate a large number of simulated values from the (possibly multidimensional) posterior of interest, sort them, then identify the k threshold of interest. Sometimes it is also difficult to sample directly from the posterior of interest, and in these cases importance sampling can by used, as described in Chapter 6.

The following pseudo-code implements this process.

```
sorted.vals = sort(rand.draw(my.posterior,n=100000))
posterior.empirical.cdf = running.sum(sorted.vals)
k = sorted.vals[ length(posterior.empirical.cdf
[posterior.empirical.cdf < 0.05]) ]
```

This code:

- Generates 100,000 values from the posterior of interest (possibly with importance sampling) and sorts these draws

- Creates an "empirical CDF" by creating a running sum along the sorted values (i.e., [1,2,3] becomes [1,3,6])
- Sorts k by the value corresponding to the vector position of the empirical CDF value of 0.05 (for a 95% HPD interval, the other size is done similarly)

The point is to determine the "altitude" whereby 5% of the values are below and 95% are above, regardless of position. The second advantage of this method is that higher-dimensional problems can be addressed with only slightly more trouble, whereas these can often be difficult analytical integration problems.

4.5.2.4 Problems with High Dimensions

Multimodal distributions present much greater challenges in higher dimensions because the important features of the likelihood surface might not "line up" with the axis when taking marginals. As a result, important or interesting characteristics of the multimodality can get lost when considering marginals alone.

As an example, consider the bimodal, three-dimensional likelihood surface in Figure 4.4, which is a mixture of two multivariate normals, $\mathcal{N}(\mu_1, \Sigma_1)$ and $\mathcal{N}(\mu_2, \Sigma_2)$, according to

$$
f(\mathbf{x}) = (2\pi)^{-\frac{1}{2}k} \left(|\Sigma_1|^{-\frac{1}{2}} p \exp\left[-\frac{1}{2}(\mathbf{x} - \mu_1)\Sigma_1^{-1}(\mathbf{x} - \mu_1)' \right] \right.
$$

$$
\left. + |\Sigma_2|^{-\frac{1}{2}}(1-p)\exp\left[-\frac{1}{2}(\mathbf{x} - \mu_2)\Sigma_2^{-1}(\mathbf{x} - \mu_2^{-1})' \right] \right), \qquad (4.31)
$$

where p is the proportion of the first component and $1 - p$ is the proportion of the second component. We consider here only a bivariate mixture, for graphic simplicity. This is a well-studied model (Fryer and Robertson 1972), and the maximum likelihood estimation process is relatively straightforward (Day 1969; Dick and Bowden 1973; Hosmer 1973a, b; Ganesalingam and McLachlan 1981). Mixture models can also be accommodated in a Bayesian setting (Smith and Makov 1978; Ferguson 1983; Titterington et al. 1985; Escobar and West 1995). Here we consider a slightly different problem where the posterior distribution is found to be bivariate normal rather than the assumed population data. The distribution in (4.31) is now a joint bivariate form between μ_1 and μ_2: $\pi(\mu_1, \mu_2)$.

How would one choose to describe this joint distribution beyond giving (4.31) with the parameters identified? What summary values are useful?[10] Clearly, the position of the highest mode along with the curvature around this model is not sufficient to inform readers fully what the distribution looks like. Notice from the contour plot in Figure 4.4 that the two high-density regions intersect at density levels between 0.01 and 0.05. This is an important feature because it describes the ridge structure evident in the wireframe plots.

---

[10]Haughton (1997) reviews standard and specialized software solutions to the problem of coefficient estimation for finite mixture distributions.

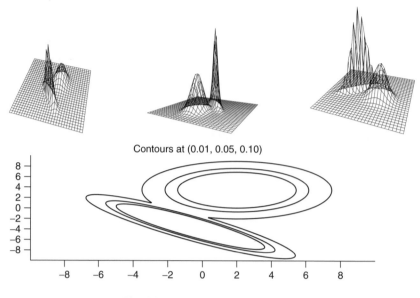

Fig. 4.4   Posterior views of $\pi(\mu_1, \mu_2)$.

Unfortunately, marginalization of this density loses some of the structural information because the axes do not line up well. In fact, if we look at the marginal posteriors from this density, much of the valley seems to disappear, particularly for μ_1. So HPD intervals of the full set of marginal densities do not contain the same information as the complete joint distribution. This is observable in Figure 4.5, where the sense of bimodality is almost gone for μ_1 and observable but indeterminant for μ_2.

There are several solutions to this problem. First, one could present the results graphically in higher dimensions such that important features are evident. Second, it is possible to use Givens rotations or the Household transformations (Thistead 1988, Chap. 3). Unfortunately, figuring out the rotations of the axes that give the most informative picture is difficult in higher dimensions and usually requires considerable trial and error. One way to get an idea of where such features exist in complex posteriors is to run the EM algorithm from many starting points and observe the location and structure of the modes.

4.5.3   Inference in the Presence of Instability

Occasionally, despite following all of the recommendations discussed in this chapter, one may not be sure whether one's answers are accurate: Perturbation tests may continue to show sensitivity to noise, or different platforms may give you substantively different answers. Despite one's best attempts to research the properties of each algorithm and software package used, and to examine all relevant diagnostics, a set of competing solutions or a range of possible answers may still exist. What can be done when faced with several competing answers?

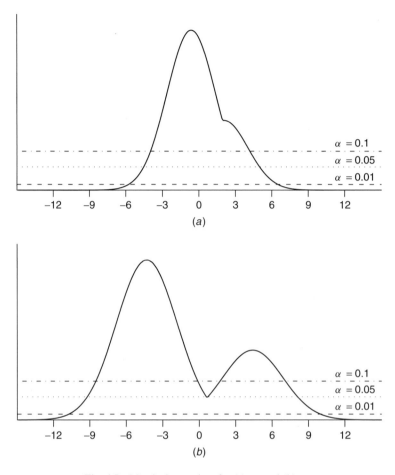

Fig. 4.5   Marginal posteriors for (a) μ_1 and (b) μ_2.

There is no simple answer to this question. Implicitly or explicitly, one must make a judgment about the tractability of the problem, the costs of further efforts put toward obtaining a better solution, and the plausibility of current solutions.

Some researchers might object (and in our experience, have objected): "If two different algorithms or implementations give different results, at least one is simply wrong. It makes no sense to use estimates based on incorrect calculations, so do more work to figure out how to do the calculations correctly."

We understand this response, and it contains some truth. If different programs yield different answers, to identify the sources of the discrepancy we recommend: examining diagnostics, applying sensitivity tests, utilizing more accurate external libraries (where possible), extending the precision of the arithmetic operators, and selecting the best algorithms available. If varying answers persist, we suggest that these variations should be reflected in the reported level of uncertainty of estimates.

The problem of selecting an implementation/algorithm is analogous to that of selecting a model. In model selection, there is simply no way of being absolutely certain that the correct model has been found. In social science, there are often a number of models that have some degree of plausibility, and some support from the data. Lacking decisive evidence for rejecting them, one has three choices: First, decline to report any estimates and collect more data, in the hope that new data will help to resolve the issue. Second, pick the most 'plausible' model and ignore the rest. Third, acknowledge that several models are plausible, and use some method to combine estimates across models.

The first approach is sometimes best but often prohibitively expensive. Worse, one may need to make decisions based on the current data immediately, before any opportunities to collect more data occur. Or, when one is trying to explain historical evidence, it may not be possible to collect more data at all.

The second approach is, we think, fairly common, but often mistaken. Ignoring other plausible models misrepresents uncertainty over the state of the world and leads to poorer estimates than incorporating the estimates from these other models (Hoeting et al. 1999).

Often, the third option is most appropriate: incorporate in your estimates a very wide range of, or even all of, the plausible models. One simple technique for doing this is to include Leamer bounds in the reported analysis (Leamer 1978, 1983; Leamer and Leonard 1983; Klepper and Leamer 1984). Leamer proposes reporting the extreme bounds for coefficient estimates observed during the specification search. In particular, coefficients with Leamer bounds that straddle zero should be regarded with some suspicion, even if they are statistically reliable in the final product. Recent use of this approach is found in Levine and Renelt (1992), Western (1995), Ades and DiTella (1997), Akhand (1998), and Smith (2003). Unfortunately, Leamer bounds require a great deal of self-regulated researcher honesty. In the process of model building, it is not uncommon to specify many different models, some of which provide wildly different coefficients for variables of prime interest. We suspect that many researchers would rationalize as follows: "Those were really just accidents, I'm not going to record." Of course, *all* aspects of the model development process require self-imposed honesty, but this is one that would not be checked by replication by others.

A second means of incorporating more widely encompassing results is the use of Bayesian model averaging. Raftery, in particular, has advocated the idea of reducing arbitrariness in social science model selection by using the Bayesian inference paradigm to average across model results, thus obtaining a metaresult that incorporates a broader perspective (Raftery 1995, 1996; Raftery et al. 1997; Volinsky et al. 1997). For an accessible introduction, see Bartels (1997) and Hoeting (1999). These methods can be applied directly to estimates produced by different packages or algorithms. If one cannot obtain a definitive answer, we recommend using these methods.

Some scholars might object that the analogy is flawed and that model selection differs from algorithm selection in important ways. They might offer two objections: that it is not possible even in theory to prove that a particular model is correct in an absolute sense but that one should be able to prove an algorithm correct; and that, in practice, distinguishing between competing models can require that more data be obtained, which may be prohibitively expensive or even impossible.

We believe that such objections have force but are not determinative. Proving an algorithm correct is often possible, but not always. Sometimes, like proving a model correct, it is impossible (or not within practical reach). Moreover, proving a complex statistical implementation correct is generally almost always beyond the ability of current software engineering practice.

In addition, global optimization of difficult nonlinear problems, which is required to solve many models, is generally computationally intractable (see Chapter 2). If the problem is big enough and is not well behaved, it is usually possible to find a local solution, but practically impossible to verify that a global solution has been reached. That is, although theoretically possible to do, it can be proved to take more time to accomplish than the expected age of the universe. In practice, we sometime have to contend with finite limits to our computational resource, just as we sometimes have to contend with limits to our ability to collect data.

Finally, it should be recognized that model uncertainty and implementation uncertainty are intertwined. Simpler statistical models will have more straightforward computational requirements and introduce less uncertainty over algorithmic and implementation choice. As Achen (2003) notes, simpler (less generalized) models may be less brittle and require significantly fewer data to estimate successfully. Complex statistical models impose additional costs due to the increased algorithmic and implementation uncertainty required in estimating them. This computationally driven uncertainty must be incorporated in the results presented using such models if the best choice is to be made among competing models.

To summarize, computational algorithm and implementation selection are analogous to model selection, although generally a milder problem. Every effort should be made to find an accurate solution, and generally, with reasonable effort, an accurate solution will be forthcoming. However, in some circumstances it will be impossible, or prohibitively expensive, to determine which of a set of solutions is correct. When this occurs, one should incorporate in the estimates the results from all plausible computations.

CHAPTER 5

Numerical Issues in Markov Chain Monte Carlo Estimation

5.1 INTRODUCTION

This chapter is focused exclusively on numerical reliability and accuracy for Markov chain Monte Carlo (MCMC) techniques used in modern Bayesian estimation. Many of the issues discussed, however, apply equally to simulation work in other areas. Notably, this chapter provides the first detailed look at the qualitative effects of pseudo-random number generation and machine rounding/truncating on the convergence and empirical reliability of Markov chain Monte Carlo estimation.

The chapter begins with a detailed review of Markov chain Monte Carlo theory and a particular focus on conditions necessary for convergence. The starting point is first principles followed by the major theoretical background required to consider problems that might occur in reaching the desired stationary distribution. We then demonstrate in detail that the period of the underlying pseudo-random number generator can have disastrous effects on the path of the Markov chain. This discussion is new to the literature. As illustrations, the example of the slice sampler and a discussion of the potential problem in WinBUGS are provided. Also highlighted is another potential pitfall: the absorbing state problem caused by machine rounding and truncating. The primary contribution of this chapter is to provide evidence that low-level numerical concerns affect MCMC analysis and that these problems are rarely discussed in either applied or theoretical work.

Thus far we have addressed important numerical issues for the standard and traditional process of computer-based estimation of linear models, descriptive statistics, and maximum likelihood functions. This is the first chapter that addresses a fundamentally different type of estimation process based on stochastic simulation. As different as these two methods are, there are some strikingly similar concerns about numerical accuracy in intermediate calculations. Some contrasts exist, however, inasmuch as standard likelihood inference tends to be mostly worried about problems with numerical derivatives (i.e., mode finding), while Bayesian inference generally seeks to estimate integral quantities based on the posterior distribution.

Numerical Issues in Statistical Computing for the Social Scientist, by Micah Altman, Jeff Gill, and Michael P. McDonald
ISBN 0-471-23633-0 Copyright © 2004 John Wiley & Sons, Inc.

Aside from the obviously contrasting basic calculus procedures, there is much more of a need for making inferences based on *simulations* in Bayesian work. Consequently, sources of inaccuracy that are relatively small and may even be ignored in a mode-finding context can be greatly magnified purely because of the large-n iterative nature of Monte Carlo procedures.

The use of modern stochastic simulation with Markov chains in Bayesian statistics begins after 1990 (Gelfand and Smith's review essay essentially initiates this era), and is therefore a relatively young subject to statisticians and others. Although a great deal of work has been done to assess numerical accuracy of calculations in standard likelihoodist statistical computing, the subject of Chapters 1 to 4, *there is currently almost no published work that analyzes the effect of well-known general numerical computing problems for MCMC.* Specifically, do unwanted round-off and truncation in register arithmetic affect the necessary theoretical characteristics of the Markov chain? Does the necessarily periodic method of generating pseudo-random numbers on computers affect the convergence of Markov chains? We seek to answer these types of questions in this chapter.

5.2 BACKGROUND AND HISTORY

A persistent and agonizing problem for those developing Bayesian models in most of the twentieth century was that it was often possible to get an unreasonably complicated posterior from multiplying realistic priors with the appropriate likelihood function. That is, the mathematical form exists, but quantities of interest such as means and quantiles cannot be calculated directly. This problem shunted Bayesian methods to the side of mainstream statistics for quite some time. What changed this state was the publication of Gelfand and Smith's 1990 review essay that described how similar problems had been solved in statistical physics with Markov chain simulation techniques.

Markov chain Monte Carlo is a set of iterative techniques where the values generated are (eventually) from the posterior of interest (Gill 2002). So the difficult posterior form can be described empirically using a large number of simulated values, thus performing difficult integral calculations through computation rather than calculation. The result of this development was a torrent of papers that solved many unresolved Bayesian problems, and the resulting effect on Bayesian statistics can easily be characterized as revolutionary.

Markov chains are a type of stochastic process where the next value depends probabilistically only on the current value of the series. In general, it is possible to set up such a chain to estimate multidimensional probability structures, the desired posterior distribution, by starting a Markov chain in the appropriate sample space and letting it run until it settles into the desired target distribution. Then, when it runs for some time confined to this particular distribution of interest, it is possible to collect summary statistics such as means, variances, and quantiles from the values simulated. The two most common procedures are the Metropolis–Hastings algorithm and the Gibbs sampler, which have been shown to possess desirable theoretical properties that lead to empirical samples from the target distribution.

There are now quite a few volumes that provide high-level descriptions of MCMC estimation, including Gelman et al. (1995), Gamerman (1997), Carlin and Louis (2000), Congdon (2001), and Gill (2002). There are also an increasing number of advanced books on MCMC, such as those of Tanner (1996), Robert and Casella (1999), Chen et al. (2000), and Liu (2001).

5.3 ESSENTIAL MARKOV CHAIN THEORY

In this section we provide the technical background necessary to understand the implications of numerical accuracy on the reliability of estimation with Markov chain Monte Carlo methods. A portion of the theory in this section is rather abstract and perhaps new to some social scientists. Consistent with earlier chapters, the primary goal here is to relate standard Markov chain theory with the numerical problems that can arise from finite register arithmetic in computer implementations.

5.3.1 Measure and Probability Preliminaries

Define H as the set defining the support of θ, a random variable of interest with individual points denoted with subscripts: $\theta_i, \theta_j, \theta_k, \ldots$. Θ is a σ-algebra of subsets of H that is assumed to be generated by a countable collection of these subsets of H, having elements (events) denoted $A, B, C, \ldots$. Thus (H, Θ) is the measurable space for θ, called the *state space*. Following standard notation we use $f, g, h, \ldots$ to denote real-valued measurable functions defined on Θ. Also define $\mathcal{M}$ as the full collection of signed measures on (H, Θ), where by convention λ and μ denote elements of the measurable space. It is also useful to be specific about the sign, where the class of positive measures is given by $\mathcal{M}^+ = \{\lambda \in \mathcal{M} : \lambda(H) > 0)$. Signed measures will be important later in this chapter when asserting Markov chain convergence.

Now define a transition kernel $K(\theta, A)$, which is the general mechanism for describing the probability structure that governs movement of the Markov chain from one state to another. This means that $K(\theta, A)$ is a defined probability measure for all θ points in the state space to the set $A \in \Theta$: a mapping of the potential transition events to their probability of occurrence (Robert and Casella 1999, p. 141; Gill 2002, p. 303). More formally, K is a nonnegative σ-finite kernel on the state space (H, Θ) that provides the mapping $H \times \Theta \to \mathcal{R}^+$ if the following conditions are met:

1. For every subset $A \in \Theta$, $K(\bullet, A)$ is measurable.
2. For every point $\theta_i \in H$, $K(\theta_i, \bullet)$ is a measure on the state space.
3. There exists a positive $\Theta \otimes \Theta$ measurable function, $f(\theta_i, \theta_j)$, $\forall \theta_i, \theta_j \in H$ such that $\int k(\theta_i, d\theta_j) f(\theta_i, \theta_j) < \infty$.

All of this is simply to assert that the transition kernel behaves nicely as a probability mechanism for determining the path of the Markov chain.

A stochastic process is a set of observed $\theta^{[t]}$ ($t \geq 0$) on the probability space $(\Omega, \mathcal{F}, P)$, where the superscript t denotes an order of occurrence and Ω is the relevant, nonempty outcome space with the associated σ-algebra $\mathcal{F}$ and probability measure P (Doob 1990; Billingsley 1995; Ross 1996). Thus the sequence of Ω-valued $\theta^{[t]}$ random elements given by $t = 0, 1, \ldots$ defines the Ω-valued stochastic process (although typically $\mathcal{R}$-valued, with some restrictions) (Karlin and Taylor 1981, 1990; Hoel et al. 1987). Note that the labeling of the sequence $T : \{\theta^{[t=0]}, \theta^{[t=1]}, \theta^{[t=2]}, \ldots\}$ implies consecutive even-spaced time intervals; this is done by convention, but not necessity.

Define at time n the *history* of the stochastic process as the increasing series of sub-σ-algebras defined by $\mathcal{F}_0 \subseteq \mathcal{F}_1 \subseteq \cdots \subseteq \mathcal{F}_n$, where θ is measurable on each. An H-valued stochastic process, θ_t; $t \geq 0$, with transition probability P and initial value θ_0 is a *Markov chain* if at the $(n + 1)$th time,

$$P(\theta_{n+1}|\mathcal{F}_n) = P(\theta_{n+1}|\theta_n), \ \forall n \geq 0. \tag{5.1}$$

[See Zhenting and Qingfeng (1978, Chap. 6) for some mathematical nuances.] In plainer words, for Markov chains the only component of the history that matters in determining movement probabilities at the current step is the current realization of the stochastic process. Therefore, for a given event A in Θ and Markov chain at time $t - 1$,

$$P(\theta^{[t]} \in A|\theta^{[0]}, \theta^{[1]}, \ldots, \theta^{[t-2]}, \theta^{[t-1]}) = P(\theta^{[t]} \in A|\theta^{[t-1]}). \tag{5.2}$$

This is the *Markovian property*, which defines a Markov chain from within the general class of stochastic processes.

5.3.2 Markov Chain Properties

Certain properties of the chain and the measure space are necessary by assumption or design to ensure useful results for statistical estimation. These are now given. The following section is somewhat more abstract and theoretical than previous discussions. This is necessary to facilitate the subsequent presentation of numerical pathologies. For more introductory treatments, see Gelman et al. (1995), Gamerman (1997), or Gill (2002).

5.3.2.1 ψ-Irreducibility

Colloquially, a set, A, is *irreducible* if every point or collection of points in A can be reached from every other point or collection of points in A, and a Markov chain is irreducible if it is defined on an irreducible set. More formally, first define ψ as a positive σ-finite measure on (H, Θ) with $A \in \Theta$ such that $\psi(A) > 0$. For the transition kernel $K(\theta, A)$, if every positive ψ-subset, $A' \subseteq A$, can be reached from every part of A, then A is called ψ-*communicating*. More important, when the full state space H is ψ-communicating, this kernel is ψ-irreducible. It is usually convenient to assume that ψ-irreducible here is *maximally* ψ-irreducible, meaning that for any other positive σ-finite measure on (H, Θ), ψ', it is necessarily true that $\psi > \psi'$. See Meyn and Tweedie (1993, pp. 88–89) for some additional but not critical details.

5.3.2.2 Closed and Absorbing Sets

We say that a nonempty set $A \in \Theta$ is *obtainable* from the state θ for the Markov chain defined at time n by the kernel K^n if

$$K^n(\theta, A) > 0 \quad \text{for some } n \geq 1, \tag{5.3}$$

and *unobtainable* if

$$K^n(\theta, A) = 0 \quad \text{for all } n \geq 1. \tag{5.4}$$

This set A is called *closed* for K if A^c is not obtainable from A:

$$K^n(\theta, A^c) = 0 \quad \text{for all } \theta \in A \text{ and all } n \geq 1. \tag{5.5}$$

The condition *absorbing* (Revuz 1975) is more restrictive than closed:

$$K(\theta, A) = K(\theta, \Theta) = 1 \quad \text{for all } \theta \in A, \tag{5.6}$$

since it is possible under the closed condition but impossible under the absorbing condition that

$$K(\theta, A) = K(\theta, \Theta) \neq 1 \quad \text{for some } \theta \in A. \tag{5.7}$$

That is, a closed set can have subsets that are unavailable but an absorbing state fully communicates with all its subsets.

5.3.2.3 Homogeneity and Periodicity

A Markov chain is said to be *homogeneous* at step n if the transition probabilities at this step do not depend on the value of n. We can also define the *period* of a Markov chain, which is simply the length of time to repeat an identical cycle of chain values. It is important to the underpinnings of the basic theories to have a chain that does not have such a defined cycle, that is, one where the only length of time for which the chain repeats some cycle of values is the trivial case with cycle length equal to 1. Such a chain is called an *aperiodic Markov chain*. Obviously, a *periodic Markov chain* is nonhomogeneous because the cycle of repetitions defines transitions, at specified points, based on time.

5.3.2.4 Recurrence

A homogeneous ψ-irreducible Markov chain on a closed set is called *recurrent* (sometimes called *persistent*) with regard to a set, A, which is a single point or a defined collection of points (required for the bounded-continuous case) if the probability that the chain occupies each subset of A infinitely often over unbounded time is 1. More informally, when a chain moves into a recurrent state, it stays there forever and visits every substate infinitely often. A recurrent Markov chain is *positive recurrent* if the mean time to return to A is bounded; otherwise, it is called *null recurrent* (Doeblin 1940).

Unfortunately, given unbounded and continuous state spaces, we have to work with a slightly more complicated version of recurrence. If there exists a σ-finite probability measure P on the measure space H such that an ψ-irreducible Markov chain, θ_n, at time n has the property $P(\theta_n \in A) = 1$, $\forall A \in H$ where $P > 0$, it is *Harris recurrent* (Harris 1956; Athreya and Ney 1978). This is necessary because in the continuous case an aperiodic, ψ-irreducible chain with an invariant distribution on an unbounded continuous state space that is *not* Harris recurrent has a positive probability of getting stuck forever in an area bounded away from convergence, given a starting point there. So the purpose of the Harris definition is to avoid worrying about the existence of a pathological null set in $\mathcal{R}^k$.

Fortunately, all of the standard MCMC algorithms implemented on (naturally) finite-state computers are Harris recurrent (Tierney 1994). From this point on we will use *recurrent* generically to mean the simpler definition of recurrent for closed or discrete sets and Harris recurrent for unbounded continuous sets.

5.3.2.5 *Transience*

First define for a set A the *expected number of visits by chain θ_n to A in the limit* $\eta_A = \sum_{n=1}^{\infty} I_{(\theta_n \in A)}$, which is just a summed indicator function that counts hits. Both transience and recurrence can be defined in terms of the expectation for η_A:

- The set A is *uniformly transient* if $\exists M < \infty \ni E[\eta_A] \leq M \ \forall \theta \in A$. A single state in the discrete state space case is *transient* if $E[\eta_\theta] < \infty$.
- The set A is *recurrent* if $E[\eta_A] = \infty \ \forall \theta \in A$. A single state in the discrete state space case is *recurrent* if: $E[\eta_\theta] = \infty$.

The important and relevant theorem is given (with proof) by Meyn and Tweedie (1993, 182-83) as well as by Nummelin (1984, p. 28):

Theorem. If θ_n is a ψ-irreducible Markov chain with transition kernel $K(\theta, A)$, it must be either transient or recurrent, depending on whether it is defined on a transient or recurrent set A.

Proof. This is a direct consequence of Kolmogorov's zero-one law. Proof details are given in Billingsley (1995, p. 120).

Thus the chain is either recurrent and we know that it will eventually settle into an equilibrium distribution, or it is transient and there is no hope that it will converge accordingly.

We can also define the *convergence parameter* of a kernel rK as the real number $0 \leq R < \infty$ on a closed set A such that $\sum_0^{\infty} r^n K^n < \infty$ for every $0 \leq$

$r < R$ and $\sum_0^\infty r^n K^n = \infty$ for every $R \geq r$. It turns out that for ψ-irreducible Markov chains there always exists a finite R that defines whether or not the kernel for this Markov chain is *R-transient* if the first condition holds, and *R-recurrent* if the second condition holds.

5.3.2.6 Stationarity

Define $\pi(\theta)$ as the stationary distribution of the Markov chain for θ on the state space H, with transition probability $P(\theta_i, \theta_j)$ to indicate the probability that the chain will move from arbitrary point θ_i to arbitrary point θ_j. The stationary distribution (sometimes called the *invariant distribution*) is then defined as satisfying

$$\sum_{\theta_i} \pi^t(\theta_i) P(\theta_i, \theta_j) = \pi^{t+1}(\theta_j) \qquad \text{discrete state space}$$

$$\int \pi^t(\theta_i) P(\theta_i, \theta_j) d\theta_i = \pi^{t+1}(\theta_j) \qquad \text{Continuous state space.} \qquad (5.8)$$

The multiplication by the transition kernel and evaluating for the current point (the summation step for discrete sample spaces and the integration step for continuous sample spaces) produces the same marginal distribution, $\pi(\theta) = \pi(\theta)P$. So the marginal distribution remains fixed when the chain reaches the stationary distribution and we can drop the superscript designation for iteration number. Once the chain reaches the stationary distribution (also called the *invariant distribution, equilibrium distribution*, or *limiting distribution*), its movement is governed purely by the marginal distribution, $\pi(\theta)$ forever. It turns out (fortunately) that an ψ-irreducible, aperiodic Markov chain is guaranteed to have exactly one stationary distribution (Häggström 2002, p. 37). This, in fact, provides the core theoretical motivation for estimation with MCMC; if the stationary distribution of the Markov chain is the posterior distribution of interest, we are certain eventually to get samples from this posterior.

5.3.2.7 Ergodicity

The fundamental result from Markov chain theory that drives the use of MCMC is the ergodic theorem. If a chain is positive (Harris) recurrent and aperiodic on some state A, it is *ergodic* (Tweedie 1975). Ergodic Markov chains have the important property

$$\lim_{n \to \infty} |P^n(\theta_i, \theta_j) - \pi(\theta_j)| = 0 \qquad (5.9)$$

for all θ_i, and θ_j in the subspace (Norris 1997, p. 53). Therefore, the chain has converged to its limiting distribution and all future draws are from the identical marginal distribution. Furthermore, once a specified chain is determined to have reached its ergodic state, sample values behave as if they were produced by the posterior of interest from the model. Ergodicity is not the only way to demonstrate convergence, but it is by far the most straightforward (Meyn and Tweedie 1994a).

Ergodicity provides a condition for convergence but does not itself provide a bound on the time required to reach convergence to the stationary distribution. It turns out that there are different "flavors" of ergodicity that provide faster convergence if the Markov chain being implemented is known to have such properties. A Markov chain with stationary distribution π at time t, positive constant $r \in (0, 1)$, and a nonnegative, real-valued function of θ, $f(\theta)$, such that there is a geometric decrease in t of the total variation distance between $P^n(\theta_i, \theta_j)$ and $\pi(\theta_j)$ [i.e., less than or equal to $f(\theta)r^t$] is *geometrically ergodic* (Nummelin and Tweedie 1978; Chan 1993; Roberts and Tweedie 1994; Meyn and Tweedie 1994b; Down et al. 1995). Mengersen and Tweedie (1996) give conditions where a simple type of Metropolis–Hastings chain is geometrically ergodic:

- **Random walk chain.** Metropolis–Hastings candidate jumping values are selected by an offset from the current state according to a simple additive scheme, $\theta' = \theta + f(\tau)$, where τ is some convenient random variable form with distribution $f(\cdot)$.

Mengersen and Tweedie also demonstrate that certain other Metropolis–Hastings derivatives are *not* geometrically ergodic (1996, p. 106). Roberts and Polson (1994) and Chan (1989, 1993) show that the Gibbs sampler is geometrically ergodic, partially justifying its widespread appeal. Detailed proofs and associated conditions are provided by Athreya et al. (1996). So ergodicity, in particular geometric ergodicity, is an attractive property of the Markov chain if it is present because one knows that convergence will occur in a "reasonable" period.

A stronger form of ergodicity is *uniform ergodicity*, which replaces the function $f(\theta)$ in geometric ergodicity with a positive constant. Only some Gibbs samplers and other MCMC implementations have been shown to have this more desirable property. Another simple Metropolis–Hastings variant that has this property is:

- **Independence chain.** Metropolis–Hastings candidate jumping values are selected from a convenient form as in the *random walk chain*, but ignoring the current position completely, $\theta' = f(\tau)$, and generating potential destinations completely on the $f(\cdot)$ distribution. This is uniformly ergodic *provided that $f(\cdot)$ is chosen such that the importance ratio $\pi(\theta)/f(\theta)$ is bounded* (Mengersen and Tweedie 1996, p. 108).

Uniform ergodicity is obviously another desirable property, but certainly not one that can be ensured.

5.3.3 The Final Word (Sort of)

So to map the logical path here. Homogeneity and ψ-irreducibility on a closed set give recurrence. Positive recurrence and aperiodicity give ergodicity: convergence

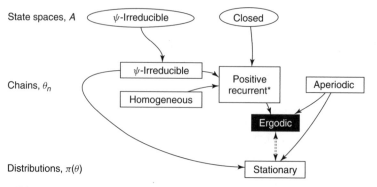

*Discrete or bounded continuous: $E[\eta_A] = \infty \; \forall \theta \in A$. Continuous: $P(\theta_n \in A) = 1, \; \forall A \in H$
where $P > 0$ (Harris).
Positive recurrence: mean return time to A is bounded.

Fig. 5.1 Conditions leading to ergodicity and stationarity.

for a Markov chain to the posterior of interest (empirical averages converge to probabilistic averages) (Athreya et al. 1996; Brémaud 1999, p. 110). This means that we can then collect samples from the simulations and calculate various posterior quantities empirically.

Figure 5.1 gives a graphic overview of these conditions where the path of arrows denotes necessary preconditions (rather than necessary implications). Note the centrality of recurrence (generally defined) in this process. This figure also distinguishes between properties associated with state spaces and properties associated with Markov chains themselves.

5.4 MECHANICS OF COMMON MCMC ALGORITHMS

In this section we review three commonly used procedures for constructing Markov chains for statistical inference. Each produces valid inferences if set up correctly, and recommended use of one over the others is usually problem dependent.

5.4.1 Metropolis–Hastings Algorithm

The Metropolis–Hastings algorithm comes from statistical physics (Metropolis et al. 1953), but has proved to be enormously useful in general statistical estimation (Hastings 1970; Peskun 1973; Chib and Greenberg 1995).

Suppose that we want to simulate θ, a J-dimensional parameter vector, from the posterior distribution $\pi(\theta)$ with support known. At the ith step in the chain, we will draw θ'_j, $j = 1 : J$ from a multivariate *candidate generating distribution* over this same support: θ'. This candidate generating distribution (also called *jumping, proposal,* or *instrumental*) is picked to be easy to sample from. One convenient possibility is a multivariate normal or t-distribution centered at the current value of the θ_j's in the process and using the empirical variance from past iterations: for each θ_j: $s^2_{\theta_j}$ (and specify $s^2_{\theta_j} = 1$ as a starting value). This requires, of course, that the posteriors of interest have support over the entire real line, but $\pi(\boldsymbol{\theta})$ can usually be transformed so that simulation occurs on the more convenient metric.

It must be possible to determine $q_t(\boldsymbol{\theta}|\boldsymbol{\theta}')$ and $q_t(\boldsymbol{\theta}'|\boldsymbol{\theta})$ for the candidate generating distribution. Under the original constraints of Metropolis et al. (1953), these two conditionals needed to be equal (symmetry), although we now know that this is not necessary and have the more flexible restriction of reversibility. That is, the *detailed balance equation* (also called *reversibility*) must be true to ensure that $\pi(\boldsymbol{\theta})$ is an invariant distribution:

$$K(\boldsymbol{\theta}', \boldsymbol{\theta})\pi(\boldsymbol{\theta}') = K(\boldsymbol{\theta}, \boldsymbol{\theta}')\pi(\boldsymbol{\theta}),$$

where $K(\boldsymbol{\theta}', \boldsymbol{\theta})$ is the kernel of the Metropolis–Hastings algorithm going from $\boldsymbol{\theta}$ to $\boldsymbol{\theta}'$. Sometimes $K(\boldsymbol{\theta}', \boldsymbol{\theta})$ is labeled as $A(\boldsymbol{\theta}', \boldsymbol{\theta})$ and called the *actual transaction function* from $\boldsymbol{\theta}$ to $\boldsymbol{\theta}'$ to distinguish it from $a(\boldsymbol{\theta}', \boldsymbol{\theta})$ below. The *acceptance ratio* is now defined as

$$a(\boldsymbol{\theta}', \boldsymbol{\theta}) = \frac{q_t(\boldsymbol{\theta}^{[t]}|\boldsymbol{\theta}')}{q_t(\boldsymbol{\theta}'|\boldsymbol{\theta}^{[t]})} \frac{\pi(\boldsymbol{\theta}')}{\pi(\boldsymbol{\theta}^{[t]})}. \tag{5.10}$$

The subsequent decision that produces the $(t+1)$st point in the chain is probabilistically determined according to

$$\boldsymbol{\theta}^{[t+1]} = \begin{cases} \boldsymbol{\theta}' & \text{with probability} & P[\min(a(\boldsymbol{\theta}', \boldsymbol{\theta}), 1)] \\ \boldsymbol{\theta}^{[t]} & \text{with probability} & 1 - P[\min(a(\boldsymbol{\theta}', \boldsymbol{\theta}), 1)]. \end{cases} \tag{5.11}$$

In the case of symmetry in the candidate generating density, $q_t(\boldsymbol{\theta}|\boldsymbol{\theta}') = q_t(\boldsymbol{\theta}'|\boldsymbol{\theta})$, the decision simplifies to a ratio of the posterior density values at the two points.

These steps can be summarized for a single scalar parameter of interest according to (5.11) with the following steps:

1. Sample θ' from $q(\theta'|\theta)$, where θ is the current location.
2. Sample u from $u[0:1]$.
3. If $a(\theta', \theta) > u$, accept θ'.
4. Otherwise, keep θ as the new point.

The algorithm described by these steps also has desirable convergence properties to the distribution of interest (Roberts and Smith 1994).

5.4.2 Hit-and-Run Algorithm

The hit-and-run algorithm is a special case of the Metropolis–Hastings algorithm that separates the move decision into a direction decision and a distance decision. This makes it especially useful in tightly constrained parameter space because we can tune the jumping rules to be more efficient. As it turns out, this directional flexibility is also helpful when there are several modes of nearly equal altitude.

The algorithm proceeds according to the following steps. From an arbitrary point θ_t at time t:

Step 1:   Generate a multidimensional direction, $\mathbf{Dr}_t$, on the surface of a J-dimensional unit hypersphere from the distribution $f(\mathbf{Dr}|\theta^{[t]})$.

Step 2:   Generate a signed distance, Ds_t, from density $g(Ds|\mathbf{Dr}_t, \theta)$.

Step 3:   Set the candidate jumping point to: $\theta' = \theta^{[t]} + Ds_t \mathbf{Dr}_t$ and calculate:

$$a(\theta', \theta^{[t]}) = \frac{\pi(\theta'|\mathbf{X})}{\pi(\theta^{[t]}|\mathbf{X})}.$$

Step 4:   Move to $\theta^{[t+1]}$ according to

$$\theta_j^{[t+1]} = \begin{cases} \theta' & \text{with probability} & P[\min(a(\theta', \theta^{[t]}), 1)] \\ \theta^{[t]} & \text{with probability} & 1 - P[\min(a(\theta', \theta^{[t]}), 1)]. \end{cases}$$

This algorithm requires the following assumptions:

- For all $\mathbf{Dr}_i$, $f(\mathbf{Dr}|\theta^{[t]}) > 0$.
- $g(Ds|\mathbf{Dr}, \theta)$ must be strictly greater than zero and have the property

$$g(Ds|\mathbf{Dr}, \theta) = g(-Ds| - \mathbf{Dr}, \theta).$$

- A new form of the detailed balance equation is required:

$$g(||\theta^{[t]} - \theta'||)a(\theta', \theta^{[t]})\pi(\theta^{[t]}|\mathbf{X}) = g(||\theta' - \theta^{[t]}||)a(\theta^{[t]}, \theta')\pi(\theta'|\mathbf{X}).$$

Subject to these conditions, *this is an ergodic Markov chain with stationary distribution $\pi(\theta|\mathbf{X})$* (Chen and Schmeiser 1993). Typically, $f(\mathbf{Dr}|\theta^{[t]})$ is chosen to be uniform, but other forms are possible if necessary. In addition, the $a(\theta', \theta^{[t]})$ criterion can be made much more general. One advantage to this algorithm over standard M-H is that $g(Ds|\mathbf{Dr}, \theta)$ is also very flexible if disengaged from the direction decision, which makes the method very tunable.

5.4.3   Gibbs Sampler

The Gibbs sampler is a transition kernel created by a series of full conditional distributions. It is a Markovian updating scheme based on these conditional probability statements. Define the posterior distribution of interest as $\pi(\theta)$, where θ is a J-length vector of coefficients to estimate, the objective is to produce a Markov chain that cycles through these conditional statements moving toward and then around this distribution.

The set of full conditional distributions for $\boldsymbol{\theta}$ are denoted $\boldsymbol{\Theta}$ and defined by $\pi(\boldsymbol{\Theta}) = \pi(\theta_i | \boldsymbol{\theta}_{-j})$ for $j = 1, \dots, J$, where the notation $\boldsymbol{\theta}_{-j}$ indicates a specific parametric form from $\boldsymbol{\Theta}$ without the θ_j coefficient. We can now define the Gibbs sampler according to the following steps:

1. Choose starting values: $\boldsymbol{\theta}^{[0]} = [\theta_1^{[0]}, \theta_2^{[0]}, \dots, \theta_J^{[0]}]$.

2. At the tth step starting at $t = 1$, complete the single cycle by drawing values from the J conditional distributions given by

$$\theta_1^{[t]} \sim \pi(\theta_1 | \theta_2^{[t-1]}, \theta_3^{[t-1]}, \dots, \theta_{J-1}^{[t-1]}, \theta_J^{[t-1]})$$

$$\theta_2^{[t]} \sim \pi(\theta_2 | \theta_1^{[t]}, \theta_3^{[t-1]}, \dots, \theta_{J-1}^{[t-1]}, \theta_J^{[t-1]})$$

$$\theta_3^{[t]} \sim \pi(\theta_3 | \theta_1^{[t]}, \theta_2^{[t]}, \dots, \theta_{J-1}^{[t-1]}, \theta_J^{[t-1]})$$

$$\vdots$$

$$\theta_{J-1}^{[t]} \sim \pi(\theta_{J-1} | \theta_1^{[t]}, \theta_2^{[t]}, \theta_3^{[t]}, \dots, \theta_J^{[t-1]})$$

$$\theta_J^{[t]} \sim \pi(\theta_J | \theta_1^{[t]}, \theta_2^{[t]}, \theta_3^{[t]}, \dots, \theta_{J-1}^{[t]}).$$

3. Increment t and repeat until convergence.

The Gibbs sampler has very attractive theoretical properties that make it extremely useful across a wide range of applications. Since the Gibbs sampler conditions only on values from its last iteration, it has the Markovian property of depending only on the current position. The Gibbs sampler is a homogeneous Markov chain: the consecutive probabilities are independent of n, the current length of the chain. This should be apparent from the algorithm above; there is nothing in the calculation of the full conditionals that is dependent on n. As mentioned previously, the Gibbs sampler has the true posterior distribution of parameter vector as its limiting distribution: $\boldsymbol{\theta}^{[i]} \xrightarrow[i=1 \to \infty]{d} \boldsymbol{\theta} \sim \pi(\boldsymbol{\theta})$ since it is *ergodic*, first proved in the original Geman and Geman (1984) paper for finite state spaces and subsequently in more general settings (Chan 1993; Roberts and Polson 1994). Furthermore, the Gibbs sampler converges very quickly relative to other algorithms since it is geometrically ergodic.

5.5 ROLE OF RANDOM NUMBER GENERATION

The purpose of Section 5.4 was to highlight and summarize the detailed workings of the most popular methods for obtaining posterior samples with MCMC. It should be obvious from the discussion that all these methods (and many others being variations on those described) rely on being able to sample (i.e., generate) random quantities from specific distributions. The Metropolis–Hastings algorithm needs to draw samples from the candidate generating distribution *and* from a random uniform distribution on [0 : 1]. The hit-and-run distribution needs to generate uniformly (typically) on the J-dimensional unit hypersphere *and* a signed distance from some specified density (often gamma) *and* from a random uniform distribution on [0 : 1]. The Gibbs sampler only needs to generate values from conditional distributions on each step, but there must be J values generated for a $1 \times J$ coefficient vector *at each step*. Therefore, random number generation is a key component of MCMC algorithms.

Random number generation is an important but understudied aspect of *applied* statistical computing, at least at the high end of implementation complexity (Marsaglia 1985), including Markov chain Monte Carlo implementations. We know from a vast foundational literature that serious problems can be caused by poorly written algorithms: Good (1957), Butcher (1961), Kronmal (1964), Coveyou and MacPherson (1967), Downham and Roberts (1967), Gorenstein (1967), Marsaglia (1968), Whittlesey (1969), Downham (1970), Toothill et al. (1971), Learmonth and Lewis (1973), Dieter (1975), Dudewicz (1975, 1976), McArdle (1976), Atkinson (1980), Morgan (1984), Krawczyk (1992), Gentle (1998), McCullough (1999a), and McCullough and Wilson (1999). In much the same way that the famously flawed but widely used RANDU algorithm from IBM was used for quite some time although it had received quite a lot of criticism in this literature (Coveyou 1960; Fishman and Moore 1982; Hellekalek 1998), it can also be difficult to generate random numbers with specialized non-random properties such as specified correlations (Hurst and Knop 1972; Kennedy and Gentle 1980; Gentle 1990; Anderson and Louis 1996; Malov 1998; Falk 1999). (We treat the issue briefly here. See Chapter 2 for a detailed discussion, and Chapter 3 for tests that can be used to evaluate the quality of random number generation.)

Random numbers generated on computers have two characteristics that make them not truly random. First, the process is wholly discrete in that they are created from a finite binary process and normalized through division. Therefore, these *pseudo-random* numbers are necessarily rational. However, truly random numbers on some defined interval are irrational with probability 1 since the irrationals dominate the continuous metric. This is a very minor technical consideration and not one worth really worrying about (especially because there is no good solution). Second, while we call these values random or more accurately pseudo-random numbers, they are not random at all since the process the generates them is completely deterministic (Jagerman 1965; Jansson 1966). The point is that the algorithms create a stream of values that is not random in the indeterminant sense but still *resembles a random process*. The utility of these deterministic streams is the degree to which they lack systematic characteristics (Coveyou 1960). These characteristics are the time it takes to repeat the stream exactly (the period) and repeated patterns in lagged sets within the stream.

5.5.1 Periodicity of Generators and MCMC Effects

The reason we care about homogeneity and aperiodicity, as described in Section 5.3, is that these conditions are required for the ergodic theorem, which establishes eventual convergence to the stationary distribution of interest. Because all MCMC computational implementations draw from target distributions using pseudo-random number generators, we should be concerned about the effect of the patterns and period of the generator, which is something that cannot be avoided. Obviously, some implementations are better (longer periods) than others, but virtually no work has been done thus far to analyze real and potential implications in the MCMC environment.

Since all pseudo-random generators are necessarily periodic, in a sense all software implementations of Markov chains produce non-ergodic Markov chains. More precisely, *a Markov chain run indefinitely in a software implementation is non-ergodic because all pseudo-random number generators have periods*. Since every underlying random generator cycles, technically the subsequent chain values are neither homogeneous nor aperiodic.

To see this in more detail, start with the transition kernel $K(\theta, A)$ that satisfies the *minorization condition*:

$$K^{t_0}(\theta, A) \geq \beta s(\theta) \nu(A), \quad \forall \theta, A \in \Theta$$

where $t_0 \in \mathbb{I} \geq 1$, β an arbitrary positive constant, $\nu \in \mathcal{M}^+$ a measure, and $s(\theta) \in \Theta$ a function. A sequence of T nonempty disjoint sets in Θ, $[A_0, A_1, \dots, A_{T-1}]$, is called a *t-cycle* for this kernel if $\forall i = 0, 1, \dots, T - 1$, and all $\theta \in \Theta$:

$$K(\theta, A_j^c) = 0 \ \forall j = i + 1 \quad (\text{mod } T) \tag{5.12}$$

(Nummelin 1984, p. 20). That is, there is actually a deterministic cycle of sets generated because the full range of alternative sets occur with probability zero: $K(\theta, A_j^c)$.

The question that remains is whether the periodicity imposed by the generator matters in finite sample implementations. Suppose that a Markov chain is run for m iterations with an underlying pseudo-random number generator that has period $N > m$. If at step m it is possible to assert convergence and collect a sample for empirical summary of the posterior equal to n such that $m + n < N$, the period of the random number generator is immaterial because no values were repeated deterministically. Two obvious complications arise from this observation: (1) it might not be possible to assert convergence yet, and (2) it might not be possible to obtain reasonable m and n iterations whereby $m + n < N$.

How real a problem is this in common implementations of MCMC algorithms? Currently, two software approaches dominate applied work: user-friendly applications coded in WinBUGS and more involved, but flexible, solutions written in C++. The pseudo-random number generator for WinBUGS is a *linear congruential generator* (discussed in Chapter 2), and this is also true for almost all system-supplied rand() functions in C++ libraries. The linear congruential generator is a simple and fast generator of uniform integers (which are then scaled for other purposes) described in Chapter 2.

The period for WinBUGS is $2^{31} = 2,147,483,648$, which seems quite large at first. The period for C++ solutions depends on the value that the user sets for

RANDMAX in the declarations, which determines the maximum possible period obtainable (although it may be lower; see Press et al. 2002). Although RANDMAX can be set as high as 2^{32}, it is often set as low as 32,767 either by the user or by the default settings of the programming libraries accompanying the compiler.

As noted previously, the PRNG periodicity problem is exacerbated in MCMC implementations because standard algorithms require more than a single random draw at each time t. For instance, the Gibbs sampler requires one for each parameter defined by the series of full conditional distributions. Therefore, for 100 parameters (not unreasonable in some literatures), the period of 2 billion given above is reduced to 20 million. Although this exceeds common use, it does not exceed all conceivable use.

Needless to say, incautious coding in C++ and other programming languages can lead to poor periodicity and other problems. [See the discussion in Park and Miller (1988).] Open-source alternatives to the default generators abound and are easy to find. These include algorithms such as *KISS*, *mother-of-all*, *RANROT*, and the *multiple recursive generator*. George Marsaglia (personal communication) notes that much can be learned about generator reliability by running the same simulation setup with several different generators to see if consistent answers result. In general, high-quality implementations are careful to pay attention to this low-level, but critical consideration. For instance, the Scythe open-source C++ library for a MCMC suite of programs, written by Andrew Martin and Kevin Quinn (<http://scythe.wustl.edu/>), uses the excellent *Mersenne twister generator* (Matsumoto and Nishimura 1998).

It should also be noted that there are two general classes of random number generators available: Standard PRNGs as discussed here and in Chapter 2, and those used by cryptographers. The latter eschew repeating patterns but at the cost of higher complexity and lower efficiency. This "crypto-strength" level of randomness is available, however, for extremely sensitive MCMC applications (see Schneier 1994).

5.5.2 Periodicity and Convergence

The question we address here is whether or not convergence to the stationary distribution is possible without asserting strict ergodicity of the Markov chain. This is a necessity since ergodicity is a property that depends on aperiodicity.

Return to the abstract measurable space (H, Θ) with events $A, B, C, \ldots$ in H, real-valued measurable functions $f, g, h, \ldots$ on Θ, and signed measure $\mathcal{M}^+$ with elements λ and μ. First, define an appropriate norm operator. The elementary form for a bounded signed measure, λ, is

$$||\lambda|| \equiv \sup_{A \in \Theta} \lambda(A) - \inf_{A \in \Theta} \lambda(A), \tag{5.13}$$

which is clearly just the total variation of λ. Second, assume that K is R-recurrent given by probability measure P, and the stationary distribution is normed such that $\pi(h) = 1$ for h on Θ. In addition, assume also that θ_n is

R-recurrent (discrete or bounded continuous space) or Harris R-recurrent (continuous unbounded space) Markov chain. If K has period $p \geq 2$, then by definition the associated Markov chain cycles between the states $\{A_0, A_1, A_2, \ldots, A_{p-1}\}$. We can also define the ψ-*null set*: $\Psi = \{A_0 \cup A_1 \cup \cdots \cup A_{p-1}\}^c$, which defines the collection of points not visited in the p-length iterations.

Let the (positive) signed measures λ and μ be any two initial distributions of the Markov chain at time zero, and therefore before convergence to any other distribution. Nummelin (1984, Chap. 6) shows that if θ_n is aperiodic, then

$$\lim_{n\to\infty} ||\lambda P^n - \mu P^n|| = 0. \tag{5.14}$$

This is essentially Orey's (1961) *total variation norm theorem* applied to an aperiodic, recurrent Markov chain. [Orey's result was more general but not any more useful for our endeavors; see also Athreya and Ney (1978, p. 498) for a proof.] But wait a minute; we know that any ψ-irreducible and aperiodic Markov chain has one and only one stationary distribution, and ψ-irreducibility is implied here by recurrence. Therefore, we can substitute into (5.14) the stationary distribution π to get

$$\lim_{n\to\infty} ||\lambda P^n - \pi|| = 0, \tag{5.15}$$

which gives ergodicity. This shows in greater detail than before the conditions by which convergence in distribution to the stationary distribution is justified.

How does this help us in the absence of aperiodicity? Suppose now that K has period $p \geq 2$, and $\lambda(A_i) = \mu(A_i) \, \forall i$, plus $\lambda(\Psi) = \mu(\Psi) = 0$. Then in this modified situation Nummelin also shows that we again get the form in (5.14). So the first price we pay for getting some form of convergence is a set of mild restrictions on the initial distributions relative to the cycled states. However, the second step above does not help us replace μP^n with π.[1] Suppose, instead, we *require* that $\mu P^n = \pi$. This is equivalent to starting the Markov chain out already in its stationary distribution. If we could do that, the periodic form improves from the (5.14) form to the (5.15) from, and the chain operates in the stationary distribution just as if it were ergodic. This works because a Markov chain *started* in its stationary distribution *remains* in this distribution.

Starting the chain in the stationary distribution is often not that difficult. In fact, its particularly easy for the Metropolis–Hastings algorithm because the stationary distribution is given explicitly. In other cases, this can be done by applying one form of the accept–reject method [a simple trial method based on testing candidate values from some arbitrary instrumental density; see Robert and Casella (1999, p. 49)] until a random variable from the stationary distribution is produced, then run the chain forward from there. Also, it is often the case that the time to get the accept–reject algorithm to produce a random variable from the stationary distribution will be shorter than any reasonable burn-in period.

---

[1] Or λP^n either, as an appeal to symmetry for that matter since the choice of λ and μ is arbitrary here.

There is a secondary issue, however. Convergence to stationarity is distinct from convergence of the empirical averages, which are usually the primary substantive interest (Gill 2002, p. 411). Consider the limiting behavior of a statistic of interest, $h(\theta)$, from an aperiodic Harris recurrent Markov chain. We typically obtain empirical summaries of this statistic using the partial sums, such as

$$\bar{h} = \frac{1}{n} \sum_{i=1}^{n} h(\theta_i). \tag{5.16}$$

The expected value of the target h is $E_f h(\theta)$, so by the established properties of Harris recurrent Markov chains (Brémaud 1999, p. 104), it is known that $\sum_{i=1}^{n} h(\theta_i)/n \to E_f h(\theta)$ as $n \to \infty$. Equivalently, we can assert that

$$\frac{1}{n} \sum_{i=1}^{n} h(\theta_i) - E_f h(\theta) \xrightarrow[n \to \infty]{} 0. \tag{5.17}$$

We can also consider the true distribution of $h(\theta)$ at time n (even if it is not observed directly) from a chain with starting point θ_0. The interest here is in $E_{\theta_0} h(\theta)$, where the expectation is with respect to the distribution of θ_n conditional on θ_0. In the next step add and subtract this term on the left-hand side of (5.17) to obtain

$$\left[\frac{1}{n} \sum_{i=1}^{n} h(\theta_i) - E_{\theta_0} h(\theta) \right] - \left[E_f h(\theta) - E_{\theta_0} h(\theta) \right] \xrightarrow[n \to \infty]{} 0. \tag{5.18}$$

The second bracketed term is obviously the difference between the expected value of the target $h(\theta)$ in the true distribution at time n and the expected value of $h(\theta)$ in the stationary distribution. Given a geometrically ergodic Markov chain, this quantity converges geometrically fast to zero since $\| E_f h(\theta) - E_{\theta_0} h(\theta) \| \le k \delta^n$ for a positive k, and a $\delta \in (0:1)$. Now the first bracketed term is the difference between the current empirical average and its expectation at time n. Except at the uninteresting starting point, these are *never* nonasymptotically equivalent, so even in stationarity, the empirical average has not converged. This is not bad news, however, since we know for certain by the central limit theorem that

$$\frac{\frac{1}{n} \sum_{i=1}^{n} h(\theta_i) - E_{\theta_0} h(\theta)}{\sigma/\sqrt{n}} \xrightarrow{d} \mathcal{N}(0, 1). \tag{5.19}$$

Therefore, as $\sqrt{n}\, \delta^n \to 0$, convergence to stationarity proceeds at a much faster rate and does not bring along convergence of empirical averages.

So what really matters for periodicity imposed by the pseudo-random number generator is whether the cycle affects the result in (5.19). It is well known that the central limit theorem holds for sequences of random variables provided that the serial dependence is "relatively weak' (Lehmann 1999) and the sequence has bounded variance. Several features of the MCMC process help this result. First, sampling truly does occur from an infinite population, even if this population exists only in the theoretical sense. Second, it is relatively easy to verify that the posterior distribution of interest does not have unbounded variance, even if there is an improper prior. Third, because of the Markovian property, the relatively weak condition should be met in all but the worst circumstances, and even in those it is easy to assess serial correlations with graphical and analytical summaries (`WinBUGS` makes this very easy).

So the net result of this section is that the long periodicity of high-quality generators does not generally affect the quality of MCMC output. There are two caveats, however. First, given models with a large number of parameters (say, $\geq$ 100) and the need for long chains, one should be particularly careful to scrutinize the output and should certainly try different generators. Second, generators with notably bad characteristics in terms of low periodicity and serial patterns (e.g., RANDU) should be avoided at all costs. The example in the following section demonstrates the problem with ignoring this last admonition.

5.5.3 Example: The Slice Sampler

Sometimes, introducing an additional variable into the Markov chain process can improve mixing or convergence. The general strategy is to augment a k-dimensional vector of coefficients, $\boldsymbol{\theta} \in \boldsymbol{\Theta}$, with (at least) one new variable, $u \in \mathbf{U}$, such that the $(k+1)$-dimensional vector on $\boldsymbol{\Theta} \times \mathbf{U}$ space has good chain properties. The slice sampler variant of this approach has many good theoretical properties, such as geometric convergence, thus making it one of the more common auxiliary variable methods.

Start with the unconditional posterior (from u) distribution of $\boldsymbol{\theta}$ given as $\pi(\boldsymbol{\theta})$, where conditioning on the data is notationally suppressed. Define $\pi(u|\boldsymbol{\theta})$ and therefore the joint posterior distribution by $\pi(\boldsymbol{\theta}, u) = \pi(u|\boldsymbol{\theta})\pi(\boldsymbol{\theta})$. It must be possible to articulate $\pi(u|\boldsymbol{\theta})$, but only convenient forms are typically considered.[2] Two conditional transition kernels are required: one that updates $\boldsymbol{\theta}$, $P[\boldsymbol{\theta} \rightarrow \boldsymbol{\theta}'|u]$, and one that updates u, $P[u \rightarrow u'|\boldsymbol{\theta}]$. In the case of Gibbs sampling these correspond to the conditionals: $\pi(\boldsymbol{\theta}'|u)$ and $\pi(u'|\boldsymbol{\theta})$, and the algorithm proceeds by cycling back and forth where the additional within-$\boldsymbol{\theta}$ steps are implied. Thus the Gibbs sampling process recovers the marginal of interest: $\pi(\boldsymbol{\theta})$. The two-step nature of the process ensures that the stationary process remains $\pi(\boldsymbol{\theta})$ since the detailed balance equation is maintained:

$$\sum_u \pi(u'|\boldsymbol{\theta})\pi(\boldsymbol{\theta})P[\boldsymbol{\theta} \rightarrow \boldsymbol{\theta}'|u)] = \sum_u \pi(\boldsymbol{\theta}'|u)\pi(u)P[\boldsymbol{\theta} \rightarrow \boldsymbol{\theta}'|u)],$$

and we can therefore run the chain in this fashion and simply discard the u values after convergence has been asserted.

---

[2]Actually, the Metropolis–Hastings algorithm is already an auxiliary variable process since the candidate generating distribution *is* auxiliary to the distribution of interest.

The *slice sampler* for a two-parameter estimation vector starts by stipulating the two marginals and the joint distribution as uniforms in the following way:

$$u^{[j+1]}|\theta^{[j]} \sim \mathcal{U}(0, \pi(\theta^{[j]})) \qquad \theta^{[j+1]}|u^{[j]} \sim \mathcal{U}(\theta : \pi(\theta^{[j+1]}) > u^{[j+1]}) \quad (5.20)$$

at time j. Here it is necessary that the target distribution be expressed as a product of the marginals: $\pi(\theta) = \prod \pi(\theta_i)$. So the auxiliary variable(s) is sampled from a uniform bounded by zero and the θ; these θ are, in turn, sampled from a uniform bounded below by the auxiliary variable(s).

As a rudimentary but instructive example (Robert and Casella 1999) suppose that we have $f(x) \propto \exp(-x^2/2)$ and wish to generate samples. Set up a uniform such that

$$u|x \sim U[0, \exp(-x^2/2)], \qquad x|u \sim U[-\sqrt{-2\log(u)}, \sqrt{-2\log(u)}]$$

to make this u the auxiliary variable with easy sampling properties. This example is interesting because although we could sample a more difficult and realistic distribution with this procedure, sampling a nonnormalized normal provides an easy post hoc test.

Now run a Gibbs sampler according to the conditionals defined, as implemented in R:

```
N <- 64000; x.vals <- 0; u.vals <- 0
for (i in 2:N)   {
    u.vals <- c(u.vals,runif(1,0,exp(-0.5*x.vals[(i-1)]^2)))
    x.vals <- c(x.vals,runif(1,-sqrt(-2*log(u.vals[i])),
                                sqrt(-2*log(u.vals[i])))))
}
```

One can see from Figure 5.2 that this process clearly produces a normal sample even though only uniform random variables are sampled. This implementation of the slice sampler can be loosely considered as an MCMC version of the Box–Müller algorithm (1958).

However, suppose that we substitute the standard pseudo-random generator in R with an intentionally poor alternative. Linear congruential generators are of the form

$$x_{n+1} = (ax_n + b) \qquad (\text{mod } m), \qquad (5.21)$$

where selection of the constants a, b, and m is critical. In fact, the period can be quite short if these are not chosen wisely. [See the survey in Park and Miller (1988) for particular choices.] Early literature on congruential generators showed that the period is maximized if (Hull and Dobell 1962; Knuth 1998):

- m and b have no common factors other than 1.
- $a - 1$ is a multiple of every prime that factors m.
- If 4 is a factor of m, then 4 is a factor of $a - 1$.

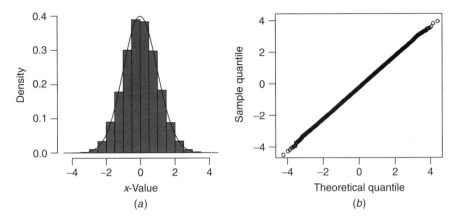

Fig. 5.2 Slice sampler without numerical problems: (*a*) histogram and density overlay; (*b*) normal Q-Q plot.

The now famously flawed `RANDU` algorithm used for decades by IBM mainframes is produced by setting $a = 65,539$, $b = 0$, and $m = 2^{31}$. The problem with `RANDU` is that it defines iterative triples which are evident when graphed in three dimensions. [See the figures of Fishman 1996, p. 620.]

Press et al. (2002) point out that a superbly bad choice for m is 32,767 (an ANSI recommendation notwithstanding!). So replace the random number generator used previously with the R code:

```
lcg.rand <- function(seed,n)  {
    m <- 32767; a <- 1103515245; c <- 12345
    out.vec <- NULL; new <- seed
    while (length(out.vec) < n)  {
        new <- (new*a + c) %% m
        out.vec <- c(out.vec, new/m)
    }
    return(out.vec)
}
```

and run the slice sampler again. This produces a sample with slightly heavier tails than the intended normal and which is only slightly noticeable with a histogram or normal quantile plot. To show this effect more clearly, we set $m = 9$, which is a stunningly bad choice since 9 is a factor of 1,103,515,245. The slice sampler then produces the output in Figure 5.3. Clearly, poor choices for the constants *can* be disastrous.

5.5.4 Evaluating `WinBUGS`

By far the most common method of MCMC estimation is the high-quality and free package from the Community Statistical Research Project at the MRC Biostatistics Unit and the Imperial College School of Medicine at St. Mary's, London, `WinBUGS` (`<http://www.mrc-bsu.cam.ac.uk/bugs/>`). Therefore, it

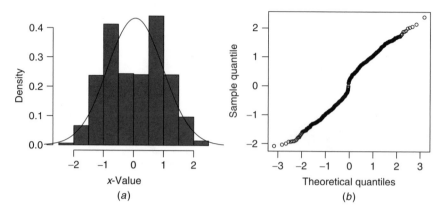

Fig. 5.3 Slice sampler with numerical problems: (*a*) histogram and density overlay; (*b*) normal Q-Q plot.

is certainly worth looking in detail at the random number generation process used by this package. The algorithm used by `WinBUGS` (and its Unix-based cousin `BUGS`) is a *multiplicative congruential generator*:

$$x_{n+1} = ax_n \quad (\text{mod } m), \tag{5.22}$$

which is a linear congruential generator with $b = 0$. Lewis et al. (1969) proposed this method with $a = 7^5 = 16,807$ and $m = 2^{31} = 2,147,483,647$, and Park and Miller (1988) showed that this is the optimal choice. Unfortunately, as Press et al. (2002) point out, the standard algorithm with this choice of m cannot be directly coded in a high-level language (`WinBUGS` is written in `MODULA-2`) since the required products would produce overflow on 32-bit registers. The authors of `WinBUGS` have used Schrage's (1979) workaround for this problem. This solution starts with defining

$$q = \lfloor m/a \rfloor \qquad r = m \quad (\text{mod } a),$$

where $\lfloor \rfloor$ indicates the floor function (integer component). Thus the Park and Miller optimal values are $q = 127,773$ and $r = 2836$. The Schrage algorithm produces new pseudo-random values according to

$$
\begin{aligned}
x_{n+1} &= ax_n \bmod(m) \\
&= \begin{cases}
a(x_n \quad (\text{mod } q)) - r\lfloor x_n/q \rfloor & \text{if greater than zero} \\
a(x_n \quad (\text{mod } q)) - r\lfloor x_n/q \rfloor + m & \text{otherwise.}
\end{cases}
\end{aligned} \tag{5.23}
$$

It turns out that this algorithm has minor problems with serial correlation. In particular, the ratio difference between a and m means that very small random numbers will by necessity produce subsequent values that are guaranteed to be less than the current sample average. This is not a big issue except that it points toward the potential for other unknown problems. Press et al. (2002) also notes that this generator also fails the χ^2 uniformity test for generated sample sizes slightly over 1 million. This generator is known to fail many of the more rigorous tests described in Chapter 3.

There is another issue with `WinBUGS` that is more operational than algorithmic. Whenever a new model is checked for syntax, the random seed is set to the default value, 9876543210. So if a user is comparing a set of subsequent models, *these models will be run with exactly the same stream of random draws*. The designers thought that this was a useful feature because it allows for direct comparison of alternative specifications, starting points, and system options. However, if there is a reason that the user needs to have different streams, the seed should be reset manually. [See Mihram and Mihram (1997) for ideas; see also Chapter 2 for a solution to this problem that uses hardware random number generators.]

Does this default seed option mean that the Markov chain moves *deterministically* in exactly the same way for a given model? The answer, surprisingly, is yes. If the same starting values are given and the same random seed is used, one can exactly replicate the path of the chain in `WinBUGS`. However, because of the Markovian property, if different starting points are given, an entirely different deterministic path is produced. Consider the Gibbs sampler in this context. The draws are from full conditional distributions, conditioned on all the other parameter values at each substep. Therefore, a different starting point produces a unique series of conditionals.

5.6 ABSORBING STATE PROBLEM

In Chapters 1 and 2 it was shown that all computational calculations necessitate rounding and truncating errors as a result of finite-precision arithmetic. Some time ago in a seminal work, Tierney (1994, p. 1722) raised an alarm about the consequences of numerical computing for the behavior of Markov chains. His concern was that machine rounding could introduce an *absorbing state* for the chain that would prevent convergence to the true stationary distribution even if it is a Markov chain algorithm with all of the appropriate theoretical properties for convergence. In other words, it is possible that due to the software implementation of the kernel, an ergodic Markov chain can be prevented from finding and exploring the posterior distribution as intended.

Robert and Casella (1999, p. 425) distinguish between *trapping sets*, which require a very large number of iterations before the chain escapes, and *absorbing sets*, which can never escape. Both of these are problematic, with the latter being far worse, of course. The central problem can be explained by the example of a simple Metropolis–Hastings algorithm. Suppose that the chain defined by kernel K enters a trapping set A such that the values produced by the instrumental distribution have the following Metropolis–Hastings characteristics:

- At an arbitrary step sample θ' from $q(\theta'|\theta)$, where $\theta \in A$.
- θ' relative to A has the following probabilities:

$$P(\theta' \in A) = 1 - \epsilon, \qquad P(\theta' \in A^c) = \epsilon.$$

- Sample u from $u[0:1]$.
- If $a(\theta', \theta) > u$, move to θ'; otherwise, remain.

Here we assume that ϵ is very small such that A is a trapping set requiring a large number of generated u values to obtain the condition that the Metropolis–Hastings acceptance ratio defined in (5.10), $a(\theta', \theta)$, is greater than a generated u. This means that the chain will be trapped in A for a long time. Now if $a(\theta', \theta)$ for all $\theta \in A$ falls below the uniform random number generator's smallest possible value, A is now an absorbing state because there cannot be a $u < a(\theta', \theta)$ giving a successful jump.

Young et al. (1998) find a related problem that is equally distressing but easily observed. In their example, if a Gibbs sampler for estimating probability parameters returns a conditional probability value outside [0 : 1], corresponding coefficient estimates will go to infinity or negative infinity. This "degeneracy" in the estimates is easy to detect and they recommend setting deliberate bounds on the parameters as part of the estimation process. This will not work well, however, if the estimates sit at the bounds cycle after cycle.

This discussion should not be interpreted as asserting that absorbing states are necessarily bad in all settings. After all, the desired stationary distribution is itself an absorbing state. The problem is actually that an absorbing state produced by machine round-off will not be the desired posterior distribution, and this problem is therefore fatal to the estimation enterprise.

Tierney (1994) also notes a related problem. It is possible to specify posterior forms that are reasonably stable near the mode but unstable in the tails. Often, this is the result of effects such as *inliers* (very small values on the original metric that go to huge negative values on the logged scale). Since it is relatively common to work with log posteriors densities, care should be exercised to ensure that this problem does not occur by analytically or diagrammatically looking at suspect regions. Fortunately, this problem is easy to detect since the unbounded values will ruin empirical averages in an obvious way.

5.7   REGULAR MONTE CARLO SIMULATION

Finally, it should be noted that problems from the pseudo-randomness of generated values do not affect MCMC procedures alone, and there is obviously quite a lot in common (Ripley 1987; Gelman 1992; Monahan 2001). In fact, regular Monte Carlo simulation for calculation of volumes and statistics are also sensitive to poor performance of the generating mechanism, albeit in a different manner. Fishman (1996) points out that "Regardless of whether or not one employs variance-reducing techniques, the statistical integrity of every Monte Carlo sampling experiment depends on the availability of an algorithm for generating random numbers, or more precisely, for generating sequences of numbers by relatively stringent standards can be regarded as indistinguishable from sequences of truly random numbers."

As Fishman's quote indicates, a central concern with MC applications is the amount of simulation error that is introduced by the numerical process and the process for accounting for it not to overstate the accuracy of simulation results. Stern (1997) emphasizes, albeit tangentially, how important the iid assumption can be in simulation studies. Obviously, the quality of the underlying PRNG affects such high-level econometric studies. So alternative generators are often evaluated by the extent to which imposed inaccuracies add to the asymptotically declining simulation error term (Rubinstein 1981).

To explain standard Monte Carlo simulation briefly, suppose that we had a probability function, $f(x)$, that was difficult to express or manipulate but for which we could generate samples on an arbitrary support of interest: $[a:b]$ (which may also be multidimensional). A typical quantity of interest is $I[A, B] = \int_A^B f(x)h(x)\,dx$. This is the expected value of some function, $h(x)$, of x distributed $f(x)$. If $h(x) = x$, $I[A, B]$ is the calculation of the mean of x over $[A:B]$. A substitute for analytically calculating this quantity is to randomly generate n values of x from $f(x)$ and calculate $\hat{I}[A, B] = (1/n)\sum_{i=1}^n h(x_i)$. So we replace analytical integration with summation from a large number of simulated values, rejecting values outside of $[A:B]$. The utility of this approach is that by the strong law of large numbers, $\hat{I}[A, B]$ converges with probability 1 to the desired value, $I[A, B]$. A second useful feature is that although $\hat{I}[A, B]$ now has the simulation error mentioned above, this error is easily measured by the empirical variance of the simulation estimate: $\text{Var}(\hat{I}[A, B]) = 1/n(n-1)\sum_{i=1}^n (h(x_i) - \hat{I}[A, B])^2$. Note that the researcher fully controls the simulation size, n, and therefore the simulation accuracy of the estimate. Also, because the central limit theorem applies here as long as $\text{Var}(I[A, B])$ is finite, credible intervals can easily be calculated by $[95\%_{\text{lower}}, 95\%_{\text{upper}}] = [\hat{I}[A, B] - 1.96\sqrt{\text{Var}(\hat{I}[A, B])},$ $\hat{I}[A, B] + 1.96\sqrt{\text{Var}(\hat{I}[A, B])}]$, or simply by reporting the 0.025 and 0.075 empirical quantiles from the simulation sample.

Although a great deal of the literature on Monte Carlo methods is concerned with variance reduction, readers often have no idea what an individual study has done with regard to simulation error and other concerns (Ripley 1987). Kennedy and Gentle (1980, p. 235) note that "Frequently one sees in articles in the statistics journals reports of 'Monte Carlo studies' that fail to supply any details of the study; only the 'results' are given. Since the results are realizations of pseudo-random processes, however, the reader of the article may be left wondering about the magnitude of the pseudo-standard errors and, indeed, about the effect of the 'pseudo' nature of the process itself." This statement is certainly no less accurate with regard to MC studies in social science journals, and the situation may even be worse.

5.8 SO WHAT CAN BE DONE?

We have identified several potential pitfalls in MCMC simulation that are substantially understudied. This is the first systematic look at the effects of numerical (in)accuracy from the computer implementation of standard algorithms. Markov chain Monte Carlo is a fairly new enterprise, and particularly so in the social sciences, so it is quite useful to understand some of the low-level issues inherent in the process.

There are just a few recommendations here that can improve the quality of MCMC estimations. In the spirit of earlier chapters, they are listed next.

- Unless you are interested in replicating a previously run Markov chain exactly, it is recommended that the random seed be changed. This is trivial in both WinBUGS and C++ implementations.

- *Never* trust the default random number generator provided with C or C++ compilers. Open-source code for alternatives is readily available (see Chapter 2).

- Be aware that periodicity matters on very long runs with many parameters. That is, the chain is not technically ergodic, and for these applications values will be repeated. Start the chain out on its stationary distribution if possible. Since the empirical averages are still a concern, run the chain from multiple starting points, all in the stationary distribution and merge.

- Problems with absorbing states can be difficult to detect and solve. With the Metropolis–Hastings algorithm, one sign of an absorbing state is a chain that rejects *many* candidate points or perhaps only accepts points that are extremely close (this can include oscillating chains). Since this could be due to scale and convergence, it is important to know where posterior modes are relative to the potential absorbing state. Gelman et al. (1995) recommend running the EM algorithm at dispersed points around the sample space as a means of detecting modes. Tierney (1994) recommends solving the problem by truncating the sample space away from the absorbing state but including the material posterior density regions. Although this obviously provides the desired numerical accuracy, it can be difficult in practice since it involves understanding the mathematical structure of the posterior.

- Finally, treat pseudo-random number generators as what they really are: just imitations of the real thing. Therefore, layering complexity through MCMC methods on top of bad simulated values creates bad inferences.

In summary, the message is not at all unlike that of other chapters in this book; do not treat this estimation process as one would linear regression on a "point-and-click" package. That is, these methods require considerably more care in setup, vigilance in process, and caution in analysis.

CHAPTER 6

Numerical Issues Involved in Inverting Hessian Matrices

Jeff Gill and Gary King

6.1 INTRODUCTION

In the social sciences, researchers typically assume the accuracy of generalized linear models by using an asymptotic normal approximation to the likelihood function or, occasionally, by using the full posterior distribution. Thus, for standard maximum likelihood analyses, only point estimates and the variance at the maximum are normally seen as necessary. For Bayesian posterior analysis, the maximum and variance provide a useful first approximation (but see Chapter 4 for an alternative).

Unfortunately, although the negative of the Hessian (the matrix of second derivatives of the posterior with respect to the parameters and named for its inventor in slightly different context, German mathematician Ludwig Hesse) must be positive definite and hence invertible so as to compute the variance matrix, invertible Hessians do not exist for some combinations of datasets and models, so statistical procedures sometimes fail for this reason before completion. Indeed, receiving a computer-generated "Hessian not invertible" message (because of singularity or nonpositive definiteness) rather than a set of statistical results is a frustrating but common occurrence in applied quantitative research. It even occurs with regularity during many Monte Carlo experiments where the investigator is drawing data from a known statistical model, due to machine effects.

The Hessian can be noninvertible for both computational reasons and data reasons. Inaccurate implementation of the likelihood function (see Chapters 2 and 3), inaccurate derivative methods (see Chapter 8), or other inappropriate choices in optimization algorithms can yield noninvertible Hessians. Where these inaccuracies cause problems with Hessians, we recommend addressing these inaccuracies directly.

If these methods aren't feasible, or don't work, which often happens, we provide an innovative new library for doing generalized inverses . *Moreover, when a Hessian is not invertible for data reasons, no computational trick can make it invertible, given the model and data chosen, because the desired inverse does*

Numerical Issues in Statistical Computing for the Social Scientist, by Micah Altman, Jeff Gill, and Michael P. McDonald
ISBN 0-471-23633-0   Copyright © 2004 John Wiley & Sons, Inc.

not exist. The advice given in most textbooks for this situation is to rethink the model, respecify it, and rerun the analysis (or in some cases get more data). For instance, in one of the best econometric textbooks, Davidson and MacKinnon (1993, pp. 185–86) write: "There are basically two options: Get more data, or estimate a less demanding model .... If it is not feasible to obtain more data, then one must accept that the data one has contain a limited amount of information and must simplify the model accordingly. Trying to estimate models that are too complicated is one of the most common mistakes among inexperienced applied econometricians." The point of this chapter is to provide an alternative to simplifying or changing the model, but the wisdom of Davidson and MacKinnon's advice is worth emphasizing in that our approach is appropriate only when the more complicated model is indeed of interest.

Respecification and reanalysis is important and appropriate advice in some applications of linear regression because a noninvertible Hessian has a clear substantive interpretation: It can only be caused by multicollinearity or including more explanatory variables than observations (although even this simple case can be quite complicated; see Searle 1971). As such, a noninvertible Hessian might indicate a substantive problem that a researcher would not be aware of otherwise. It is also of interest in some nonlinear models, such as logistic regression, where the conditions of noninvertibility are also well known. In nonlinear models, however, noninvertible Hessians are related to the shape of the posterior density, but how to connect the problem to the question being analyzed can often be extremely difficult.

In addition, for some applications, the textbook advice is disconcerting, or even misleading, because the same model specification may have worked in other contexts and really is the one from which the researcher wants estimates. Furthermore, one may find it troubling that dropping variables from the specification substantially affects the estimates of the remaining variables and therefore the interpretation of the findings (Leamer 1973).

The point developed in this chapter is that although a noninvertible Hessian means the desired variance matrix does not exist, the likelihood function may still contain considerable information about the questions of interest. As such, discarding data and analyses with this valuable information, even if the information cannot be summarized as usual, is an inefficient and potentially biased procedure.

In situations where one is running many parallel analyses (say, one for each U.S. state or population subgroup), dropping only those cases with noninvertible Hessians, as is commonly done, can easily generate selection bias in the conclusions drawn from the set of analyses. Here, restricting all analyses to the specification that always returns an invertible Hessian risks other biases. Similarly, Monte Carlo studies that evaluate estimators risk severe bias if conclusions are based (as usual) on only those iterations with invertible Hessians.

Rather than discarding information or changing the questions of interest when the Hessian does not invert, we discuss some methods that are sometimes able to extract information in a convenient format from problematic likelihood functions or posterior distributions without respecification.[1] This has always been possible within Bayesian analysis, by using algorithms that enable one to draw directly from the posterior of interest. However, the algorithms, such as those based on Monte Carlo Markov chains or higher-order analytical integrals, are normally much more involved to set up than calculating point estimates and asymptotic variance approximations to which social scientists have become accustomed, and so they have not been adopted widely. Our approach can be thought of as Bayesian, too, although informative prior distributions need not be specified; we focus only on methods that are relatively easy to apply. Although a sophisticated Bayesian analyst could figure out how to elicit information from a posterior with a noninvertible Hessian without our methods in particular instances, we hope that our proposals will make this information available to many more users and may even make it easier for those willing to do the detailed analysis of particular applications. In fact, the methods we discuss are appropriate even when the Hessian does invert and in many cases may be more appropriate than classical approaches. We begin in Section 6.2 by providing a summary of the posterior that can be calculated, even when the mode is uninteresting and the variance matrix is nonexistent. The road map to the rest of the chapter concludes that motivating section.

6.2 MEANS VERSUS MODES

When a posterior distribution contains information but the variance matrix cannot be computed, all hope is not lost. In low-dimensional problems, plotting the posterior is an obvious solution that can reveal all relevant information. In a good case, this plot might reveal a narrow plateau around the maximum, or collinearity between two relatively unimportant control variables (as represented by a ridge in the posterior surface). Unfortunately, most social science applications have enough parameters to make this type of visualization infeasible, so some summary is needed. [Indeed, this was the purpose of maximum likelihood estimates, as opposed to the better justified likelihood theory of inference, in the first place; see King 1989].

We propose an alternative strategy. We do not follow the textbook advice by asking the user to change the *substantive question* they ask, but instead, ask the researcher to change their *statistical summary* of the posterior so that useful information can still be elicited without changing their substantive questions, statistical specification, assumptions, data, or model. All available information from the model specified can thus be extracted and presented, at which point one may wish to stop or instead respecify the model on the basis of substantive results.

In statistical analyses, researchers collect data, specify a model, and form the posterior. They then summarize this information, essentially by posing a question

---

[1]For simplicity, we refer to the objective function as the posterior distribution from here on, although most of our applications will involve flat priors, in which case, of course, the posterior is equivalent to a likelihood function.

about the posterior distribution. The question answered by the standard maximum likelihood (or maximum posterior) estimates is: What is the mode of the posterior density and the variance around this mode?

In cases where the mode is on a plateau or at a boundary constraint, or the posterior's surface has ridges or saddlepoints, the curvature will produce a noninvertible Hessian. In these cases, the Hessian also suggests that the mode itself may not be of use even if a reasonable estimate of its variability were known. That is, when the Hessian is noninvertible, the mode may not be unique and is, in any event, not an effective summary of the full posterior distribution. In these difficult cases, we suggest that researchers pose a different but closely related question: What is the mean of the posterior density and the variance around the mean?

When the mode and mean are both calculable, they often give similar answers. If the likelihood is symmetric, which is guaranteed if n is sufficiently large, the two are identical, so switching questions has no cost. Indeed, the vast majority of social science applications appeal to asymptotic normal approximations for computing the standard errors and other uncertainty estimates, and for these the mode and the mean are equal. As such, for these analyses, our proposals involve no change of assumptions.

If the maximum is not unique, or is on a ridge or at the boundary of the parameter space, the mean and its variance can be found, but a unique mode and its variance cannot. At least in these difficult cases, when the textbook suggestion of substantive respecification is not feasible or if it is not desirable, we propose switching from the mode to the mean.

Using the mean and its variance seems obviously useful when the mode or its variance do not exist, but in many cases when the two approaches differ and both exist, the mean would be preferred to the mode. For an extreme case, suppose that the posterior for a parameter θ is truncated normal with mean 0.5, standard deviation 10, and truncation is on the [0, 1] interval (cf. Gelman et al. 1995, p. 114, Prob. 4.8). In this case, the posterior, estimated from a sample of data, will be a small segment of the normal curve. Except when the unit interval captures the mode of the normal posterior (very unlikely given the size of the variance), the mode will almost always be a corner solution (0 or 1). In contrast, the mean posterior will be some number within (0,1). In this case, it seems clear that 0 or 1 does not make good single-number summaries of the posterior, whereas the mean is likely to be much better.

In contrast, when the mean is not a good summary, the mode is usually not satisfactory either. For example, the mean will not be very helpful when the likelihood provides little information at all, in which case the result will effectively return the prior. The mean will also not be a very useful summary for a bimodal posterior, since the point estimate would fall between the two humps in an area of low density. The mode would not be much better in this situation, although it does at least reasonably characterize one part of the density.

In general, when a point estimate makes sense, the mode is easier to compute, but the mean is more likely to be a useful summary of the full posterior. We believe that if the mean were as easy to compute as the mode, few would choose the mode. We thus hope to reduce the computational advantage of the mode over the mean by proposing some procedures for computing the mean and its variance.

6.3 DEVELOPING A SOLUTION USING BAYESIAN SIMULATION TOOLS

When the inverse of the negative Hessian exists, we compute the mean and its variance by importance resampling. That is, we take random draws from the exact posterior in two steps. We begin by drawing a large number of random numbers from a normal distribution, with mean set at the vector of maximum posterior estimates and variance set at the estimated variance matrix. Then we use a probabilistic rejection algorithm to keep only those draws that are close enough to the correct posterior. These draws can then be used directly to study some quantity of interest, or they can be used to compute the mean and its variance.

When the inverse of the negative Hessian does not exist, we suggest two separate procedures to choose from. One is to create a *pseudovariance matrix* and use it, in place of the inverse, in our importance resampling scheme. In brief, applying a generalized inverse (when necessary, to avoid singularity) and generalized Cholesky decomposition (when necessary, to guarantee positive definiteness) together often produce a pseudovariance matrix for the mode that is a reasonable summary of the curvature of the posterior distribution. (The generalized inverse is a commonly used technique in statistical analysis, but to our knowledge, the generalized Cholesky has not been used before for statistical purposes.) Surprisingly, the resulting matrix is not usually ill conditioned. In addition, although this is a "pseudo" rather than an "approximate" variance matrix (because the thing that would be approximated does not exist), the calculations change the resulting variance matrix as little as possible to achieve positive definiteness. We then take random draws from the exact posterior using importance resampling as before, but using two diagnostics to correct problems with this procedure.[2]

Our solution is nothing more than a way to describe the difficult posterior form using importance sampling, which is a standard tool for Bayesians because they often end up with posterior forms that are difficult to describe analytically. This method of using a convenient candidate distribution and then accepting or rejecting values depending on their resemblance to those produced by the real posterior is supported by a large body of theoretical work starting with Ott (1979), Rubin (1987a), and Smith and Gelfand (1992). Recent discussions of the theoretical validity as well as properties of importance sampling are given by Geweke (1989), Gelman et al. (1995), Robert and Casella (1999), and Tanner (1996). Before continuing, it is also important to note that this proposed solution uses simulation but is not estimation based on Markov chain Monte Carlo analysis.

---

[2]This part of our method is what most separates it from previous procedures in the literature that sought to find a working solution based on the generalized inverse alone (Riley 1955; Marquardt 1970; Searle 1971).

6.4   WHAT IS IT THAT BAYESIANS DO?

We are certainly "borrowing" from the Bayesian perspective: mean summaries and statistical summary through simulation. However, philosophically we are not requiring that one subscribe to the tenants of Bayesian inference: stipulation of prior distributions for unknown parameters, a belief that these parameters should be described distributionally conditional on the data observed and posteriors based on updating priors with likelihoods.

The essence of Bayesian inference is encapsulated in three general steps:

1. Specify a probability model for unknown parameter values that includes some prior knowledge about the parameters if available.
2. Update knowledge about the unknown parameters by conditioning this probability model on observed data.
3. Evaluate the fit of the model to the data and the sensitivity of the conclusions to the assumptions.

The second step constitutes the core of this process and is accomplished through Bayes' law:

posterior probability $\propto$ prior probability $\times$ likelihood function

$$\pi(\theta|\mathbf{D}) = \frac{p(\theta)L(\theta|\mathbf{D})}{\displaystyle\int_{\Theta} p(\theta)L(\theta|\mathbf{D})\,d\theta}$$

$$\propto p(\theta)L(\theta|\mathbf{D}),$$

where $\mathbf{D}$ is a generic symbol denoting the observed data at hand. A consequence is that $\pi(\theta|\mathbf{D})$ is a model summary that obviously retains its distributional sense. This is useful because it allows a more general look at what the model is asserting about parameter location and scale. It also pushes one away from simply describing this posterior with a point estimate and standard error for each parameter since this could miss some of the important features of the posterior shape. These additional features can include multimodality, skewness, and flat regions.

The Bayesian reporting mechanisms include the credible interval (computed exactly like the non-Bayesian confidence interval) and the highest posterior density (HPD) interval. The HPD interval contains the $100(1-\alpha)\%$ highest posterior density and therefore meets the criteria $C = \{\theta : \pi(\theta|\mathbf{x}) \geq k\}$, where k is the largest number assuring that $1 - \alpha = \int_{\theta:\pi(\theta|\mathbf{x})>k} \pi(\theta|\mathbf{x})\,d\theta$. This is the region where the probability that θ is in the region is maximized at $1 - \alpha$, regardless of modality.

Bayesian statistical methods have some distinct advantages over conventional approaches in modeling social science data (Poirer 1988; Western 1998, 1999), including overt expression of model assumptions, an exclusive focus on probability-based statements, direct and systematic incorporation of prior knowledge, and the ability to "update" inferences as new data are observed. Standard Bayesian statistical references include Box and Tiao (1973), Berger (1985), Bernardo and Smith (1994), and Robert (2001).

Our solution to the noninvertible Hessian problem is technically not at all Bayesian since there is no stipulation of priors and no treatment of posteriors as general conditional distributions in this Bayesian sense. We do, however, use this distributional treatment as an interim process since the importance sampling step samples from the difficult posterior as a complete distribution. Since the point estimate and subsequent standard errors are reported, it is essentially back to a likelihoodist result in summary. The key point from this discussion is that researchers do not need to subscribe to the Bayesian inference paradigm to find our techniques useful.

We next describe in substantive terms what is "wrong" with a Hessian that is noninvertible (Section 6.5), describe how we create a pseudovariance matrix (in Section 6.7), with algorithmic details and numerical examples, outline the concept of importance resampling to compute the mean and variance (in Section 6.9). We give our alternative procedure in Section 6.11.1, an empirical example (Section 6.10), and other possible approaches (in Section 6.11).

6.5 PROBLEM IN DETAIL: NONINVERTIBLE HESSIANS

Given a joint probability density $f(\mathbf{y}|\boldsymbol{\theta})$ for an $n \times 1$ observed data vector $\mathbf{y}$ and unknown $p \times 1$ parameter vector $\boldsymbol{\theta}$, denote the $n \times p$ matrix of first derivatives with respect to $\boldsymbol{\theta}$ as

$$g(\boldsymbol{\theta}|\mathbf{y}) = \partial \ln[f(\mathbf{y}|\boldsymbol{\theta})]/\partial\boldsymbol{\theta},$$

and the $p \times p$ matrix of second derivatives as

$$\mathbf{H} = \mathbf{H}(\boldsymbol{\theta}|\mathbf{y}) = \partial^2 \ln[f(\mathbf{y}|\boldsymbol{\theta})]/\partial\boldsymbol{\theta}\,\partial\boldsymbol{\theta}'.$$

Then the Hessian is $\mathbf{H}$, normally considered to be the estimate

$$E[g(\boldsymbol{\theta}|\mathbf{y})g(\boldsymbol{\theta}|\mathbf{y})'] = E[\mathbf{H}(\boldsymbol{\theta}|\mathbf{y})].$$

The standard maximum likelihood or maximum posterior estimate, which we denote as $\hat{\boldsymbol{\theta}}$, is obtained by setting $g(\boldsymbol{\theta}|\mathbf{y})$ equal to zero and solving, analytically or numerically. When $-\mathbf{H}$ is positive definite in the neighborhood of $\hat{\boldsymbol{\theta}}$, the theory is well known and no problems arise in application. This occurs the vast majority of the time.

The problem described as "a noninvertible Hessian" can be decomposed into two distinct parts. The first problem is *singularity*, which means that $(-\mathbf{H})^{-1}$ does not exist. The second is *nonpositive definiteness*, which means that $(-\mathbf{H})^{-1}$ may exist but its contents do not make sense as a variance matrix. (A matrix that is positive definite is nonsingular, but nonsingularity does not imply positive definiteness.) Statistical software normally describes both problems as noninvertibility because their inversion algorithms take computational advantage of the fact that the negative of the Hessian must be positive definite if the result is to be a variance matrix. This means that these programs do not bother to invert nonsingular matrices (or even to check whether they are nonsingular) unless it is established first that they are also positive definite.

We first describe these two problems in single-parameter situations, where the intuition is clearest but where our approach does not add much of value (because the full posterior can easily be visualized). We then move to more typical multiple-parameter problems, which are more complicated but where we can help more. In one dimension, the Hessian is a single number measuring the degree to which the posterior curves downward on either side of the maximum. When all is well, $\mathbf{H} < 0$, which indicates that the mode is indeed at the top of the hill. The variance is then the reciprocal of the negative of this degree of curvature, $-1/\mathbf{H}$, which, of course, is a positive number, as a variance must be.

The first problem, singularity, occurs in the one-dimensional case when the posterior is flat near the mode—so that the posterior forms a plateau at best or a flat line over $(-\infty, \infty)$ at worst. Thus, the curvature is zero at the mode and the variance does not exist, since $1/0$ is not defined. Intuitively, this is as it should be since a flat likelihood indicates the absence of information, in which case any point estimate is associated with an (essentially) infinite variance (to be more precise, $1/\mathbf{H} \to \infty$ as $\mathbf{H} \to 0$).

The second problem occurs when the "mode" identified by the maximization algorithm is at the bottom of a valley instead of the top of a hill [$\mathbf{g}(\theta|y)$ is zero in both cases], in which case the curvature will be positive. (This is unlikely in one dimension, except for seriously defective maximization algorithms, but the corresponding problem in high-dimensional cases of *saddlepoints*, where the top of the hill for some parameters may be the bottom for others, is more common.) The difficulty here is that $-1/\mathbf{H}$ exists, but it is negative (or in other words, is not positive definite), which obviously makes no sense as a variance.

A multidimensional variance matrix is composed of variances, which are the diagonal elements and must be positive, and correlations that are off-diagonal elements divided by the square root of the corresponding diagonal elements. Correlations must fall within the $[-1, 1]$ interval. Although invertibility is an either/or question, it may be that information about the variance or covariances exist for some of the parameters but not for others.

In the multidimensional case, singularity occurs whenever the elements of $\mathbf{H}$ that would map to elements on the diagonal of the variance matrix, $(-\mathbf{H})^{-1}$, combine in such a way that the calculation cannot be completed because they would involve divisions by zero. Intuitively, singularity indicates that the variances to be calculated would be (essentially) infinite. When $(-\mathbf{H})^{-1}$ exists, it is a valid variance matrix only if the result is positive definite. Observe that $(-\mathbf{H})^{-1}$ is a positive definite matrix if for any nonzero $p \times 1$ vector $\mathbf{x}$, $\mathbf{x}'(-\mathbf{H})^{-1}\mathbf{x} > 0$. Nonpositive definiteness occurs in simple cases either because the variance is negative or the correlations are exactly -1 or 1.

6.6   GENERALIZED INVERSE/GENERALIZED CHOLESKY SOLUTION

The alternative developed here uses a generalized inverse, then a generalized Cholesky decomposition [if necessary when the generalized inverse of $(-\mathbf{H})$ is not positive definite], and subsequent refinement with importance sampling. The generalized inverse is produced by changing the parts of $-\mathbf{H}$ that get mapped to the variances so that they are no longer infinities. The generalized Cholesky adjusts inappropriate terms that would get mapped to the correlations (by slightly increasing variances in their denominator) to keep them within the required range of $[-1, 1]$. So the pseudovariance matrix is calculated as $\mathbf{V}'\mathbf{V}$, where $\mathbf{V} = \text{GCHOL}(\mathbf{H}^-)$, $\text{GCHOL}(\cdot)$ is the generalized Cholesky, and $\mathbf{H}^-$ is the generalized inverse of the Hessian.

The result of this process is a pseudovariance matrix that is in most cases well conditioned in that it is not nearly singular. Actually, this generalized inverse/generalized Cholesky approach is closely related to, but distinct from, the quasi-Newton *Davidson–Fletcher–Powell* (DFP) *method*. The difference is that the DFP method uses iterative differences to converge on an estimate of the negative inverse of a nonpositive definite Hessian. [See Greene (2003) for details.] However, the purpose of the DFP method is computational rather than statistical and therefore does not include our importance sampling step. Note that this method includes a default such that if the Hessian is really invertible, the pseudovariance matrix is the usual inverse of the negative Hessian.

6.7   GENERALIZED INVERSE

The literature on the theory and application of the generalized inverse is vast and spans several fields. Here we summarize some of the fundamental principles. [See Harville (1997) for further details.] The procedure begins with a generalized inverse procedure to address singularity in the $-\mathbf{H}$ matrix. This process resembles a standard matrix inversion to the greatest extent possible. The standard inverse $\mathbf{A}^{-1}$ of $\mathbf{A}$ meets five well-known conditions:

1. $\mathbf{H}\mathbf{A}^{-1}\mathbf{A} = \mathbf{A}$
2. $\mathbf{A}^{-1}\mathbf{A}\mathbf{A}^{-1} = \mathbf{A}^{-1}$
3. $(\mathbf{A}\mathbf{A}^{-1})' = \mathbf{A}^{-1}\mathbf{A}$
4. $(\mathbf{A}^{-1}\mathbf{A}) = \mathbf{A}\mathbf{A}^{-1}$
5. $\mathbf{A}^{-1}\mathbf{A} = \mathbf{I}$

(where conditions 1 to 4 are implied by condition 5). However, the *Moore–Penrose generalized inverse matrix*, $\mathbf{A}^-$ of $\mathbf{A}$, meets only the first four conditions listed above. Any matrix, $\mathbf{A}$, can be decomposed as

$$\underset{(p \times q)}{\mathbf{A}} = \underset{(p \times p)(p \times q)(q \times q)}{\mathbf{L} \quad \mathbb{D} \quad \mathbf{U}} \quad \text{where} \quad \mathbb{D} = \begin{bmatrix} \mathbf{D}_{r \times r} & 0 \\ 0 & 0 \end{bmatrix}, \tag{6.1}$$

and both $\mathbf{L}$ (lower triangular) and $\mathbf{U}$ (upper triangular) are nonsingular (even given a singular $\mathbf{A}$). The diagonal matrix $\mathbf{D}_{r \times r}$ has dimension and rank r corresponding to the rank of $\mathbf{A}$. When $\mathbf{A}$ is nonnegative definite and symmetric, the diagonals of $\mathbf{D}_{r \times r}$ are the eigenvalues of $\mathbf{A}$. If $\mathbf{A}$ is nonsingular, positive definite, and symmetric, as in the case of a proper invertible Hessian, $\mathbf{D}_{r \times r} = \mathbb{D}$ (i.e., $r = q$) and $\mathbf{A} = \mathbf{L}\mathbb{D}\mathbf{L}'$. The matrices $\mathbf{L}$, $\mathbb{D}$, and $\mathbf{U}$ are all nonunique unless $\mathbf{A}$ is nonsingular.

By rearranging (6.1) we can diagonalize any matrix as

$$\mathbb{D} = \mathbf{L}^{-1}\mathbf{A}\mathbf{U}^{-1} = \begin{bmatrix} \mathbf{D}_{r \times r} & 0 \\ 0 & 0 \end{bmatrix}. \tag{6.2}$$

Now define a new matrix, $\mathbb{D}^{-}$, created by taking the inverses of the nonzero (diagonal) elements of $\mathbb{D}$:

$$\mathbb{D}^{-} = \begin{bmatrix} \mathbf{D}_{r \times r}^{-} & 0 \\ 0 & 0 \end{bmatrix}. \tag{6.3}$$

If $\mathbb{D}\mathbb{D}^{-} = \mathbf{I}_{q \times q}$, we could say that $\mathbb{D}^{-}$ is *the* inverse of $\mathbb{D}$. However, this is not true:

$$\mathbb{D}\mathbb{D}^{-} = \begin{bmatrix} \mathbf{D}_{r \times r} & 0 \\ 0 & 0 \end{bmatrix} \begin{bmatrix} \mathbf{D}_{r \times r}^{-} & 0 \\ 0 & 0 \end{bmatrix} = \begin{bmatrix} 1 & 0 \\ 0 & 0 \end{bmatrix}.$$

Instead, we notice that

$$\mathbb{D}\mathbb{D}^{-}\mathbb{D} = \begin{bmatrix} 1 & 0 \\ 0 & 0 \end{bmatrix} \begin{bmatrix} \mathbf{D}_{r \times r} & 0 \\ 0 & 0 \end{bmatrix} = \begin{bmatrix} \mathbf{D}_{r \times r} & 0 \\ 0 & 0 \end{bmatrix} = \mathbb{D}.$$

So $\mathbb{D}^{-}$ is *a* generalized inverse of $\mathbb{D}$ because of the extra structure required. Note that this is *a* generalized inverse, not *the* generalized inverse, since the matrices on the right side of (6.1) are nonunique. By rearranging (6.1) and using (6.3) we can define a new $q \times p$ matrix: $\mathbf{G} = \mathbf{U}^{-1}\mathbb{D}^{-}\mathbf{L}^{-1}$. The importance of the *generalized inverse* matrix $\mathbf{G}$ is revealed in the following theorem.[3]

Theorem. (Moore 1920). $\mathbf{G}$ is a generalized inverse of $\mathbf{A}$ since $\mathbf{AGA} = \mathbf{A}$.

The new matrix $\mathbf{G}$ necessarily has rank r since the product rule states that the result has rank less than or equal to the minimum of the rank of the factors, and $\mathbf{AGA} = \mathbf{A}$ requires that $\mathbf{A}$ must have rank less than or equal to the lowest rank of itself or $\mathbf{G}$. Although $\mathbf{G}$ has infinitely many definitions that satisfy the Theorem, any one of them will do for our purposes: for example, in linear regression, the

---

[3]The generalized inverse is also sometimes referred to as the *conditional inverse, pseudo inverse,* and *g-inverse.*

fitted values, defined as $\mathbf{XGX'Y}$, with $\mathbf{G}$ as the generalized inverse of $\mathbf{X'X}$, $\mathbf{X}$ as a matrix of explanatory variables, and $\mathbf{Y}$ as the outcome variable, are invariant to the definition of $\mathbf{G}$. In addition, we use our pseudovariance only as a first approximation to the surface of the true posterior, and we will improve it in our importance resampling stage. Note, in addition, that $\mathbf{AG}$ is always idempotent [$\mathbf{GAGA} = \mathbf{G}(\mathbf{AGA}) = \mathbf{GA}$], and rank($\mathbf{AG}$) = rank($\mathbf{A}$). These results hold whether or not $\mathbf{A}$ is singular.

Moore (1920) and (apparently unaware of Moore's work) Penrose (1955) reduced the infinity of generalized inverses to the one unique solution given above by imposing four reasonable algebraic constraints, all met by the standard inverse. This G matrix is unique if the following hold:

1. *General condition:* $\mathbf{AGA} = \mathbf{A}$
2. *Reflexive condition:* $\mathbf{GAG} = \mathbf{G}$
3. *Normalized condition:* $(\mathbf{AG})' = \mathbf{GA}$
4. *Reverse normalized condition:* $(\mathbf{GA})' = \mathbf{AG}$

The proof is lengthy, and we refer the interested reader to Penrose (1955). There is a vast literature on generalized inverses that meet some subset of the Moore–Penrose condition. A matrix that satisfies the first two conditions is called a *reflexive* or *weak generalized inverse* and is order dependent. A matrix that satisfies the first three conditions is called a *normalized generalized inverse*. A matrix that satisfies the first and fourth conditions is called a *minimum norm generalized inverse*.

Because the properties of the Moore–Penrose generalized inverse are intuitively desirable, and because of the invariance of important statistical results to the choice of generalized inverse, we follow standard statistical practice by using this form from now on. The implementations of the generalized inverse in Gauss and Splus are both the Moore–Penrose version.

The Moore–Penrose generalized inverse is also easy to calculate using QR factorization. QR factorization takes the input matrix, $\mathbf{A}$, and factors it into the product of an orthogonal matrix, $\mathbf{Q}$, and a matrix, $\mathbf{R}$, which has a triangular leading square matrix ($\mathbf{r}$) followed by rows of zeros corresponding to the difference in rank and dimension in $\mathbf{A}$:

$$\mathbf{A} = \begin{bmatrix} \mathbf{r} \\ \mathbf{0} \end{bmatrix}.$$

This factorization is implemented in virtually every professional-level statistical package. The Moore–Penrose generalized inverse is produced by

$$\mathbf{G} = \begin{bmatrix} \mathbf{r}^{-1}\mathbf{0} \end{bmatrix} \mathbf{Q}',$$

where $\mathbf{0}$ is the transpose of the zeros' portion of the $\mathbf{R}$ matrix required for conformability.

6.7.1 Numerical Examples of the Generalized Inverse

As a means of motivating a simple numerical example of how the generalized inverse works, we develop a brief application to the linear model where the $\mathbf{X}'\mathbf{X}$ matrix is noninvertible because $\mathbf{X}$ is singular. In this context, the generalized inverse provides a solution to the normal equations (Campbell and Meyer 1979, p. 94), and both the fitted values of $\mathbf{Y}$ and the residual error variance are invariant to the choice of $\mathbf{G}$ (Searle 1971, pp. 169–71). We use the Moore–Penrose generalized inverse.

Let

$$
\mathbf{X} = \begin{bmatrix} 5 & 2 & 5 \\ 2 & 1 & 2 \\ 3 & 2 & 3 \\ 2.95 & 1 & 3 \end{bmatrix} \qquad \mathbf{Y} = \begin{bmatrix} 9 \\ 11 \\ -5 \\ -2 \end{bmatrix}
$$

(Our omission of the constant term makes the numerical calculations cleaner but is not material to our points.) Applying the least squares model to these data ($\mathbf{X}$ is of full rank) yields the coefficient vector

$$
\hat{\mathbf{b}} = (\mathbf{X}'\mathbf{X})^{-1}\mathbf{X}'\mathbf{Y} = (222.22, -11.89, -215.22)',
$$

fitted values,

$$
\hat{\mathbf{Y}} = \mathbf{X}\hat{\mathbf{b}} = (11.22, 2.11, -2.78, -2.00)',
$$

and variance matrix

$$
\Sigma = \begin{bmatrix} 57283.95 & -1580.25 & -56395.06 \\ -1580.25 & 187.65 & 1491.36 \\ -56395.06 & 1491.36 & 55550.62 \end{bmatrix}.
$$

What we call the *standardized correlation matrix*, a correlation matrix with standard deviations on the diagonal, is then

$$
\mathbf{C}_s = \begin{bmatrix} 239.34 & -0.48 & -0.99 \\ -0.48 & 13.69 & 0.46 \\ -0.99 & 0.46 & 235.69 \end{bmatrix}.
$$

Now suppose that we have a matrix of explanatory effects that is identical to $\mathbf{X}$ except that we have changed the bottom left number from 2.95 to 2.99:

$$
\mathbf{X}_2 = \begin{bmatrix} 5 & 2 & 5 \\ 2 & 1 & 2 \\ 3 & 2 & 3 \\ 2.99 & 1 & 3 \end{bmatrix}.
$$

Using the same $\mathbf{Y}$ outcome vector and applying the same least squares calculation now gives

$$\hat{\mathbf{b}}_2 = (1111.11, -11.89, -1104.11)'$$

and

$$\hat{\mathbf{Y}} = (11.22, 2.11, -2.78, -2.00)'.$$

However, the variance–covariance matrix reacts sharply to the movement toward singularity as seen in the standardized correlation matrix:

$$\mathbf{C}_s = \begin{bmatrix} 1196.70 & -0.48 & -0.99 \\ -0.48 & 13.70 & 0.48 \\ -0.99 & 0.48 & 1193.00 \end{bmatrix}.$$

Indeed, if $\mathbf{X}_3 = 2.999$, $\mathbf{X}'\mathbf{X}$ is singular (with regard to precision in Gauss and Splus) and we must use the generalized inverse. This produces

$$\tilde{\mathbf{b}}_3 = \mathbf{GX}'\mathbf{Y} = (1.774866, -5.762093, 1.778596)'$$

and

$$\hat{\mathbf{Y}} = \mathbf{XGX}'\mathbf{Y} = (11111.11, -11.89, -11104.11)'.$$

The resulting pseudovariance matrix (calculated now from $\mathbf{G}\sigma^2$) produces larger standard deviations for the first and third explanatory variables, reflecting greater uncertainty, again displayed as a standardized correlation matrix:

$$\mathbf{C}_s = \begin{bmatrix} 11967.0327987 & -0.4822391 & -0.9999999 \\ -0.4822391 & 13.698 & 0.4818444 \\ -0.9999999 & 0.4818444 & 11963.3201730 \end{bmatrix}.$$

6.8 GENERALIZED CHOLESKY DECOMPOSITION

We now describe the classic Cholesky decomposition and recent generalizations designed to handle nonpositive definite matrices. A matrix $\mathbf{C}$ is positive definite if for any $\mathbf{x}$ vector except $\mathbf{x} = \mathbf{0}$, $\mathbf{x}'\mathbf{Cx} > 0$, or in other words, if $\mathbf{C}$ has all positive eigenvalues. Symmetric positive definite matrices are nonsingular, have only positive numbers on the diagonal, and have positive determinants for all principal leading submatrices. The Cholesky matrix is defined as $\mathbf{V}$ in the decomposition $\mathbf{C} = \mathbf{V}'\mathbf{V}$. We thus construct our pseudovariance matrix as $\mathbf{V}'\mathbf{V}$, where $\mathbf{V} = \mathrm{GCHOL}(\mathbf{H}^-)$, $\mathrm{GCHOL}(\cdot)$ is the generalized Cholesky described below, and $\mathbf{H}^-$ is the Moore–Penrose generalized inverse of the Hessian.

6.8.1   Standard Algorithm

The classic Cholesky decomposition algorithm assumes a positive definite matrix and symmetric variance matrix ($\mathbf{C}$). It then proceeds via the matrix decomposition

$$\underset{(k \times k)}{\mathbf{C}} = \underset{(k \times k)(k \times k)(k \times k)}{\mathbf{L} \quad \mathbf{D} \quad \mathbf{L}'} . \tag{6.4}$$

The basic Cholesky procedure is a one-pass algorithm that generates two output matrices which can then be combined for the desired "square root" matrix. The algorithm moves down the main diagonal of the input matrix determining diagonal values of $\mathbf{D}$ and triangular values of $\mathbf{L}$ from the current column of $\mathbf{C}$ and previously calculated components of $\mathbf{L}$ and $\mathbf{C}$. Thus the procedure is necessarily sensitive to values in the original matrix and previously calculated values in the $\mathbf{D}$ and $\mathbf{L}$ matrices. There are k stages in the algorithm corresponding to the k-dimensionality of the input matrix. The jth step ($1 \le j \le k$) is characterized by two operations:

$$\mathbf{D}_{j,j} = \mathbf{C}_{j,j} - \sum_{\ell=1}^{j-1} \mathbf{L}_{j,\ell}^2 \mathbf{D}_{\ell,\ell} \tag{6.5}$$

and

$$\mathbf{L}_{i,j} = \left[\mathbf{C}_{i,j} - \sum_{\ell=1}^{j-1} \mathbf{L}_{j,\ell} \mathbf{L}_{i,\ell} \mathbf{D}_{\ell,\ell} \right] \Big/ \mathbf{D}_{j,j}, \qquad i = j+1, \ldots, k, \tag{6.6}$$

where $\mathbf{D}$ is a positive diagonal matrix so that on completion of the algorithm, its square root is multiplied by $\mathbf{L}$ to give the Cholesky decomposition. From this algorithm it is easy to see why the Cholesky algorithm cannot tolerate singular or nonpositive definite input matrices. Singular matrices cause a divide-by-zero problem in (6.6), and nonpositive definite matrices cause the sum in (6.5) to be greater than $\mathbf{C}_{j,j}$, causing negative diagonal values. Furthermore, these problems exist in other variations of the Cholesky algorithm, including those based on svd and qr decomposition. Arbitrary fixes have been tried to preserve the mathematical requirements of the algorithm, but they do not produce a useful result (Fiacco and McCormick 1968, Matthews and Davies 1971; Gill et al. 1974).

6.8.2   Gill–Murray Cholesky Factorization

Gill and Murray (1974) introduced, and Gill et al. (1981) refined, an algorithm to find a nonnegative diagonal matrix, $\mathbf{E}$, such that $\mathbf{C} + \mathbf{E}$ is positive definite and the diagonal values of $\mathbf{E}$ are as small as possible. This could easily be done by taking the greatest negative eigenvalue of $\mathbf{C}$, λ_1, and assigning $\mathbf{E} = -(\lambda_1 + \epsilon)I$, where ϵ is a small positive increment. However, this approach (implemented in various computer programs, such as the Gauss "maxlike" module) produces $\mathbf{E}$

values that are much larger than required, and therefore the $\mathbf{C}+\mathbf{E}$ matrix is much less like $\mathbf{C}$ than it could be.

To see Gill et al.'s (1981) approach, we rewrite the Cholesky algorithm provided as (6.5) and (6.6) in matrix notation. The jth submatrix of its application at the jth step is

$$\mathbf{C}_j = \begin{bmatrix} c_{j,j} & \mathbf{c}'_j \\ \mathbf{c}_j & \mathbf{C}_{j+1} \end{bmatrix}, \tag{6.7}$$

where $c_{j,j}$ is the jth pivot diagonal, $\mathbf{c}'_j$ is the row vector to the right of $c_{j,j}$, which is the transpose of the $\mathbf{c}_j$ column vector beneath $c_{j,j}$, and $\mathbf{C}_{j+1}$ is the $(j+1)$th submatrix. The jth row of the $\mathbf{L}$ matrix is calculated by: $L_{j,j} = \sqrt{c_{j,j}}$, and $\mathbf{L}_{(j+1):k,j} = \mathbf{c}_{(j+1):k,j}/L_{j,j}$. The $(j+1)$th submatrix is then updated by

$$\mathbf{C}^*_{j+1} = \mathbf{C}_{j+1} - \frac{\mathbf{c}_j \mathbf{c}'_j}{L^2_{j,j}}. \tag{6.8}$$

Suppose that at each iteration we defined $L_{j,j} = \sqrt{c_{j,j} + \delta_j}$, where δ_j is a small positive integer sufficiently large so that $\mathbf{C}_{j+1} > \mathbf{c}_j \mathbf{c}'/L^2_{j,j}$. This would obviously ensure that each of the j iterations does not produce a negative diagonal value or divide-by-zero operation. However, the size of δ_j is difficult to determine and involves trade-offs between satisfaction with the current iteration and satisfaction with future iterations. If δ_j is picked such that the new jth diagonal is just barely bigger than zero, subsequent diagonal values are greatly increased through the operation of (6.8). Conversely, we don't want to be adding large δ_j values on any given iteration.

Gill et al. (1981) note the effect of the $\mathbf{c}_j$ vector on subsequent iterations and suggest that minimizing the summed effect of δ_j is equivalent to minimizing the effect of the vector maximum norm of $\mathbf{c}_j$, $\|\mathbf{c}_j\|_\infty$, at each iteration. This is done at the jth step by making δ_j the smallest nonnegative value satisfying

$$\|\mathbf{c}_j\|_\infty \beta^{-2} - c_{j,j} \leqslant \delta_j \tag{6.9}$$

where

$$\beta = \max \begin{cases} \max(\text{diag}(\mathbf{C})) \\ \max(not\,\text{diag}(\mathbf{C}))\sqrt{k^2 - 1} \\ \epsilon_m, \end{cases}$$

where ϵ_m is the smallest positive number that can be represented on the computer used to implement the algorithm (normally called the *machine epsilon*) (see Chapter 4). This algorithm always produces a factorization and has the advantage of not modifying already positive definite $\mathbf{C}$ matrices. However, the bounds in (6.9) have been shown to be nonoptimal and thus provide $\mathbf{C} + \mathbf{E}$ that is again farther from $\mathbf{C}$ than necessary.

6.8.3   Schnabel–Eskow Cholesky Factorization

Schnabel and Eskow (1990) improve on the $C+E$ procedure of Gill and Murray by applying the Gerschgorin circle theorem to reduce the infinity norm of the E matrix. The strategy is to calculate δ_j values that reduce the *overall* difference between C and $C+E$. Their approach is based on the following theorem (stated in the context of our problem):

Theorem. Suppose that $C \in \mathbb{R}^k$ with eigenvalues $\lambda_1, \ldots, \lambda_k$, and define the ith Gerschgorin bound as

$$
G_i\,(\text{lower,upper}) = \left[C_{i,i} - \sum_{\substack{j=1 \\ j \neq i}}^{n} |C_{i,j}|, \; C_{i,i} + \sum_{\substack{j=1 \\ j \neq i}}^{n} |C_{i,j}| \right].
$$

Then $\lambda_i \in \left[G_1 \cup G_2 \cup \cdots \cup G_k \right], \forall \lambda_{1 \leq i \leq k}$.

But we know that λ_1 is the largest negative amount that must be corrected, so the process suggested by the theorem simplifies to the following decision rule:

$$
\delta_j = \max \left(\epsilon_m, \max_i (G_i\,(\text{lower})) \right). \tag{6.10}
$$

In addition, we do not want any δ_j to be less than δ_{j-1} since this would cause subsequent submatrices to have unnecessarily large eigenvalues, so a smaller quantity is subtracted in (6.8). Adding this condition to (6.10) and protecting the algorithm from problems associated with the machine epsilon produces the following determination of the additional amount in $L_{j,j} = \sqrt{c_{j,j} + \delta_j}$:

$$
\delta_j = \max \left(\epsilon_m, -C_{j,j} + \max(\|a_j\|, (\epsilon_m)^{1/3} \max(\text{diag}(C)), E_{j-1,j-1} \right). \tag{6.11}
$$

The algorithm follows the same steps as that of Gill–Murray except that the determination of δ_j is done by (6.11). The Gerschgorin bounds, however, provide an order-of-magnitude improvement in $\|E\|_\infty$. We refer to this Cholesky algorithm based on Gerschgorin bounds as the generalized Cholesky since it improves the common procedure, accommodates a more general class of input matrices, and represents the "state of the art" with regard to minimizing $\|E\|_\infty$.

6.8.4   Numerical Examples of the Generalized Cholesky Decomposition

Suppose that we have the positive definite matrix

$$
\Sigma_1 = \begin{bmatrix} 2 & 0 & 2.4 \\ 0 & 2 & 0 \\ 2.4 & 0 & 3 \end{bmatrix}.
$$

This matrix has the Cholesky decomposition:

$$\text{chol}(\Sigma_1) = \begin{bmatrix} 1.41 & 0 & 1.70 \\ 0 & 1.41 & 0 \\ 0 & 0 & 0.35 \end{bmatrix}.$$

Now suppose that we have a very similar but nonpositive definite matrix that requires the generalized Cholesky algorithm. The only change from the input matrix above is that the values on the corners have been changed from 2.4 to 2.5:

$$\Sigma_2 = \begin{bmatrix} 2 & 0 & 2.5 \\ 0 & 2 & 0 \\ 2.5 & 0 & 3 \end{bmatrix}.$$

This matrix has the generalized Cholesky decomposition

$$\text{GCHOL}(\Sigma_2) = \begin{bmatrix} 1.41 & 0 & 1.768 \\ 0 & 1.41 & 0 \\ 0 & 0 & 0.004 \end{bmatrix}$$

So the generalized Cholesky produces a very small change here so as to obtain a positive definite input matrix. This reflects the fact that this nonpositive definite matrix is actually very close to being positive definite. Now suppose that we create a matrix that is deliberately very far from positive definite status:

$$\Sigma_3 = \begin{bmatrix} 2 & 0 & 10 \\ 0 & 2 & 0 \\ 10 & 0 & 3 \end{bmatrix}$$

This matrix has the Cholesky decomposition

$$\text{GCHOL}(\Sigma_3) = \begin{bmatrix} 1.41 & 0 & 7.071 \\ 0 & 1.41 & 0 \\ 0 & 0 & 0.008 \end{bmatrix}$$

The effects are particularly evident when we square the Cholesky result:

$$\text{GCHOL}(\Sigma_3)'\text{GCHOL}(\Sigma_3) = \begin{bmatrix} 2 & 0 & 10 \\ 0 & 2 & 0 \\ 10 & 0 & 50 \end{bmatrix},$$

so the diagonal of the **E** matrix is very large: $[8, 6, 11]$.

6.9   IMPORTANCE SAMPLING AND SAMPLING IMPORTANCE RESAMPLING

The algorithm called *sampling importance resampling* (SIR) or simply *importance resampling* is a Monte Carlo simulation technique used to draw random numbers directly from an exact (finite sample) posterior distribution. The original idea comes from Rubin (1987a, pp. 192–94), but see also Wei and Tanner (1990), Tanner (1996), and Gill (2002). For social science applications, see King (1997) and King et al. (1998). The primary requirement for effective implementation of the algorithm is the specification of a reasonable approximation to the exact (but inconvenient) posterior. If this requirement is not met, the procedure can take excessively long to be practical or can miss features of the posterior distribution. Also, while the approximating distribution is required, it need not be normalized. So there is a lot of flexibility in this choice.

A common choice for the approximation distribution, based on flexibility and convenience, is the multivariate normal distribution. Sometimes the multivariate *t* distribution is substituted when the sample size is small or there is general concern about the tails. Using the normal or *t*-distribution should be relatively uncontroversial for our purposes here, since the algorithm in applied cases for which the asymptotic normal approximation was assumed appropriate from the start, and for most applications it probably would have worked except for the failed variance in the original matrix calculation. So this first approximation retains as many of the assumptions of the original model as possible. However, other distributions can easily be used if that seems necessary.

Using either the normal or *t*-distribution, the mean is set at $\hat{\boldsymbol{\theta}}$, the vector of maximum likelihood or maximum posterior estimates. Recall that this vector of point estimates was reported by the computer program that failed before it failed the variance calculation. For the normal this is simple: Set the variance equal to our pseudovariance matrix. For the *t*, the pseudovariance is required that there be an adjustment by the degrees of freedom to yield the appropriate scatter matrix.

6.9.1   Algorithm Details

The basic idea of importance resampling is to draw a large number of simulations from the approximation distribution, decide how close each is to the target posterior distribution, and keep those close with higher probability than for those farther away. The main difficulty is in determining an approximation distribution that somewhat resembles the difficult posterior. So we use normal or *t*-distributions centered at the posterior mean and the pseudovariance matrix calculated as $\mathbf{V'V}$, where $\mathbf{V} = \text{GCHOL}(\mathbf{H}^-)$, $\text{GCHOL}(\cdot)$ is the generalized Cholesky, and $\mathbf{H}^-$ is the generalized inverse of the Hessian.

Denote $\tilde{\boldsymbol{\theta}}$ as one random draw of $\boldsymbol{\theta}$ from the approximating distribution, and use it to compute the *importance ratio*: the ratio of the posterior $P(\cdot)$ to the normal approximation, where both are evaluated at $\tilde{\boldsymbol{\theta}}$: $P(\tilde{\boldsymbol{\theta}}|y)/N(\tilde{\boldsymbol{\theta}}|\hat{\boldsymbol{\theta}}, \mathbf{V'V})$. Then keep $\tilde{\boldsymbol{\theta}}$, as if it where a random draw from the posterior, with probability proportional to this ratio. The procedure is repeated until the desired (generally large) number of simulations have been accepted.

Suppose that we wish to obtain the marginal distribution for some parameter θ_1 from a joint distribution: $f(\theta_1, \theta_2|\mathbf{X})$. If we actually knew the parametric form for this joint distribution, it would be straightforward to integrate out the second parameter analytically over its support as shown in basic texts:

$$f(\theta_1|\mathbf{X}) = \int f(\theta_1, \theta_2|\mathbf{X}) \, d\theta_2. \qquad (6.12)$$

However, in many settings this is not possible, and more involved numerical approximations are required. Suppose that we could posit a *normalized* conditional posterior approximation density of θ_2, $\hat{f}(\theta_2|\theta_1, \mathbf{X})$, that would often be given a normal or t form, as mentioned above. The trick that this approximation gives is that an expected value formulation can be substituted for the integral and repeated draws used for numerical averaging. Specifically, the form for the marginal distribution is developed as

$$f(\theta_1|\mathbf{X}) = \int f(\theta_1, \theta_2|\mathbf{X}) \, d\theta_2$$

$$= \int \frac{f(\theta_1, \theta_2|\mathbf{X})}{\hat{f}(\theta_2|\theta_1, \mathbf{X})} \hat{f}(\theta_2|\theta_1, \mathbf{X}) \, d\theta_2$$

$$= E_{\theta_2} \left[\frac{f(\theta_1, \theta_2|\mathbf{X})}{\hat{f}(\theta_2|\theta_1, \mathbf{X})} \right]. \qquad (6.13)$$

The fraction

$$\frac{f(\theta_1, \theta_2|\mathbf{X})}{\hat{f}(\theta_2|\theta_1, \mathbf{X})}, \qquad (6.14)$$

called the *importance weight*, determines the probability of accepting sampled values of θ_2. This setup provides a rather simple procedure to obtain the estimate of $f(\theta_1|\mathbf{X})$. The steps are summarized as follows:

1. Divide the support of θ_1 into a grid with the desired level of granularity determined by k: $\theta_1^{(1)}, \theta_1^{(2)}, \ldots, \theta_1^{(k)}$.
2. For each of the $\theta_1^{(i)}$ values along the k-length grid, determine the density estimate at that point by performing the following steps:

   (a) Simulate N values of $\hat{\theta}_2$ from $\hat{f}(\theta_2|\theta_1^{(i)}, \mathbf{X})$.

   (b) Calculate $f(\theta_1^{(i)}, \hat{\theta}_{2n}|\mathbf{X})/\hat{f}(\hat{\theta}_{2n}|\theta_1^{(i)}, \mathbf{X})$ for $i = 1$ to N.

   (c) Use (6.13) to obtain $f(\theta_1^{(i)}|\mathbf{X})$ by taking the means of the N ratios just calculated.

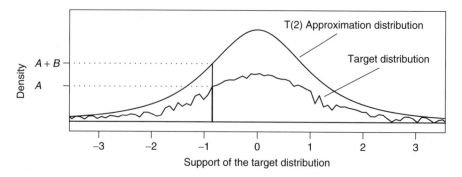

Fig. 6.1   Importance sampling illustration.

The user controls the level of accuracy of this estimate by increasing the granularity of the grid and the number of draws per position on that grid. In addition, this procedure can also be used to perform standard numerical integration, provided that a suitable normalized approximation function can be found (albeit somewhat less efficiently than standard algorithms; see Gill 2002, Chap. 8). These considerations make importance sampling a very useful and very common tool in applied mathematics.

The importance sampling algorithm is illustrated in Figure 6.1, where the importance ratio calculation is shown for an arbitrary point along the x-axis. The approximation distribution is t with two degrees of freedom and the target distribution is a contrived problematic form. The point indicated is accepted into the sample with probability $A/(A + B)$, which can be viewed as the quality of the approximation at this point.

6.9.2   SIR Output

The resulting simulations can easily be displayed with a histogram to give the full marginal distribution of a quantity interest (see Tanner 1996; King et al. 2000) or just a parameter of the model. Taking the sample average and sample standard deviation of the simulations can be used to compute the mean and standard error or full variance matrix of the parameters if these common summaries are desired. The computed variance matrix of the means will almost always be positive definite, as long as enough simulations are drawn such that there are sufficient elements of the mean vector and variance matrix (normally, one would want at least one order of magnitude more than that number). It is also possible, however, that the resulting variance matrix will be singular even when based on many simulations if the likelihood or posterior contains exact dependencies among the parameters. But in this case, singularity in the variance matrix (as opposed to the Hessian) poses no problem, since it is already on the variance–covariance metric (inverted), and the only problem is that some of the correlations will be exactly 1 or −1, which can actually be very informative substantively, and standard errors, for example, will still be available.

One diagnostic often used to detect a failure of importance resampling is when many candidate values of $\tilde{\theta}$ are rejected due to low values of the importance ratio. In this case the procedure will take a very long time, and to be useful a better approximation is certainly needed. Here, the long run time indicates a problem, and letting it run longer may eventually yield sufficient sample size. However, this can be very frustrating and time consuming from a practical point of view. There is a danger here, though: if the approximation distribution entirely misses a range of values of θ that have posterior density systematically different from the rest. Since the normal has support over $(-\infty, \infty)$, the potential for this problem to occur vanishes as the number of simulations grows. Therefore, one check is to compute a very large number of simulations with an artificially large variance matrix, such as the pseudovariance matrix multiplied by a positive factor, which we label F. This works since obviously the coverage is more diffuse. Like all related simulation procedures, it is impossible to cover the full continuum of values that θ can take, and the procedure can miss subtle features such as pinholes in the surface, very sharp ridges, or other eccentricities.

6.9.3 Relevance to the Generalized Process

The importance sampling procedure cannot be relied on *completely* in our case, since we know that the likelihood surface is nonstandard by definition of the problem. The normal approximation requires an invertible Hessian. The key to extracting at least some information from the Hessian via the derived pseudovariance matrix is determining whether the problems are localized or, instead, affect all the parameters. If they are localized, or the problem can be reparameterized so that they are localized, some parameters effectively have infinite standard errors, or pairs of parameters have perfect correlations. The suggestion here is to perform two diagnostics to detect these problems and to alter the reported standard errors or covariances accordingly. For small numbers of parameters, using profile plots of the posterior can be helpful, and trying to isolate the noninvertibility problem in a distinct set of parameters can be very valuable in trying to understand the problem.

To make the normal or t-approximation work more efficiently, it is generally advisable to reparameterize so that the parameters are unbounded and approximately symmetric. This strategy is pretty standard in this literature and normally makes the maximization routine work better. This can be broadly used; for example, instead of estimating $\sigma^2 > 0$ as a variance parameter directly, one could estimate γ, where $\sigma^2 = e^\gamma$, since γ can take on any real number.

6.10 PUBLIC POLICY ANALYSIS EXAMPLE

This real-data example looks at public policy data focused on poverty and its potential causes, measured by state at the county level (FIPS). The data highlight a common and disturbing problem in empirical model fitting. Suppose that a researcher seeks to apply a given model specification to multiple datasets for the purpose of comparison: comparing models across 50 U.S. states, 25 OECD countries, 15 EU countries, or even the same unit in some time series. Normally, if the Hessian fails to invert for a small number of the cases, generally the researcher respecifies the model for nonsubstantive, technical reasons, even though some other specification may be preferred for substantive reasons. If the researcher respecifies only the problem cases, differences among the results are contaminated by investigator-induced omitted variable bias. Otherwise, all equations are respecified in an effort to get comparable results, in which case the statistical analyses differs from the original substantive question posed. Obviously, neither approach is satisfactory from a substantive research perspective.

It is important to note, prior to giving the empirical example, *that we do not extract, fabricate, or simulate information from the likelihood function that does not exist.* That is, the culpable dimension will be given an infinite variance posterior, reflecting a complete lack of information about its form. What the algorithm does accomplish is the recovery of information on the other dimensions that otherwise would not be available to researchers. Therefore, a model that would have been dismissed as nonidentified for purely data reasons can now be partially recovered.

6.10.1 Texas

The example here uses data from the 1989 county-level economic and demographic survey for all 2276 nonmetropolitan U.S. counties ("ERS Typology") organized hierarchically by state such that each state is a separate unit of analysis with counties as cases. The U.S. Bureau of the Census, U.S. Department of Agriculture, and state agencies collect these data to provide policy-oriented information about conditions leading to high levels of rural poverty. The dichotomous outcome variable indicates whether 20% or more of the county's residents live in poverty (a standard measure in this field). The specification includes the following explanatory variables:

- Govt: a dichotomous factor indicating whether various government activities contributed a weighted annual average of 25% or more labor and proprietor income over the three preceding years.
- Service: a dichotomous factor indicating whether service-sector activities contributed a weighted annual average of 50% or more labor and proprietor income over the three preceding years.
- Federal: a dichotomous factor indicating whether federally owned lands make up 30% or more of a county's land area.
- Transfer: a dichotomous factor indicating whether income from transfer payments (federal, state, and local) contributed a weighted annual average of 25% or more of total personal income over the preceding three years.
- Population: the log of the county population total for 1989.
- Black: the proportion of black residents in the county.
- Latino: the proportion of Latino residents in the county.

This model provides the results given in Table 6.1.

Table 6.1 Logit Regression Model: Nonsingular Hessian, Texas

Parameter	Standard Results		Without `Federal`		Importance Sampling	
	Est.	Std. Err.	Est.	Std. Err.	Est.	Std. Err.
`Black`	15.91	3.70	16.04	3.69	15.99	3.83
`Latino`	8.66	1.48	8.73	1.48	8.46	1.64
`Govt`	1.16	0.78	1.16	0.78	1.18	0.74
`Service`	0.17	0.62	0.20	0.63	0.19	0.56
`Federal`	−5.78	16.20	—	—	−3.41	17.19
`Transfer`	1.29	0.71	1.17	0.69	1.25	0.63
`Population`	−0.39	0.22	−0.39	0.22	−0.38	0.21
`Intercept`	−0.47	1.83	−0.46	1.85	−0.51	1.68

A key substantive question is whether the black fraction predicts poverty levels even after controlling for governmental efforts and the other control variables. Since the government supposedly has a lot to do with poverty levels, it is important to know whether they are succeeding in a racially fair manner or whether there is more poverty in counties with larger fractions of African Americans. That is, whether the hypothesized effect is due to more blacks being in poverty or more whites and blacks in heavily black counties being in poverty would be interesting to know but is not material for our substantive purposes.

We analyze these data with a standard logistic regression model, so $P(Y_i = 1|X_i) = [1 + \exp(X_i\beta)]^{-1}$, where X_i is a vector of all our explanatory variables for case i. Using this specification, 43 of the U.S. states produce invertible Hessians and therefore available results. Rather than alter our theory and search for a new specification driven by numerical and computational considerations, we apply our approach to the remaining state models. From this 43:7 dichotomy, a matched pair of similar states is chosen for discussion here, where one case produces a (barely) invertible Hessian with the model specification (Texas) and the other is noninvertible (Florida). These states both have large rural areas, similar demographics, and similar levels of government involvement in the local county economies, and we would like to know whether the black fraction predicts poverty in similar fashions.

The logit model for Texas counties ($n = 196$) produces the results in the first pair of columns in Table 6.1. The coefficient on the black fraction is very large, and statistically reliable, thus supporting the racial bias hypothesis. It turns out that the variable `Federal` is problematic in these models and as noted below, actually prevents estimation using the Florida data. The second pair of columns reestimates the Texas model without the `Federal` variable, and the results for the black fraction (and the other variables) are mostly unchanged. In contrast to the *modes* and their standard deviations in the first two sets of results, the final pair of columns gives the *means* and their standard deviations by implementing our importance resampling but without the need for a pseudovariance matrix calculation. The means here are very close to the modes, and the standard errors in the two cases are very close as well, so the importance resampling in this (invertible) case did not generate important differences.

Below is the Hessian from this estimation, which supports the claim that the variable Federal is a problematic component of the model. Note the zeros and very small values in the fourth row and column of H.

$$
H =
\begin{vmatrix}
0.13907100 & 0.00971597 & 0.01565632 & 0.00000000 \\
0.00971597 & 0.00971643 & 0.00000000 & 0.00000000 \\
0.01565632 & 0.00000000 & 0.01594209 & 0.00000000 \\
0.00000000 & 0.00000000 & 0.00000000 & 0.00000003 \\
0.01165964 & 0.00022741 & 0.00305369 & 0.00000000 \\
1.27113747 & 0.09510282 & 0.14976776 & 0.00000044 \\
0.01021141 & 0.00128841 & 0.00170421 & -0.00000001 \\
0.03364064 & 0.00211645 & 0.00246767 & 0.00000000 \\
\end{vmatrix}
$$

$$
\begin{vmatrix}
0.01165964 & 1.27113747 & 0.01021141 & 0.03364064 \\
0.00022741 & 0.09510282 & 0.00128841 & 0.00211645 \\
0.00305369 & 0.14976776 & 0.00170421 & 0.00246767 \\
0.00000000 & 0.00000044 & -0.00000001 & 0.00000000 \\
0.01166205 & 0.10681518 & 0.00136332 & 0.00152559 \\
0.10681518 & 11.77556446 & 0.09904505 & 0.30399224 \\
0.00136332 & 0.09904505 & 0.00161142 & 0.00131032 \\
0.00152559 & 0.30399224 & 0.00131032 & 0.01222711 \\
\end{vmatrix}
$$

To see this near singularity implied by this Hessian, Figure 6.2 provides a matrix of the bivariate profile contour plots for each pair of coefficients from the Texas data, with contours at $0.05, 0.15, \ldots, 0.95$ where the 0.05 contour line bounds approximately 0.95 of the data, holding all other parameters constant at their maxima. These easy-to-compute profile plots are distinct from the more desirable but harder-to-compute marginal distributions: Parameters not shown are held constant in the former but integrated out in the latter. In these data, the likelihood is concave at the global maximum, although the curvature for Federal is only slightly greater than zero. This produces a near-ridge in the contours for each variable paired with Federal, and although it cannot be seen in the figure, the ridge is gently sloping around the maximum value in each profile plot, thus allowing estimation.

The point estimates and standard errors correctly pick up the unreliability of the Federal coefficient value by giving it a very large standard error, but as is typically the case, the graphed profile contours reveal more information. In particular, the plot indicates that distribution of the coefficient on Federal is quite asymmetric, and indeed, very informative in the manner by which the probability density drops as we come away from the near ridge. The modes and their standard errors, in the first pair of columns in Table 6.1, cannot reveal this additional information. In contrast, the importance resampling results reveal the richer set of information. For example, to compute the entries in the last two columns of Table 6.1, we first took many random draws of the parameters from

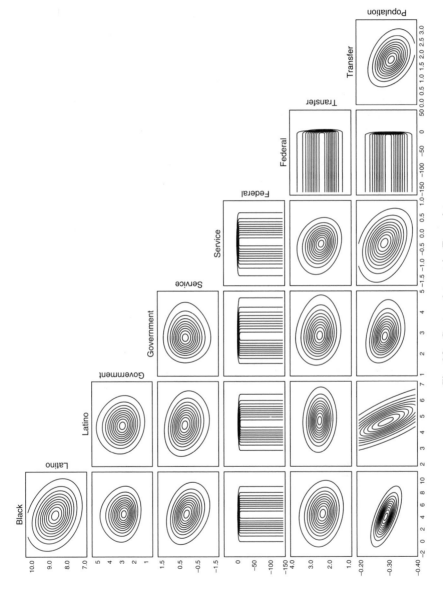

Fig. 6.2 Contourplot matrix, Texas data.

their exact posterior distribution. If instead of summarizing this information with their means and standard deviations, as in Table 6.1, we presented univariate or bivariate histograms of the draws, we would reveal all the information in Figure 6.2. In fact, the histograms would give the exact marginal distributions of interest (the full posterior, with other parameters integrated out) rather than merely the profile contours as shown in the figures, so the potential information revealed, even in this case where the Hessian is invertible, could be substantial. We do not present the histograms in this example because they happen to be similar to the contours in this particular dataset.

Note that although logit is known to have a globally concave likelihood surface in theory, actual estimates are not strictly concave, due to numerical imprecision. In the present data, the Hessian is barely invertible, making the likelihood surface sensitive to numerical imprecision. As it turns out, there are at least two local maxima on the marginal likelihood for Federal and thus potential attractors. The statistical package Gauss found a solution at −11.69 and the package R at −5.78 (reported). This discrepancy is typical of software solutions to poorly behaved likelihood functions, as algorithmic differences in the applied numerical procedures have different intermediate step locations. The difference in the results here is not particularly troubling, as no reasonable analyst would place faith in either coefficient estimate for Federal, given the large reported standard error. Note also that Govt and Service fall below conventional significance threshold levels as well. Our primary concern with Federal is that it alone prevents the Florida model (Section 6.10.2) from producing conventional results.

6.10.2 Florida

We ran the same specification used in Texas for Florida (33 counties), providing the maximum likelihood parameter estimates in Table 6.2 and the following Hessian, *which is now noninvertible*. The standard errors are represented in the table with question marks since standard estimates are not available.

Table 6.2 Logit Regression Model: Singular Hessian, Florida

Parameter	Standard Results		Without Federal		Importance Sampling	
	Est.	Std. Err.	Est.	Std. Err.	Est.	Std. Err.
Black	5.86	???	5.58	5.34	5.56	2.66
Latino	4.08	???	3.21	8.10	3.97	2.73
Government	−1.53	???	−1.59	1.24	−1.49	1.04
Service	−2.93	???	−2.56	1.69	−2.99	1.34
Federal	−21.35	???			−20.19	∞
Transfer	2.98	???	2.33	1.29	2.98	1.23
Population	−1.43	???	−0.82	0.72	−1.38	0.47
Intercept	12.27	???	6.45	6.73	11.85	4.11

$$
H = \begin{vmatrix}
0.13680004 & 0.04629599 & 0.01980602 & 0.00000001 & 0.05765988 & 1.32529504 & 0.02213744 & 0.00631444 \\
0.04629599 & 0.04629442 & 0.00000000 & -0.00000004 & 0.03134646 & 0.45049457 & 0.00749867 & 0.00114495 \\
0.01980602 & 0.00000000 & 0.01980564 & 0.00000000 & 0.01895061 & 0.19671280 & 0.00234865 & 0.00041155 \\
0.00000001 & -0.00000004 & 0.00000000 & 0.00000000 & 0.00000000 & 0.00000000 & 0.00000000 & 0.00000002 \\
0.05765988 & 0.03134646 & 0.01895061 & 0.00000000 & 0.05765900 & 0.57420212 & 0.00817570 & 0.00114276 \\
1.32529504 & 0.45049457 & 0.19671280 & 0.00000000 & 0.57420212 & 12.89475788 & 0.21458995 & 0.06208332 \\
0.02213744 & 0.00749867 & 0.00234865 & 0.00000000 & 0.00817570 & 0.21458995 & 0.00466134 & 0.00085111 \\
0.00631444 & 0.00114495 & 0.00041155 & 0.00000002 & 0.00114276 & 0.06208332 & 0.00085111 & 0.00088991
\end{vmatrix}
$$

Consider first Figure 6.3, which provides the same type of matrix of the bivariate profile plots for each pair of coefficients for the Florida data, like Texas with contours at $0.1, 0.2, \ldots, 0.9$. The problematic profile likelihood is clearly for Federal, but in this case the modes are not unique, so the Hessian is not invertible. Interestingly, except for this variable, the posteriors are very well behaved and easy to summarize. If one was forced to abandon the specification at this point, this is exactly the information that would be lost forever. The loss is especially problematic when contrasted with the Texas case, for which the contours do not look a lot more informative, but we were barely able to get an estimate.

Here is the key trap. A diligent data analyst using classical procedures with our data might reason as follows:

- The Texas data clearly suggest racial bias, but no results are available in Florida with the same specification.
- Follow the textbook advice and respecify by dropping Federal and rerunning the model for both Texas and Florida (these results are in both Tables 6.1 and 6.2).
- Note that the new results for black reveal a coefficient for Florida that is only a third of the size it was in Texas and only slightly larger than its standard error.

Now the contrast with the previous results is striking: a substantial racial bias in Texas and no evidence of such in Florida. However, with this approach it is impossible to tell whether these interesting and divergent substantive results in Florida are due to omitted variable bias rather than true political and economic differences between the states.

What can our analysis do? One reasonable approach is to assume that the (unobservable) bias that resulted from omitting Federal in the Florida specification would be of the same degree and direction as the (observable) bias that

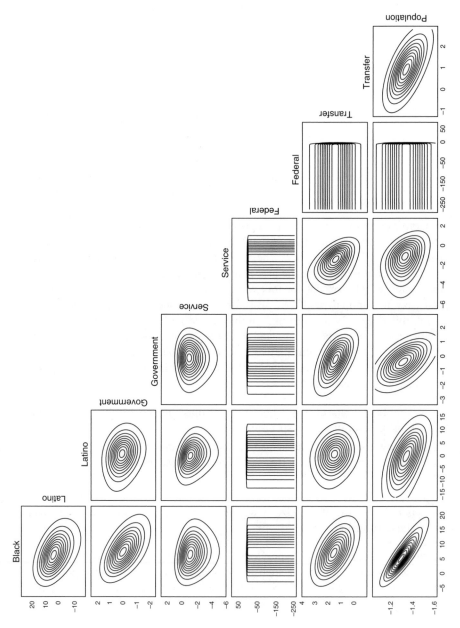

Fig. 6.3 Contourplot matrix, Florida data.

would occur by omitting the variable in the Texas data. Therefore, one can easily estimate the bias in Texas by omitting `Federal`. This is done in the second pair of columns in Table 6.1, and the results suggest that there is no bias introduced since the results are nearly unchanged from the first two columns. Although this seems like a reasonable procedure (and one that most analysts have no doubt tried at one time or another), it is of course based on the completely unverifiable assumption that the biases are the same in the two states. With the present data, this assumption is false, as our procedure now shows.

We now recover the information lost in the Florida case by first applying our generalized inverse and generalized Cholesky procedures to the singular Hessian to create a pseudovariance matrix. We then perform importance resampling using the multivariate normal, with the mode and pseudovariance matrix, as the first approximation. We use a t-distribution with three degrees of freedom as the approximation distribution so as to be as conservative as possible since we know from graphical evidence that one of the marginal distributions is problematic. The last two columns of Table 6.2 give the means and standard deviations of the marginal posterior for each parameter. We report ∞ for the standard error of `Federal` to emphasize the lack of information. Although the data and model contain no useful information about this parameter, the specification did control for `Federal`, so any potentially useful information about the other parameters and their standard errors are revealed with our procedure without the potential for omitted variable bias that would occur by dropping the variable entirely.

The results are indeed quite informative. They show that the effect of `Black` is indeed smaller in Florida than in Texas, but the standard error for Florida is now almost a third of the size of the coefficient. Thus, the racial bias is clearly large in both states, although larger in Texas than Florida. This result thus precisely reverses the conclusion from the biased procedure of dropping the problematic `Federal` variable. Of course, without the generalized inverse/generalized Cholesky technique, there would be no results to evaluate for Florida at all.

6.11 ALTERNATIVE METHODS

6.11.1 Drawing from the Singular Normal

In this section we describe another procedure for drawing the random numbers from a different approximating density: the truncated singular normal. The key idea is to draw directly from the singular multivariate density with a noninvertible Hessian. It should be true that the generalized Cholesky procedure will work better if the underlying model is identified, but numerical problems lead to apparent nonidentification. However, the singular normal procedure will perform better when the underlying model would have a noninvertible Hessian even if one were able to run it on a computer with infinite precision.

Again consider the matrix of second derivatives, $\mathbf{H}$, along with a $k \times 1$ associated vector of maximum likelihood estimates, $\hat{\boldsymbol{\theta}}$. Again, the matrix $(-\mathbf{H})^{-1}$ does not exist due either to nonpositive definiteness or to singularity ($r \leq k$). Suppose that one can actually set some reasonable bounds on the posterior distribution of each of the k coefficient estimates in $\hat{\boldsymbol{\theta}}$. These bounds may be set according to empirical observation with similar models, as a Bayes-like prior assertion (Hathaway 1985; O'Leary and Rust 1986; McCullagh and Nelder 1989; Geyer 1991; Wolak 1991; Geyer and Thompson 1992; Dhrymes 1994, Sec. 5.11). Thus, we assume that $\boldsymbol{\theta} \in [\mathbf{g}, \mathbf{h}]$, where $\mathbf{g}$ is a $k \times 1$ vector of lower bounds and $\mathbf{h}$ is a $k \times 1$ vector of upper bounds.

The goal now is to draw samples from the distribution of $\hat{\boldsymbol{\theta}} : \hat{\boldsymbol{\theta}} \sim N(\boldsymbol{\theta}, (-H)^{-1}) \propto e^{-T/2}$, truncated to be within $[\mathbf{g}, \mathbf{h}]$, and where $T = (\hat{\boldsymbol{\theta}} - \boldsymbol{\theta})'\mathbf{H}(\hat{\boldsymbol{\theta}} - \boldsymbol{\theta})$. Note that the normal density does not include an expression for the variance–covariance matrix—only the inverse (i.e., the negative of the Hessian), which exists here. We thus decompose T as follows:

$$T = (\hat{\boldsymbol{\theta}} - \boldsymbol{\theta})'\mathbf{H}(\hat{\boldsymbol{\theta}} - \boldsymbol{\theta})$$
$$= (\hat{\boldsymbol{\theta}} - \boldsymbol{\theta})'\mathbf{U}'\mathbf{L}\mathbf{U}(\hat{\boldsymbol{\theta}} - \boldsymbol{\theta}), \tag{6.15}$$

where $\mathbf{U}'\mathbf{L}\mathbf{U}$ is the *spectral decomposition* of $\mathbf{H}$; $\text{rank}(\mathbf{H}) = r \leq k$; $\mathbf{H}$ has r non-zero eigenvalues, denoted $d_1, \ldots, d_r$; $\mathbf{U}$ is $k \times k$ and orthogonal and hence $(\mathbf{U})^{-1} = \mathbf{U}'$; and $\mathbf{L} = \text{diag}(\mathbf{L}_1, 0)$, where $\mathbf{L}_1 = \text{diag}(d_1, \ldots, d_r)$. Thus, the $\mathbf{L}$ matrix is a diagonal matrix with r leading values of eigenvalues and $n - r$ trailing zero values.

Now make the transformation $\mathbf{A} = \mathbf{U}(\hat{\boldsymbol{\theta}} - [\mathbf{h} + \mathbf{g}]/2)$, the density for which would normally be $\mathbf{A} \sim N(\mathbf{U}(\boldsymbol{\theta} - [\mathbf{h} + \mathbf{g}]/2), (-\mathbf{L})^{-1})$. This transformation centers the distribution of $\mathbf{A}$ at the middle of the bounds, and since $\mathbf{L}$ is diagonal, it factors into the product of independent densities. But this expression has two problems:

- $(-\mathbf{L})^{-1}$ does not always exist.
- $\mathbf{A}$ has complicated multivariate support (a hypercube not necessarily parallel with the axes of the elements of $\mathbf{A}$), which is difficult to draw random numbers from.

We now address these two problems. First, in place of $\mathbf{L}$, we use $\mathbf{L}^*$ defined such that $L_i^* = L_i$ if $L_i > 0$ and L_i^* is equal to some small positive value otherwise (where the subscript refers to the row and column of the diagonal element). Except for the specification of the support of A that we consider next, this transforms the density into

$$\mathbf{A} \sim N(\mathbf{U}'\boldsymbol{\theta}, (-\mathbf{L}^*)^{-1})$$
$$= \prod_i N(U_i(\theta_i - [h_i + g_i]/2, -1/L_i^*). \tag{6.16}$$

Second, instead of trying to draw directly from the support of **A**, we draw from a truncated density with support that is easy to compute and encompasses the support of **A** (but is larger than it), transform back via $\hat{\boldsymbol{\theta}} = \mathbf{U}'\mathbf{A} + (\mathbf{h}+\mathbf{g})/2$, and accept the draw only if $\hat{\boldsymbol{\theta}}$ falls within its (easy to verify) support, $[\mathbf{g}, \mathbf{h}]$. The encompassing support we use for each element in the vector **A** is the hypercube $[-Q, Q]$, where the scalar Q is the maximum Euclidean distance from $\boldsymbol{\theta}$ to any of the 2^k corners of the hyperrectangle defined by the bounds. Since by definition $\boldsymbol{\theta} \in [\mathbf{g}, \mathbf{h}]$, we should normally avoid the sometimes common pitfall of rejection sampling—having to do an infeasible number of draws from **A** to accept each draw of $\hat{\boldsymbol{\theta}}$.

Now the principle of rejection sampling is satisfied here: that we can sample from any space (in our case, using support for **A** larger than its support) as long as it fully encompasses the target space and the standard accept–reject algorithm operates appropriately. If $-\mathbf{H}$ were positive definite, this algorithm would return random draws from a truncated normal distribution. When $-\mathbf{H}$ is not positive definite, it returns draws from a singular normal, but truncated as indicated.

So now we have draws of $\hat{\boldsymbol{\theta}}$ from a singular normal distribution. We then repeat procedure m, which serves to provide draws from the enveloping distribution that is used in the importance sampling procedure. That is, we take these simulations of $\hat{\boldsymbol{\theta}}$ and accept or reject according to the importance ratio. We keep going until we have enough simulated values.

6.11.2 Aliasing

The problem of computational misspecification and covariance calculation is well studied in the context of generalized linear models, particularly in the case of the linear model (Albert 1973). McCullagh and Nelder (1989) discuss this problem in the context of generalized linear models where specifications that introduce overlapping subspaces due to redundant information in the factors produce *intrinsic aliasing*. This occurs when a linear combination of the factors is reduced to fewer terms than the number of parameters specified. McCullagh and Nelder solve the aliasing problem by introducing suitable constraints, which are linear restrictions that increase the dimension of the subspace created by the factors specified. A problem with this approach is that the suitable constraints are necessarily an arbitrary and possibly atheoretical imposition. In addition, it is often difficult to determine a minimally affecting, yet sufficient set of constraints.

McCullagh and Nelder also identify *extrinsic aliasing*, which produces the same modeling problem but as a result of data values. The subspace is reduced below the number of factors because of redundant case-level information in the data. This is only a problem, however, in very low sample problems atypical of political science applications.

6.11.3 Ridge Regression

Another well-known approach to this problem in linear modeling is *ridge regression*, which essentially trades the multicollinearity problem for introduced bias.

Suppose that the $\mathbf{X'X}$ matrix is singular or nearly singular. Then specify the smallest scalar possible, ζ, that can be added to the characteristic roots of $\mathbf{X'X}$ to make this matrix nonsingular. The linear estimator is now defined as

$$\hat{\boldsymbol{\beta}}(\theta) = (\mathbf{X'X} + \zeta \mathbf{I})^{-1} \mathbf{X'y}.$$

There are two very well known problems with this approach. First, the coefficient estimate is by definition biased, and there currently exists no theoretical approach that guarantees some minimum degree of bias. Some approaches have been suggested that provide reasonably small values of ζ based on graphical methods (Hoerl and Kennard 1970a,b), empirical Bayes (Efron and Morris 1972; Amemiya 1985), minimax considerations (Casella 1980, 1985), or generalized ridge estimators based on decision-theoretical considerations (James and Stein 1961; Berger 1976; Strawderman 1978). Second, because ζ is calculated with respect to the smallest eigenvalue of $\mathbf{X'X}$, it must be added to every diagonal of the matrix: $\mathbf{X'X} + \zeta \mathbf{I}$. So by definition the matrix is changed more than necessary (in contrast to the Schnabel–Eskow method). For a very important critique, see Smith and Campbell (1980) along with the comments that follow.

6.11.4 Derivative Approach

Another alternative was proposed by Rao and Mitra (1971). Define $\delta\boldsymbol{\theta}$ as an unknown correction that has an invertible Hessian. Then (ignoring higher-order terms in a Taylor series expansion of $\delta\boldsymbol{\theta}$)

$$f(\mathbf{x}|\boldsymbol{\theta}) = H(\boldsymbol{\theta})\, \delta\boldsymbol{\theta}. \tag{6.17}$$

Since $H(\boldsymbol{\theta})$ is singular, a solution is available only by the generalized inverse:

$$\delta\boldsymbol{\theta} = H(\boldsymbol{\theta})^{-} f(\mathbf{x}|\boldsymbol{\theta}). \tag{6.18}$$

When there exists a parametric function of $\boldsymbol{\theta}$ that is estimable and whose first derivative is in the column space of $H(\boldsymbol{\theta})$, there exists a unique, maximum likelihood estimate of this function, $\phi(\hat{\boldsymbol{\theta}})$, with asymptotic variance–covariance matrix:

$$\phi(\hat{\boldsymbol{\theta}}) H(\boldsymbol{\theta}_0)^{-} \phi(\hat{\boldsymbol{\theta}}). \tag{6.19}$$

The difficulty with this procedure is finding a substantively reasonable version of $\phi(\hat{\boldsymbol{\theta}})$. Rao and Mitra's point is nevertheless quite useful since it points out that any generalized inverse has a first derivative in the column space of $H(\boldsymbol{\theta})$.

6.11.5 Bootstrapping

An additional approach is to apply bootstrapping to the regression procedure so as to produce empirical estimates of the coefficients, which can then be used to obtain subsequent values for the standard errors. The basic procedure (Davidson and MacKinnon 1993, pp. 330–31; Efron and Tibshirani 1993, pp. 111–12) is to bootstrap from the residuals of a model where coefficients estimates are obtained but where the associated measures of uncertainty are unavailable or unreliable.

The steps for the linear model are given by Freedman (1981):

1. For the model $\mathbf{y} = \mathbf{X}\boldsymbol{\beta} + \boldsymbol{\epsilon}$, obtain $\hat{\boldsymbol{\beta}}$ and the centered residuals $\boldsymbol{\epsilon}^*$.
2. Sample size n with replacement m times from $\boldsymbol{\epsilon}^*$ and calculate m replicates of the outcome variable by $\mathbf{y}^* = \mathbf{X}\hat{\boldsymbol{\beta}} + \boldsymbol{\epsilon}^*$.
3. Regress the m iterates of the $\mathbf{y}^*$ vector on $\mathbf{X}$ to obtain m iterates of $\hat{\boldsymbol{\beta}}$.
4. Summarize the coefficient estimates with the mean and standard deviation of these bootstrap samples.

The generalized linear model case is only slightly more involved since it is necessary to incorporate the link function and the (Pearson) residuals need to be adjusted (see Shao and Tu 1995, pp. 341–43).

Applying this bootstrap procedure to the problematic Florida data where the coefficient estimates are available but the Hessian fails, we obtain the standard error vector: [9.41, 9.08, 1.4, 2.35, 25.83, 1.43, 11.86, 6.32] (in the same order as Table 6.2). These are essentially the same standard errors as those in the model dropping `Federal` except that the uncertainty for `Population` is much higher. This bootstrapping procedure does not work well in non-iid settings (it assumes that the error between $\mathbf{y}$ and $\mathbf{X}\hat{\boldsymbol{\beta}}$ is independent of $\mathbf{X}$) and it is possible that spatial correlation that is likely to be present in FIPS-level population data is responsible for this discrepancy.

An alternative bootstrapping procedure, the paired bootstrap, generates m samples of size n directly from (y_j, x_j) together to produce $\mathbf{y}^*, \mathbf{X}^*$ and then generates $\hat{\boldsymbol{\beta}}$ values. While the paired bootstrap is less sensitive to non-iid data, it can produce simulated datasets (the $\mathbf{y}^*, \mathbf{X}^*$) that are very different from the original data (Hinkley 1988).

6.11.6 Respecification (Redux)

Far and away the most common way of recovering from computational problems resulting from forms of collinearity is respecification. Virtually every basic and intermediate textbook on linear and nonlinear regression techniques gives this advice. The respecification process can vary from ad hoc trial error strategies to more sophisticated approaches based on principal components analysis (Krzanowski 1988). Although these approaches often work, they force the user to change their research question due to technical concerns. As the example in Section 6.10 shows, we should not be forced to alter our thinking about a research question as a result of computational issues.

6.12   CONCLUDING REMARKS

The purpose of this chapter is twofold. The first objective is to illuminate the central role of the Hessian in standard likelihood-based estimation. The second objective is to provide a working solution to noninvertible Hessian problems that might otherwise cause researchers to discard their substantive objectives. This method is based on some established theories, but is new as a complete method. Currently, the generalized inverse/generalized Cholesky procedure is implemented in King's EI software (<http://gking.harvard.edu/stats.shtml>), and the R and Gauss procedures are freely available at <http://www.hmdc.harvard. edu/numerical_issues/>.

So what can a frustrated practitioner do? We have given several alternatives to the standard (albeit eminently practical) recommendation to respecify. Our new method is intended to provide results even in circumstances where it is not usually possible to invert the Hessian and obtain coefficient standard errors. The usual result is that the problematic coefficient has huge posterior variance, indicating statistical unreliability. This is exactly what should happen: A model is produced and poor contributors are identified.

Although the likelihood estimation process from a given dataset may have imposing problems, the data may still contain revealing information about the question at hand. The point here is therefore to help researchers avoid giving up the question they posed originally and instead, to extract at least some of the remaining information available. The primary method we offer here is certainly not infallible, nor are the listed alternatives. Therefore, considerable care should go into their use and interpretation.

CHAPTER 7

Numerical Behavior of King's EI Method

7.1 INTRODUCTION

The ecological inference (EI) problem occurs when one attempts to make inferences about the behavior of individuals (or subaggregates) from aggregate data describing group behavior. Ecological inference problems abound in the social sciences and occur frequently in voting rights litigation, where the voting patterns of individual members of racial groups are inferred from election returns aggregated into voting precincts. The problem is that information is lost in aggregation, and patterns observed in aggregate data may not correspond to patterns in subaggregate (individual) data (Robinson 1950). As King puts it, "... there is no way to make certain inferences about individual level behavior from aggregate data alone" (King 1997, p. 36).

Ecological inference is claimed to be the "longest standing, hitherto unsolved problems in quantitative social science" (King 1997). The problem is easy to define but difficult to solve. In the simplest case, a 2×2 table represents the unknown quantities one wishes to estimate, wherein only the table marginals are known. In voting rights litigation, for example, the election returns and racial characteristics of a given voting precinct are known. Because of the secret ballot, however, the number (or percentage) of black and white voters that voted for the black and white candidates are unknown. Ecological inference is to estimate these unknown quantities, which are used in this context to determine the degree of racially polarized voting—how often voters of a racial group vote only for candidates of their same racial type.

Interest is not just confined to voting and race. Other social scientists have used ecological regression methods to make claims about individual-level behavior. For instance, Gove and Hughes (1980) test a variant of Durkheim's hypothesis by asking whether living alone is related to suicide and alcoholism. That is, aggregate statistics exist on living arrangements, suicides, and (slightly less reliably) alcoholism. However, individual data can be difficult to obtain (for physiological as well as psychological reasons). Historians have a parallel interest in ecological

Numerical Issues in Statistical Computing for the Social Scientist, by Micah Altman, Jeff Gill, and Michael P. McDonald
ISBN 0-471-23633-0   Copyright © 2004 John Wiley & Sons, Inc.

inference because they often want to conjecture about individual behavior in the past that can only be observed as aggregates. Smith (1972) wants to understand the linkage between social class status and voting for Perón in the 1946 election. Obviously, one cannot go back in time and survey people during or immediately after the election, and surviving voters may not be a random sample (or voters may not even remember their vote correctly). So here there is no other alternative but to make some ecological claim.

Two approaches to solving the ecological inference problem were proposed in the 1950s. Goodman (1953) was the first to propose a linear regression solution to the ecological inference problem. "Goodman's regression" was adopted by the U.S. Supreme Court in *Thornburg* v. *Gingles* (1986) as a standard for evaluating racially polarized voting, and continues to be used in voting rights litigation. The drawback of linear regression is that it may produce nonsensical results outside the logical bounds of the table marginals, such as turnout rates above 100%. A second approach is to bound quantities of interest with information from the table marginals (Duncan and Davis 1953). This approach provides 100% confidence intervals (assuming zero measurement error) since the bounds are deterministic mathematical results, not statistical findings. Unfortunately, these intervals are sometimes too wide to be practically informative.

In 1997, King provided a new method that incorporates both the statistical information in Goodman's regression and the deterministic information in Duncan and Davis's bounds. This method has come to be known as EI (after King's popular software program of the same name). The method continues to be extended, both statistically and in its practical applications (e.g., King et al. 1999).

Early solutions to the ecological inference problem were relatively simple. In contrast, recent solutions to the ecological inference problem are theoretically detailed and computationally elaborate. King (1997) has proposed a solution to the ecological inference problem that purport to produce more accurate (and realistic) estimates of the true individual values but with considerably more elaborate computations.

Although much of the attention has been positive, King's solution has been criticized by a number of authors (Tam Cho 1998; Ferree 1999; Freedman et al. 1999; Herron and Shotts 2003a & b; McCue 2001). It is unfortunate but not unusual— in social science that computational issues tend to be mentioned in passing. For example, although the software that King distributes to compute his EI model contains many numerically sophisticated features, King devotes only two out of the nearly 350 pages in his book to computational details (see King 1997, App. F).

Debate over computational details are not uncommon when new statistical methods are introduced, but such debate is less often documented. The history of statistical and mathematical approaches to ecological inference is particularly noteworthy for generating controversy. From the early debates between those advocating the *method of bounds* (Duncan and Davis 1953) and those supporting Goodman's (1953) linear regression solution, to recent exchanges between King (1999) and Freedman et al. (1998, 1999), strongminded controversy over assumptions and implications has been common.

The focus here is on computation; this work explicitly examines the numerical properties of the leading proposed solution to the EI problem. In this chapter we provide a detailed examination of the numerical behavior of King's (1997) approach to ecological inference. We analyze the performance of this approaches in solving the ecological inference problem through sensitivity analysis. A data perturbation technique is used to evaluate the pseudostability of this technique across identical datasets. The results illuminate the trade-offs among correctness complexity and numerical sensitivity.

7.2 ECOLOGICAL INFERENCE PROBLEM AND PROPOSED SOLUTIONS

Formally, the ecological inference problem is typically described using a 2×2 table. More general $R \times C$ tables may also be constructed, but in practice the estimation problem is sufficiently complex that solutions are not reliable unless more information can be brought to bear on the problem.

The notation follows that of King (1997, p. 31) in formalizing the ecological inference problem in terms of an ecological inference problem of inferring from aggregate data in an election precinct, i, the proportion of blacks and whites who voted or did not vote. In King's notation:

- T_i: proportion of voting-age population turning out to vote
- X_i: proportion of voting-age population who are black
- β_i^b: proportion of voting-age blacks who vote
- β_i^w: proportion of voting-age whites who vote

These quantities are presented in a 2×2 table in Table 7.1.

The first wave of proposed solutions to the ecological inference problem occurred in the 1950s, with research by Duncan and Davis (1953) and Goodman (1953). Goodman (1953) proposed a simple linear regression model to solve the ecological inference problem, that generalized the accounting identity implied by Table 7.1, and which is still used frequently today:

$$T_i = \beta_i^b X_i + \beta_i^w (1 - X_i). \tag{7.1}$$

Contemporaneously, but independently, Duncan and Davis (1953) proposed that the logical bounds of the allocation of aggregate data to the subaggregates provides a range of possible solutions. For example, suppose that 75 blacks and 25 whites live in an election precinct (all of whom vote), and that the black candidate receives 75 votes and the white candidate 25. It is not necessarily true that all blacks voted for the black candidate, even though the numbers are equivalent. We know that black voters cast anywhere from [50, 75] of the votes for the black candidate, and that white voters cast anywhere from [0, 25] of the votes for the black candidate. Both black and white voters may have cast anywhere

Table 7.1   Simplified Notation for Ecological Inference Problem

Race of Voting-Age Person	Voting Decision		
	Vote	No Vote	
Black	β_i^b	$1 - \beta_i^b$	X_i
White	β_i^w	$1 - \beta_i^w$	$1 - X_i$
	T_i	$1 - T_i$	

from [0, 25] of the votes for the white candidate. Also important, the allocation of votes to one candidate depends on the number of votes allocated to the other candidate. Intuitively, these feasible values lie on a line implied by the solution for β_i^w and β_i^b in (7.1). King (1997, p. 81) summarizes the feasible values for all precincts into a *tomography plot*, which portrays these values of allocated votes within all precincts (as percentages rather than absolute numbers) as lines that crisscross the unit square.

Neither approach produces entirely satisfactory results. Goodman's regression ignores the obvious proportion bounds, [0, 100]%, and those logically deduced by Duncan and Davis. Duncan and Davis bounds provide only a narrower possible range of solutions, without inferential analysis in accompaniment.

King (1997) proposes *combining* the normality assumptions of Goodman's regression with the Duncan and Davis method of bounds. Strictly speaking, King assumes that β_i^w and β_i^b are distributed as truncated bivariate normal distribution over the unit square, conditional on up to two vectors of covariates. King's solution is also computationally complicated by its use of a generous amount of Bayesian-style simulation in lieu of closed-form solutions to transform variables.

7.3   NUMERIC ACCURACY IN ECOLOGICAL INFERENCE

In this section the heuristic sensitivity tests discussed in Chapter 4 are applied to applications of King's solution to the ecological inference problem. Examples drawn from King's (1997) book, *A Solution to the Ecological Inference Problem*, and from a study of split-ticket voting appearing in *American Political Science Review* (Burden and Kimball 1998) are employed.

Accuracy is a concern for all statistical algorithms and software, and the solutions to the ecological inference problem that we investigate in this chapter are no exception. Like most computer-intensive statistical programs, King's solution is potentially vulnerable in four areas. First, the program uses floating point arithmetic, so is potentially susceptible to floating point inaccuracies. Second, King's solution makes extensive use of the cumulative bivariate normal distribution, which can be difficult to compute accurately, particularly in the tails of the distribution. Third, the solution relies on solving a nonlinear constrained maximum likelihood problem, and despite assurances that convergence occurs "almost every time" and that the results are robust to local minima if any exist (King 1997, pp. 310–11), no formal proof is provided that the likelihood function is globally concave. So one cannot rule out the possibility that the search algorithm will settle at a local optimum. Fourth, King's solution uses simulation to reparameterize some of the variables, which requires a generous quantity of Pseudo-random numbers.

Fortunately, because King supplies the source code to his solution, it is possible to substitute alternative methods for computing some aspects of it (such as PRNGs) and to automate the replication of analysis under numerous combinations of input, options, and methods. The latter feature is of particular importance when the analysis required running thousands of replications, which at times took months of continuous computer time to complete.

The purpose is to test the sensitivity of King's solution to floating point error, to errors in the calculation of the bivariate normal distribution function, to search algorithm choice, and to Pseudo-random number generation (see Chapter 2). This discussion also applies the perturbation and sensitivity analyses described in Chapter 4, which highlight numerical problems and sensitivity to measurement error. To examine platform dependence, we ran different versions of the program on HP-Unix and Windows, as these operating systems may vary the underlying binary arithmetic computations. To test the accuracy of the bivariate normal distribution function used by EI, we substitute EI's function with a quadruple-precision function based on an extension of Drezner and Wesolowsky (1989). An expert in this area, Allan Genz, supplied a quadruple-precision function based on an extension of Drezner and Wesolowsky (1989). After porting this function to Gauss and integrating it into King's program, the areas of previous instability were tested. These were greatly improved, although not eliminated. The more accurate function has now been incorporated into a new version of King's programs as an option. This approach to removing numerical inaccuracies may prove fruitful for sophisticated consumers of statistical software.

To test the sensitivity of the estimation to the maximum likelihood algorithm, automation of the variation of the numerous algorithm options available in Gauss was constructed. Finally, a test for sensitivity to variations of default maximum likelihood search options was performed. Throughout, the sensitivity of the estimates was observed in the various circumstances as indications of numerical inaccuracies severe enough to caution inference from the examples investigated.

During the course of the analysis, and in part because of it, the programs distributed by King have undergone revision. The replication runs sometimes required long periods of time to complete, and it has happened that a new version of the program became available even as computers were incrementing through replications. New versions of the program will certainly be available after this chapter is in print. Readers should therefore understand that many, if not all of the problems identified herein have already been fixed and incorporated in EI. The purpose at hand is not to critique EI but to demonstrate the techniques used to identify and solve problems with implementations of algorithms.

7.3.1 Case Study 1: Examples from King (1997)

The analysis begins by evaluating the computational behavior of King's EI program using examples from King's (1997) book, *A Solution to the Ecological Inference Problem*. The replication files that King archived also contained many of the details of the original analysis that he performed using the original version of the EI program. These data and Gauss programs used to replicate the tables in King's book are available as Inter-university Consortium for Political and Social Research (ICPSR) publication replication archive as Study 1132. The discussion follows naming conventions in the replication files to identify the seven examples analyzed here: **CEN1910, FULTON, KYCK88, LAVOTE, MATPROII, NJ,** and **SCSP**. Selected results for these replications are presented in the first column of Table 7.2: the log-likelihood of the estimated solution and the estimated coefficients and standard errors for β^b and β^w. In turn, the tests and results of the accuracy of King's ecological inference solution algorithm to floating point arithmetic and platform sensitivity are presented.

Also included here are the effects of variations in Pseudo-random number generation that necessarily occur. For the record, runs under "HP-Unix" were created under HP-Unix 10.10 K900, running Gauss v.3.2.43, with CML v.1.0.41, plus EI v.1.0 and EI v.1.63 installed. The "Windows" system is a Pentium II running Windows 95 Sr2 and EzI v.2.23.

In this section we provide only an overview of the tests, for a full discussion of these and other sensitivity tests that we employ here, see Chapter 4. In interpreting the results of the analysis, it is important to caution readers that these data perturbations are diagnostic sensitivity tests. They are not classical statistical tests, although many can be interpreted within a statistical framework.

7.3.1.1 Perturbation Analysis

As described in Chapter 4, the intuition behind this test is straightforward: one should be wary of knife-edge results. If a small amount of noise introduced to the inputs of an analysis cause dramatic variations in the resulting estimates, one should suspect that the software used is particularly sensitive to numerical error or that the model is ill-conditioned with respect to these data, or both. Since this use of perturbations represents a sensitivity analysis and not a statistical test, there is no significance threshold or confidence interval directly associated with variations in estimates. As a rule of thumb, however, one should exercise caution when small amounts of noise lead to estimates that are outside the confidence intervals one would otherwise report.

To test for the sensitivity of EI to input perturbations for each of King's seven examples, we performed 100 perturbed replications of EI. For each precinct, X_i

Table 7.2 Analysis of Numeric Computation Properties of King's Ecological Inference Solution

Replication	Case	Original Analysis	Perturbation Analysis EI v1.63 Min–Max	Perturbation Analysis EI v1.63 Mean	Option Variation Analysis EI v1.63 Min–Max	Option Variation Analysis EI v1.63 Mean	PRNG Analysis EI v1.63	Platform and Version Variation Analysis EI v1.0	Platform and Version Variation Analysis EI v1.63	Platform and Version Variation Analysis EzI v2.23
CEN1910 (n = 1040)	LL	2449	2316–2402	2359	2448–2448	2448	2448	2448	2448	2448
	β^b	0.6383	0.6365–0.6547	0.6435	0.6348–0.6364	0.6357	0.6355	0.6363	0.636	0.6357
	Std. error	0.0082	0.0041–0.0550	0.0088	0.0038–0.0050	0.0044	0.005	0.0043	0.0047	0.0042
	β^w	0.9497	0.9405–0.9508	0.9477	0.9506–0.9512	0.9508	0.9509	0.9506	0.957	0.9508
	Std. error	0.0034	0.0017–0.0229	0.0037	0.0016–0.0021	0.0018	0.0021	0.0018	0.0019	0.0017
FULTON (n = 289)	LL	591.5	365.2–373.4	368.9	365.0–369.4	369.0	370.3	369.37	370.3	589.2
	β^b	0.5705	0.4070–0.4174	0.4126	0.3524–0.4141	0.4070	0.4262	0.4149	0.4266	0.5723
	Std. error	0.0052	0.0109–0.0153	0.0129	0.0118–0.0332	0.0147	0.0201	0.0128	0.0194	0.0046
	β^w	0.0428	0.4079–0.4228	0.4149	0.4115–0.5570	0.4283	0.3829	0.4096	0.3821	0.0385
	Std. error	0.0122	0.0257–0.0361	0.0304	0.0278–0.0782	0.0346	0.0474	0.0302	0.0458	0.0108
KYCK88 (n = 118)	LL	198.6	22.54–64.12	47.87	Fatal errors		73.7	44.9	73.7	214.8
	β^b	0.4216	0.4967–0.5433	0.5201	—		0.5158	0.5258	0.523	0.5212
	Std. error	0.1	0.0863–0.1269	0.1057	—		0.1144	0.1103	0.1198	0.1581
	β^w	0.7703	0.7556–0.7718	0.7635	—		0.7634	0.7628	0.763	0.7632
	Std. error	0.0071	0.0060–0.0095	0.0076			0.0082	0.0079	0.0085	0.0113
LAVOTE (n = 3262)	LL	6487	3203–3283	3237	3294–6486	5361	3621	3295	3621	6812
	β^b	0.6259	0.6230–0.6282	0.6259	0.6247–0.6307	0.6265	0.6266	0.6261	0.6268	0.6251
	Std. error	0.0016	0.0062–0.0093	0.0075	0.0016–0.0137	0.0046	0.0094	0.008	0.0091	0.0016
	β^w	0.7068	0.7055–0.7074	0.7065	0.7047–0.7069	0.7062	0.7062	0.7063	0.7061	0.7067
	Std. error	0.0006	0.0023–0.0034	0.0027	0.0006–0.0050	0.0017	0.0034	0.0029	0.0033	0.0006

(*Continued overleaf*)

Table 7.2 (Continued)

Replication	Case	Original Analysis	Perturbation Analysis EI v1.63		Option Variation Analysis EI v1.63		PRNG Analysis EI v1.63	Platform and Version Variation Analysis		
			Min-Max	Mean	Min-Max	Mean		EI v1.0	EI v1.63	EzI v2.23
MATPROII (n = 268)	LL	412.6	393.9–412.0	403.3	411.2–412.6	412.0	411.2	411.2	411.2	412.6
	β^b	0.5922	0.3309–0.8344	0.6022	0.5901–0.6085	0.5997	0.6115	0.6071	0.6123	0.5924
	Std. error	0.044	0.0443–0.2334	0.0766	0.0384–0.0553	0.0447	0.0541	0.0468	0.054	0.037
	β^w	0.8143	0.7473–0.8873	0.8113	0.8097–0.8149	0.8122	0.8088	0.8101	0.8086	0.8142
	Std. error	0.0125	0.0126–0.0673	0.0217	0.0109–0.0157	0.0127	0.0154	0.0133	0.0153	0.024
NJ (n = 493)	LL	1043	1007–1034	1020	1043–1043	1043	1043	1043	1043	1043
	β^b	0.0627	0.0559–0.0761	0.0667	0.0643–0.0695	0.0663	0.0611	0.0642	0.0616	0.0629
	Std. error	0.0097	0.0081–0.0111	0.0096	0.0122–0.0175	0.0144	0.0095	0.0142	0.0095	0.0105
	β^w	0.3791	0.3742–0.3808	0.3782	0.3775–0.3787	0.3783	0.3795	0.3787	0.3794	0.379
	Std. error	0.0023	0.0019–0.0026	0.0022	0.0028–0.0041	0.0033	0.0022	0.0033	0.0022	0.0024
SCSP (n = 3185)	LL	5339	5222–5305	5270	5094–1748000	430900	5657	5337	5657	5688
	β^b	0.1267	0.0358–0.1281	0.1139	0.0198–0.2572	0.0946	0.1162	0.1197	0.1156	0.1139
	Std. error	0.0041	0.0038–0.0529	0.0113	0.0015–0.1340	0.0263	0.0288	0.0109	0.0278	0.0424
	β^w	0.1786	0.1771–0.2622	0.1908	0.0548–0.2770	0.2082	0.1883	0.1841	0.1888	0.1904
	Std. error	0.0038	0.0035–0.0487	0.0104	0.0014–0.1234	0.0242	0.0265	0.01	0.0256	0.0391

and T_i are perturbed a small amount, distributed uniformly in $[-0.025, 0.025]$. (These runs were performed under HP-Unix.) Typically, the form of the perturbation in this type of analysis is usually either uniform noise, as in Beaton et al. (1976), Gill et al. (1981), and Parker et al. (2000), or normal, as in St. Laurent and Cook (1993). (For a detailed discussion of these tests, see Chapter 4.) However, the proportional data used as the input to an EI analysis complicates the perturbation. Both types of perturbations can yield proportions outside of the legal [0, 1] interval.

The easiest way to avoid perturbing values outside their legal range is to truncate any illegal value to the constraint, and this was the initial approach. This approach accumulates mass at the boundary points, although in the analysis only a handful of observations were affected by this truncation scheme. In later analyses two alternative methods of treating this problem were implemented. One approach is a resampling scheme, which rejects and resamples any draws that result outside the legal bounds, The second approach is a set of "shrinking" truncated noise distributions, made symmetric to avoid biasing the data, and shrunk to fall within the bounds for all noise draws. A consequence of the latter approach is that observations closest to the [0,1] constraint are effectively subject to less noise—those at the boundaries being subject to no noise at all. Six additional combinations of uniform and normal, truncated, resampled, and shrinking noise are compared, at levels both above and below that in the initial run. There is no substantive difference among the results of these six schemes. Although this is gratifying, in that the results are not sensitive to the exact form of noise used, it still warrants caution that the form of noise may matter for *other* problems.

The degree of perturbation sensitivity is gauged by comparing the results of these replications with the original published results found in King's replication files. The maximum, minimum, and mean values of the log-likelihood and the coefficients and standard errors of β^b and β^w across the 100 perturbations of these data for each example are presented in the second column of Table 7.2. In Chapter 4 there is a discussion on how introducing noise induces attenuation bias and may lower the magnitude of estimates coefficients. Despite this issue, for the replications of **CEN1910**, **LAVOTE**, and **NJ**, perturbations have a small effect on β^b and β^w, causing variations of only approximately 1% from the results published in King (1997). For **FULTON** the perturbation range is larger but is within the original confidence intervals for the parameters.

In three cases, **MATPROII**, **KYCK88**, and **SCSP**, results seem overly sensitive to perturbations. **MATPROII**'s β^b varies by much as 50% of its possible range. The same is true for the coefficient β^b in **KYCK88**. The sensitivity to perturbations observed suggests that numeric stability, data issues, or both are present in these examples.

MATPROII and **KYCK88** are instructive examples of the difficulty in parsing numeric stability issues with data issues. For both examples, the Duncan–Davis bounds for β^b are mostly uninformative, as presented in tomography plots in King (1997, pp. 204, 229). The values of β^b for individual precincts can fall practically anywhere within the [0, 1] bound. In addition, the likelihood function is not sharply peaked in β^b for these two examples. In particular, for **KYCK88**, King shows (p. 389) that a ridge exists on the likelihood surface for β^b. Perturbing these data will have a much greater effect than other examples on the shape of the likelihood function, and thus one cannot say how much of the variation observed across replication runs is a matter of numeric stability or a feature of the data.

Numeric issues are more pronounced in the presence of data issues. As discussed in Chapters 4 and 6, numerical accuracy along a flat likelihood surface requires accurate calculations to a high level of precision. In situations such as these, it is recommended that researchers use the most accurate algorithms and software available. There is evidence that King attempted alternative computational strategies, as reflected by the fact that the original log-likelihood, reported in King's replication files, exceeds that of the perturbation runs. Oddly, as the platform analysis below shows, it was often not possible to replicate, without perturbation, the original log-likelihood using the original version of the software, or the original platform.

SCSP is given in Chapter 11 of King's book (1997, pp. 215–25). King's diagnostics again point toward the presence of aggregation bias, citing the negative coefficient on β^b estimated by Goodman's regression and further analysis of "truth" (which, of course, would not normally be available to researchers). The sensitivity of β^b and β^w to perturbations is less than with the other problematic examples, but the estimates for the two beta parameters still vary by about 10% of the possible range.

These results show that for some, but not all, examples there was sensitivity of results to data perturbations. This is perhaps the most troubling case, since if EI always worked, there would be little need for alarm, and if it always failed, researchers would probably be aware of the potential problems and be on guard. These results show that for cases where EI diagnostics indicate potential difficulties, such as the presence of aggregation bias, researchers should be cautious, not only because the assumptions of the model may be violated, but also because the EI program itself may be prone to numerical inaccuracies.

7.3.2   Nonlinear Optimization

Standard mathematical solutions to numerical problems are analytical abstractions and therefore invariant to the optimization method used. In practice, this is often not the case. Nonlinear optimization can be performed using a number of different algorithms. Although the use of BFGS (Broyden, Fletcher, Goldfarb, Shannon) is recommended as a good initial choice (see Chapter 4) for the types of problems that social science researchers are most likely to encounter, no optimization algorithm uniformly dominates all others, and most of those offered by statistics packages are not presumptively invalid. However, some implementation options, such as forward differences, are clearly less accurate than the alternatives.

Here, we make an extensive investigation into the sensitivity of EI estimates to optimization method. The intuition behind this sensitivity test is to determine how dependent estimates are on the estimation process used. If the results of an analysis are *strongly* dependent on the choice of method, the author should provide evidence defending that choice.

To explore EI's sensitivity to changes in optimization method, we return to the examples, varying only the options to the *constrained maximum likelihood* (CML) Gauss library to vary across 40 combinations of optimization (secant) method, step (line search) method, and numerical derivative methods. Note that the claim here is not that all of these optimization methods are necessarily equally appropriate for EI but that there is *no justification given* for using a particular one. If a particular method is most appropriate, it should be documented, along with the reasons for its choice and diagnostics for confirming its success. Note also that for the most part, the versions of EI examined herein simply adopted the default options of the CML package and that many of these defaults (such as the use of forward differences in earlier versions) are not set solely with accuracy and robustness in mind.

As with the previous perturbation analyses, the results include the minimum, maximum, and mean values across the 40 replications. In a small number of runs for each of the datasets, CML failed to converge properly, almost invariably because the maximum number of iterations had been exceeded. These results are excluded in the analysis presented in the third column of Table 7.2. In two cases, **MATPROII** and **NJ**, EI estimates are robust to the optimization algorithm used to perform the constrained maximum likelihood estimation. In two other cases, **CEN1910** and **LAVOTE**, the ranges of the estimates of standard errors do not cover the published results, even through the estimated coefficients and log-likelihood values are generally near the published results.

In the remaining three cases, the estimates are sensitive to the optimization method. For **KYCK88**, the CML algorithm encountered fatal errors and the search algorithm was halted prematurely. For **FULTON**, every variation of the search algorithm does not find the published maximum; the log-likelihood never reaches the value of the published results and the range of the coefficients does not cover the published results. For **SCSP**, EI's estimate of β^b and β^w varied by roughly 25% of the possible $[0:1]$ range. Even more dramatically, the estimate of log-likelihood varied by a factor of over 1000.

Again, there is sensitivity to the optimization method for some but not all of King's examples. The recommendation is again for researchers to pay special attention to optimization methods by trying different configurations, especially when EI diagnostics point to problems with the model itself.

7.3.3 Pseudo-Random Number Generation

King performs CML estimation on a transformation of the variables and then uses simulation rather than analytical methods to transform estimated coefficients into the coefficients of interest to ecological inference (see King 1997, pp. 145–49).

These simulations require a number of pseudo-random numbers to be generated, on the order of the number of simulations requested times the number of precincts in the analysis. In addition, Gauss's CML solver may make use of pseudo-random numbers to add randomness to the search algorithm.

At the time of the analysis, Gauss's PRNG failed a number of diagnostic tests for randomness. Subsequently, a new version of Gauss was released that addressed all the failures we noted (Altman and McDonald 2001). King's example cases require hundreds of thousands of random numbers to produce final estimates, which far exceeds the maximum suggested for Gauss's updated PRNG. To test if these theoretical failings had an observable effect, EI was reconfigured here to use a better generator, Marsaglia and Zaman (1993) KISS generator, and then replicated King's results using this new generator. These results are presented in the fourth column of Table 7.2. The table reports some good news: The change of generators had no discernible effect on the estimates (except for **FULTON**, which apparently is a consequence of the search algorithm settling on a local optima), despite Gauss's poor native PRNG.

7.3.4 Platform and Version Sensitivity

At the time of writing, the most recent versions of EI and EzI were used. We also obtained the original version of King's EI program, which was used to produce the results in his book, and to replicate, as far as possible, the original computing environment, although the exact version of Gauss used by King is no longer available. We tested the sensitivity of the results to version and platform by replicating King's original analysis on these different versions and platforms. To be precise, for the HP-Unix replications we use versions 1.0 and 1.63 of EI, Gauss v3.2.43, and its constrained maximum likelihood (CML) algorithm v1.0.41, all on a system running the current version of HP-Unix. For the Windows replications, the EzI v2.23 package was run on Windows 2000.

The results of the analysis are presented in the fourth, fifth, and sixth columns of Table 7.2. First, the "null" results: For **NJ**, the three replications across versions and platform are consistent with the published results. The log-likelihood for **CEN1910** and two of the **MATPROII** EI replications are slightly smaller than the published results, which may explain the small differences in parameter estimates in these cases. Sizable differences in the log-likelihood for **LAVOTE** EI replications do not appear to greatly affect the parameter estimates. Neither do those replications for SCSP, where EzI and EI v1.63 find solutions with higher log-likelihoods than published.

Enough version and platform sensitivity is observed to change substantive results in **FULTON**. Recall that this example is particularly tricky, as the perturbation, optimization option, and PRNG analyses all encountered difficulty in replication. Here EzI is able to replicate the published results closely, but not exactly. CML in the GaussEI replications settles at a local optimum and fails to replicate.

These results lead to more than mild concern. As the **FULTON** and **LAVOTE** examples demonstrate, these discrepancies are not problems with EI or EzI per se, but rather, are more likely to be a result of differences between Gauss's implementation on HP-Unix and Windows operating systems than a direct result of the small differences in EI code between these two platforms. Ironically, although the published results are from the EI version of the program, the EzI replications are more faithful to the published results in these two examples, whereas the Gauss replications are not. In four examples; **FULTON, KYCK88, LAVOTE**, and **SCSP**, note that the CML search algorithm converged at a higher local optimum in the EI v1.63 replications than in the v1.0 replications, indicating, at least in some small part, improvement of the overall EI algorithm over time. However, these results show evidence of implementation dependence across the platform and version of King's ecological inference solution. Users of the program should use the most current version of EI and all programs in general, and if concerned about implementation dependence, should run the analysis on the EI and EzI versions of the program.

7.4   CASE STUDY 2: BURDEN AND KIMBALL (1998)

In the December 2001 issue of the *American Political Science Review*, the editor publicly withdrew an offer to publish an accepted manuscript written by Tam Cho and Gaines (2001) that critiqued the article published by Burden and Kimball (1998), focusing especially on the numerical accuracy of King's (1997) EI program, used in Burden and Kimball's analysis:[1] "Because of inaccuracies discovered during the prepublication process, 'Reassessing the Study of Split-Ticket Voting,' by Wendy K. Tam Cho and Brian J. Gaines, previously listed as forthcoming, has been withdrawn from publication." (p. iv). This unusual decision by the *American Political Science Review* editor was prompted by flawed replication of Burden and Kimball's analysis by Tam Cho and Gaines, who claim that rerunning Burden and Kimball's EzI analysis yields substantively different results. Unfortunately, they unwittingly changed the case selection rule across runs and did not provide evidence for numerical inaccuracy. In the replication of both Burden and Kimball's analysis and Tam Cho and Gaines's reanalysis, it was discovered that there were execution bugs and usability issues related to option parameters in EzI, which probably contributed further to the errors made by Tam Cho and Gaines. This is an interesting case study because it shows that even the most careful and thoughtful of authors can inadvertently make serious numerical computing errors.

More recently, King's iterative procedure for applying ecological inference to $R \times C$ tables, and Burden and Kimball's use of it, have been criticized for logical

---

[1]Tam Cho and Gaines's original unpublished paper, along with responses and replication data, are available in the replication archive of the Inter-university Consortium for Political and Social Research, as replication data set 1264, available as ftp://ftp.icpsr.umich.edu/pub/PRA/outgoing/s1264/.

and statistical inconsistency (Ferree 1999; Herron and Shotts 2002). It is true also that casual use of ecological inference coefficients in second-stage regression analyses can lead to serious problems.[2]

In this chapter we do not provide a theoretical or practical justification for using larger than 2×2 tables, which King states should only be used in special circumstances not established by Burden and Kimball (Adolph and King 2003), but focus only on the computational issues involved in producing such inferences. The investigation confirms that Burden and Kimball ignore computational issues, especially with respect to simulation-induced variation of estimates, and as a consequence underestimate the uncertainty of the estimates in their analysis.

Burden and Kimball (1998), using King's method to study split-ticket voting, is one of the first applications of King's method to appear independently in print. Burden and Kimball's (1998) aggregate data represent the percentage of people who voted for the two major party presidential and congressional candidates in 1988 congressional districts. From this, the authors wish to estimate the percentage that split their tickets among candidates from different parties, which they refer to as Bush and Dukakis splitters. Since typically fewer people vote in congressional than in presidential races (i.e., roll-off), Burden and Kimball must estimate a two-stage EI model, which essentially produces ecological inference estimates for a 2×3 table.

In the first stage, Burden and Kimball estimate the percentage of roll-off among the two types of ticket splitters. This estimate of the number of people who did not vote in the congressional election is removed from the analysis through an iterative application of EI . In the second stage, Burden and Kimball use the first-stage estimates to estimate rates of ticket splitting. Burden and Kimball (1998) seek to estimate a 2×3 table of aggregate data, whereas King's examples are 2×2 tables. King's solution is designed for a 2×2 table, but he recommends and provides in his EI program a multistage approach to solving the more general $2 \times C$ case (row $\times$ column), which involves iteratively applying the estimation to data analyzed through a previous 2×2 estimation (see King 1997, Sec. 8.4 and Chap. 15).

The algorithm differs for one- and two-stage models in a way that has important consequences for the current analysis. In the discussion of PRNGs above, note that King's method uses simulation to solve otherwise intractable analytic computations when transforming the maximum likelihood estimates into quantities of interest. Unless the PRNG seed is held constant, no two runs of the program will produce the same precinct-level beta parameters at the first stage, even though the maximum likelihood optimizer should settle on the same optimum. This is called *simulation variance* (i.e., from Monte Carlo error) and should not be confused with numerical inaccuracies of the program. Furthermore, because PRNGs are deterministic processes, they are not equivalent to true random draws from a

---

[2]Herron and Shotts also have a forthcoming paper (2003a) whose main point is a proof that using point estimates from the EI model as outcome variables in a second-stage regression model is wrong, usually having logically incompatible residual assumptions. The authors also provide a simple test based on bivariate linear regression.

distribution. As a consequence, increasing the number of simulations only reduces simulation variance up to a point and does not guarantee that the results will not be slightly biased. (See Chapter 2 for a discussion of these issues.)

These random perturbations of the first stage are carried through to the second stage, where the beta parameters of the first stage are used to estimate the beta parameters of the second stage. This is, in effect, similar to the previous perturbation analysis. In the second stage, the maximum likelihood optimum should vary as a consequence of the perturbations introduced through simulation at the first stage. The beta parameters of the second stage will be affected by both the simulation variance at the first stage which affects the maximum likelihood optimization, and the simulation variance at the second stage. This method of estimation suggests that a simple heuristic test of numeric accuracy of the two-stage method is to run the program multiple times and note large variations in the second-stage results.

A total of 195 replications of Burden and Kimball's analysis on HP-Unix (EI) and Windows (EzI) platforms are summarized in Table 7.3. The original results were run on the Windows version of King's program, EzI , Version 1.21. Although replication data are available for Burden and Kimball's research at the ICPSR publication archive as Study 1140, special EI replication files are not available, so one is unable to view some of the details of the original runs, as done here with King's examples. The published results presented in Table 7.3 are drawn directly from the Burden and Kimball (1999) article.

As with the previous case study of examples from King (1997), we employ heuristic tests to examine the stability of results to perturbing data, changing default options, and running the analysis on different operating platforms. In this case-study we pay further special attention to two options that may explain why the original version of EI produces results different from the current version. One option controls the type of cumulative bivariate normal distribution algorithm used by EI , and the other controls the method EI uses to invert the Hessian. This additional analysis illustrates the benefit of changing options when available to identify sources of numerical inaccuracy.

7.4.1 Data Perturbation

The analysis is presented in Table 7.3. At a first blush, EI produces approximately deterministic first-stage estimates; all first-stage replications, where the analysis only varies the version and some important default settings, the CML search algorithm consistently finds the same optimum for each configuration. Point estimates of the beta parameters fluctuate around the point estimate of the standard error, as is expected with simulation variance.

The test for the numeric stability of the original and current versions of EI is accomplished by adding 1% simulated measurement error to T_i, which in this case is the House election turnout divided by the presidential turnout. The results are presented in Table 7.3 in the section "Perturbation Analysis". Apparently, the CML algorithm finds different optima and greater variation in the point estimates,

Table 7.3 Burden and Kimball Replication

	First-Stage Estimates			Second-Stage Estimates		
	β^b	β^b	Log-Likelihood	Dukakis Splitters	Bush Splitters	Log-Likelihood
Burden and Kimball published results						
Windows EzI						
Windows EzI v1.21	Not reported			0.198 (0.007)	0.331 (0.006)	Not reported
HP-Unix						
(25 replications)						
EzI v1.0						
(v1.0) cdfbvn setting	[0.9711–0.9713]	[0.8664–0.8666]	−671.9	[0.2185–0.2214]	[0.3094–0.3121]	[0494.4–496.8]
(v1.0) Hessian setting	[0.0008–0.0010]	[0.0006–0.0008]		[0.0060–0.0082]	[0.0057–0.0078]	
EzI v1.63						
(v1.63) cdfbvn setting	[0.9304–0.9314]	[0.9006–0.9013]	1057	[0.2125–0.2181]	[0.3220–0.3269]	[0494.4–496.8]
(v1.63) Hessian setting	[0.0024–0.0032]	[0.0021–0.0027]		[0.0068–0.0088]	[0.0059–0.0078]	
(v1.0) cdfbvn setting	[0.9271–0.9452]	[0.8886–0.9041]	1034	[0.2074–0.2134]	[0.3145–0.3222]	[0495.8–507.9]
(v1.63) Hessian setting	[0.0344–0.0488]	[0.0294–0.0416]		[0.0065–0.0092]	[0.0063–0.0085]	
(v1.63) cdfbvn setting	[0.8984–0.9158]	[0.9138–0.9286]	1057	[0.2091–0.2178]	[0.3290–0.3374]	[0490.4–498.9]
(v1.0) Hessian setting	[0.0186–0.0274]	[0.0159–0.0234]		[0.0072–0.0099]	[0.0059–0.0089]	
(v1.0) cdfbvn setting	[0.9404–0.9411]	[0.8922–0.8928]	1034	[0.2129–0.2170]	[0.3178–0.3230]	[0494.3–499.9]
(v1.0) Hessian setting	[0.0020–0.0026]	[0.0017–0.0023]		[0.0061–0.0088]	[0.0056–0.0080]	

Perturbation analysis

(25 replications)					
EI v1.0	[0.8271–0.9943]	[0.8469–0.9892]	[−645.3–1045]	[0.1791–0.2211]	[0.2971–0.3467]	[490.7–505.1]
	[0.0012–0.0451]	[0.0011–0.0385]		[0.0064–0.0094]	[0.0057–0.0084]	
EI v1.63	[0.9016–0.9555]	[0.8797–0.9255]	[939.4–996.5]	[0.2075–0.2170]	[0.3118–0.3298]	[492.8–503.6]
	[0.0022–0.0635]	[0.0019–0.0542]		[0.0072–0.0203]	[0.0067–0.0176]	

Windows EzI

(5 replications)					
EzI v2.3						
(v2.3) cdfbvn setting	[0.9026–0.9122]	[0.9169–0.9250]	1060	[0.2015–0.2094]	[0.3240–0.3294]	[494.6–502.0]
(v2.3) Hessian setting	[0.0366–0.0391]	[0.0312–0.0334]		[0.0080–0.0092]	[0.0071–0.0080]	
(v1.0) cdfbvn setting	[0.9060–0.9169]	[0.9169–0.9250]	1057	[0.2048–0.2388]	[0.3246–0.3534]	[496.5–501.5]
(v2.3) Hessian setting	[0.0338–0.0401]	[0.0289–0.0343]		[0.0074–0.0787]	[0.0058–0.0666]	
(v2.3) cdfbvn setting	[0.8837–0.8842]	[0.9407–9412]	1060	[0.1917–0.2000]	[0.3298–0.3319]	[500.7–501.7]
(v1.0) Hessian setting	[0.0020–0.0026]	[0.0017–0.0023]		[0.0069–0.0086]	[0.0055–0.0069]	
(v1.0) cdfbvn setting	[0.8838–0.8841]	[0.9408–0.9411]	1057	[0.1975–0.1991]	[0.3298–0.3315]	[501.2–501.7]
(v1.0) Hessian setting	[0.0015–0.0017]	[0.0013–0.0014]		[0.0074–0.0087]	[0.0059–0.0070]	

as expected when the likelihood function is perturbed. The analysis also indicates that changes to EI have reduced the sensitivity of the first-stage estimates to perturbations of data. The original version of EI v1.0 shows considerable sensitivity to perturbations, with large swings in the value of the log-likelihood and point estimates of the coefficients, while EI v1.63 shows roughly a threefold decrease in the sensitivity of EI's point estimates to perturbation. Further, for no replication did the constrained maximum likelihood procedure fail to converge or find an optimum far from that found in the unperturbed replications of EI v1.63.

The value of the second-stage log-likelihood varies across the perturbed datasets, but the optima are close to one another, again indicating stability. The second-stage point estimates show stability within the range of their simulation variance, even where the first-stage estimates show greater sensitivity to perturbations. Yet again, perturbation analysis of the second-stage estimates suggests that improvements in King's program have led to a threefold decrease in the sensitivity of the second-stage estimates to perturbations.

7.4.2 Option Dependence

In this section we analyze option dependence by investigating two changes in EI that are potential sources of it: changes to the cumulative bivariate normal distribution algorithm (cdfbvn) and method of inverting the Hessian. The procedure is to force the program to use the methods from EI Version 1.0 by changing global variables and assess how these changes affect estimation. Here 25 replications are run for each setting, to account for variation due to simulation variance.

The cumulative bivariate normal distribution algorithm is an important factor in determining the shape of the likelihood function for the EI method. The "shape" of the likelihood function determines not only the location of the mode of posterior, but also the Hessian matrix, which is the curvature around this mode as measured by the second derivative of the likelihood function at the modal value. The importance of the Hessian matrix is that it produces, by inversion, the variance–covariance matrix of the coefficient estimate. King recognizes that this process is not always straightforward and provides options for users to choose six different methods of calculating the cumulative bivariate normal distribution with the _Ecdfbvn option, which is referred to as cdfbvn (i.e., cumulative density function, bivariate normal). The original default cdfbvn is a fast algorithm, but subject to inaccuracies for small values, while the current default represents a trade-off between accuracy and speed. King once recommended the use of the current default (King 1998, p. 8), although he now provides a more accurate version as a consequence of the investigation into the accuracy of the function.

A second source of option dependence is the method used to represent the inverse of a nonpositive definite Hessian. (The version of this routine that became available after this chapter was drafted is both more accurate and faster; see

Chapter 6.) In all of these replications the first-stage estimation results in a Hessian that is not positive definite. In these circumstances the program uses specialized methods to find a "close" Hessian that is invertible. The program attempts a number of methods in sequence and exits on execution of the first successful method. As new versions of the program have been developed, new techniques have been devised to handle situations when the Hessian is not positive definite. The sequence of methods applied when the default method fails has also changed. In early versions of the program, such as the one used by Burden and Kimball, the first specialized method that the program will attempt is documented as a "wide-step procedure" or "quadratic approximation with falloff" (King 1998, p. 10). In later versions of EI, the program attempts a generalized inverse Cholesky alternative proposed by Gill and King (see Chapter 6).

A summary of the results are presented in Table 7.3. There is ample evidence that Tam Cho and Gaines were correct in their suspicions: Replication of Burden and Kimball is indeed dependent on the version and option settings of EI. The range of first- and second-stage beta parameters do not overlap between EI v1.0 and v1.63 using their default settings. It is possible to tease out what has changed between the two versions by changing the option settings for cdfbvn and the method of inverting the Hessian. Changing the defaults alters the first- and second-stage estimates outside the range of simulation variance. For example, using EI (v1.63) default settings, the range of β^b across the 25 replications is [0.9304–0.9314], whereas when using EI v1.63 set at the original (v1.0) method of inverting the Hessian, β^b is [0.8984–0.9158]. The range of the second-stage estimates does not overlap for either Dukakis or Bush splitters.

EI v1.63 estimates come closest to agreeing with EI v1.0 estimates when cdfbvn option is set to the v1.0 setting. The value of the log-likelihood is also different depending on whether cdfbvn option is set to v1.0 or v1.63, suggesting that the shape of the likelihood function is dependent on the cumulative bivariate normal distribution algorithm.

It is also observed that the cdfbvn option setting interacts strangely with the Hessian option setting. When both are set to either v1.0 or v1.63, the range of the simulation variance narrows to fluctuations at the third significant digit. When the options are mismatched across versions, fluctuations are greater and occur at the second significant digit.

7.4.3   Platform Dependence

Burden and Kimball performed their original research using EzI v1.21. Unfortunately, EzI v1.21 is no longer available, even from the author, so here EzI v2.3 is used in platform dependence analysis. EzI uses a Windows interface that cannot be automated, so the analysis performs only 5 EzI replications, compared to the 25 EI replications in Section 7.4.2.

When EzI is run on Windows, the results are within simulation variance of Burden's and Kimball's original results. However, when we run a purportedly identical analysis using a corresponding version of EI on Linux, using the same algorithmic options, the results are significantly different. These results are presented in the bottom four rows of Table 7.3. Again note option dependence in EzI, just as done with the analysis of EI in Section 7.4.2. Estimates for Bush splitters overlap only slightly using EzI v2.3 with the EI v1.0 and EI v1.63 settings. Only using the EI v1.0 setting is it possible to replicate the published Burden and Kimball results within simulation variance.

7.4.4 Discussion: Summarizing Uncertainty

Burden and Kimball take the EzI estimates from one run of the program and use these point estimates as an outcome variable in regression analysis. This procedure is known as EI-R, and Burden and Kimball's use is the subject of debate (Ferree 1999; Adolph and King 2003; Herron and Shotts 2003b) centering on the uncertainty of EI point estimates and the resulting inconsistency of using these estimates as a dependent variable in a subsequent regression. (See also Chapter 4 and for a discussion of this issue in the context of data perturbations.) Here, focus is on another source of error in the estimates, that of implementation dependence, and the consequence on Burden and Kimball's inference.

Table 7.4 replicates Table 6 of Burden and Kimball (1999). The uncertainty of simulation variance, along with the uncertainty of platform dependence, is seen by replicating two runs of EzI v2.3 on Windows and two runs of EI v1.63 on both HP-Unix and Linux operating systems. The table shows that even though the standard errors for the two EzI runs are in general agreement with the published results, the coefficients vary simply across the two Windows runs. This variation is due to the aforementioned simulation variance that occurs within both stages of the EI analysis performed by Burden and Kimball. On this issue alone, Burden and Kimball (and all researchers who use point estimates from one model as a dependent variable in another) underestimate the uncertainty of their coefficients (Adolph and King 2003; Herron and Shotts 2003b).

Table 7.4 Summarizing Uncertainty in Burden and Kimball

	Democratic Incumbent	Spending Ratio	Ballot Format	South
Windows				
Run 1	0.111 (0.015)	0.331 (0.020)	−0.027 (0.008)	0.056 (0.009)
Run 2	0.106 (0.015)	0.337 (0.121)	−0.031 (0.008)	0.062 (0.009)
HP-Unix				
Run 1	0.106 (0.014)	0.324 (0.020)	−0.025 (0.008)	0.056 (0.008)
Run 2	0.106 (0.014)	0.322 (0.020)	−0.024 (0.007)	0.053 (0.009)
Linux				
Run 1	0.089 (0.026)	0.374 (0.039)	−0.021 (0.013)	0.055 (0.015)
Run 2	0.089 (0.026)	0.377 (0.038)	−0.021 (0.012)	0.051 (0.015)
Published values	0.107 (0.015)	0.350 (0.021)	−0.032 (0.008)	0.065 (0.008)

When looking at variation across platforms, there is even greater variation in coefficients and standard errors. For example, the coefficient for *Democratic incumbent* ranges from 0.089 in the Linux runs, to 0.106 in the HP-Unix runs, to a maximum of 0.111 in the Windows runs. The estimates for standard errors also vary: from 0.026 in the Linux runs to 0.015 in the Windows runs. In practice, these standard errors are normally related to the degree of simulation variance. Here they are not, which suggests to us that these differences across statistical packages are due to something more than simulation variance. On this issue of implementation dependence, Burden and Kimball again underestimate the uncertainty of their EI-R regression estimates.

How might one summarize the uncertainty of the estimates across computing platforms? It is possible to approach the problem by concluding that one set of estimates are the true estimates and that the others are wrong, caused by numerical inaccuracies. The problem is that lacking the truth, it is not possible to assess which set of estimates are the "true" values, so this approach is unfruitful.

A second approach is that one could assume that the estimates derived from the different platforms come from "draws" from some underlying distribution of implementation dependence, perhaps a normal distribution, and then use this distribution to characterize the uncertainty of the estimates. This approach is more attractive, and if one knew the distribution, perhaps one could construct an adjustment to the coefficients and standard errors to account for the uncertainty. [See Adolph and King (2003) and Herron and Shotts (2003a) for a method in the context of EI-R.] Unfortunately, this distribution is unavailable as well, so this approach is also unfruitful.

The best that one can do is to provide ranges, as done here, to the coefficients and standard errors, and hold that any of these are equally valid measures. This approach is similar to one involving Leamer bounds (Leamer 1978, 1983), described in Chapter 4, and is equally valid when observing varied estimates across different statistical software packages. It is doubtful that most researchers will go the extra step of capturing platform and software implementation dependence in their analysis in the manner done here. Still, because such dependence was observed, EI analysis may be susceptible to additional uncertainty due to platform dependence, and more generally, complex models may be susceptible to similar implementation dependence. When observed, it is recommended that researchers include results from different implementations in their analyses and report ranges so that this uncertainty can be better measured.

7.5 CONCLUSIONS

The ecological inference problem stands out in social science methodological research because it has shown to be substantively important and there is still no definitive solution. Given the complexity and popularity to King's solution, detailed evaluation of its numerical properties is easily justified. Numerical sensitivity analysis gives us the opportunity to take specific case studies and test their sensitivity to numeric and non-numeric noise, and to implementation and algorithmic specific options.

The extensive analysis of the EI solutions in this chapter serves as an example of the battery of diagnostic tests that researchers may put their own programs through. We draw from our experience some practical recommendations. Using the recommended, most numerically accurate options may help reduce computational problems, but does not guarantee success. King's admonishments to users of his EI program should be well-heeded: Researchers should carefully scrutinize all available diagnostic tools available to them, and apply common sense. However, although researchers can avoid some of these problems by carefully scrutinizing standard statistical diagnostics, additional diagnostic sensitivity tests, and an understanding of how statistical computations work may still be needed to avoid problems in complex estimations.

The results suggest that both data and numerical issues play a role in enabling reliable estimates. Estimated results will be less stable when a problem is ill-conditioned, and where there are numerical inaccuracies in implementation. In these cases, researchers should pay special attention to numerical issues, as correct inference may be dependent on the choice of platform, implementation, and options used.

CHAPTER 8

Some Details of Nonlinear Estimation

B. D. McCullough

8.1 INTRODUCTION

The traditional view adopted by statistics and econometrics texts is that in order to solve a nonlinear least squares (NLS) or nonlinear maximum likelihood (NML) problem, a researcher need only use a computer program. This point of view maintains two implicit assumptions: (1) one program is as good as another, and (2) the output from a computer is always reliable. Both of these assumptions are false.

Recently, the National Institute of Standards and Technology (NIST) released the "Statistical Reference Datasets" (StRD), a collection of numerical accuracy benchmarks for statistical software. It has four suites of tests: univariate summary statistics, one-way analysis of variance, linear regression, and nonlinear least squares. Within each suite the problems are graded according to level of difficulty: low, average, and high. NIST provides certified answers to 15 digits for linear problems and 11 digits for nonlinear problems. Complete details can be found at <http://www.nist.gov/itl/div898/strd>. [For a detailed discussion of these tests, see Chapter 2.]

Nonlinear problems require starting values, and all the StRD nonlinear problems come with two sets of starting values. Start I is far from the solution and makes the problem difficult to solve. Start II is near the solution and makes the problem easy to solve. Since the purpose of benchmark testing is to say something useful about the underlying algorithm, start I is more important, since the solver is more likely to report false convergence from start I than from start II. To see this, consider two possible results: (A) that the solver can solve an easy problem correctly, and (B) that the solver can stop at a point that is not a solution and nonetheless declare that it has found a solution. Clearly, (B) constitutes more useful information than (A).

The StRD nonlinear suite has been applied to numerous statistical and econometric software packages, including SAS, SPSS, S-Plus, Gauss, TSP, LIMDEP, SHAZAM, EViews, and several others. Some packages perform very well on these tests; others exhibit a marked tendency to report false convergence

Numerical Issues in Statistical Computing for the Social Scientist, by Micah Altman, Jeff Gill, and Michael P. McDonald
ISBN 0-471-23633-0   Copyright © 2004 John Wiley & Sons, Inc.

(i.e., the nonlinear solver stops at a point that is not a minimum and nevertheless reports that it has found a minimum). Thus, the first point of the traditional view is shown to be false: some packages are very good at solving nonlinear problems, and other packages are very bad. Even the packages that perform well on the nonlinear suite of tests can return false convergence if not used carefully, so the second point of the traditional view is shown to be false. Users of nonlinear solvers need some way to protect themselves against false results. In this chapter we offer useful guidance on the matter.

In Section 8.2 we present the basic ideas behind a nonlinear solver. In Section 8.3 we consider some details on the mechanics of nonlinear solvers. In Section 8.4 we analyze a simple example where a nonlinear solver produces several incorrect answers to the same problem. In Section 8.5 we offer a list of ways that a user can guard against incorrect answers and Monte Carlo evidence on profile likelihood. Wald and likelihood inference are compared in Section 8.6. In Section 8.7 we offer conclusions, including what to look for in a nonlinear solver.

8.2 OVERVIEW OF ALGORITHMS

On the kth iteration, gradient-based methods for finding the set of coefficients $\hat{\beta}$ that minimize a nonlinear least squares function[1] take the form

$$\hat{\beta}^{k+1} = \hat{\beta}^k + \lambda_k d_k, \tag{8.1}$$

where $\beta = [\beta_1, \beta_2, \dots, \beta_m]'$ is the vector of parameters to be estimated. The objective function is the sum of squared residuals, denoted $S(\hat{\beta})$. From some selected vector of starting values, β^0, the iterative process proceeds by taking a step of length λ in some direction d. Different choices for λ and d give rise to different algorithms. The *gradient methods* are based on $d = Wg$, where g is the gradient of the objective function and W is some matrix. (See Chapter 4 for a discussion of other algorithms, including nongradient methods.)

One approach to choosing λ and d is to take a linear approximation to the objective function

$$F(\hat{\beta}_k + \lambda) \equiv F_k + g_k' \lambda. \tag{8.2}$$

This leads to the choice $W = I$, where I is the identity matrix and yields $d = -g_k$, thus producing the algorithm known as *steepest descent*. While it requires numerical evaluation of only the function and the gradient, it makes no use of the curvature of the function (i.e., it makes no use of the Hessian). Thus the steepest descent method has the disadvantage that it is very slow: The steepest descent method can require hundreds of iterations to do what other algorithms can do in just several iterations.

Another approach to choosing λ and d is to take a quadratic approximation to the objective function. Then a first-order Taylor expansion about the current

---

[1] In the remainder of this section we address nonlinear least squares directly, but much of the discussion applies, *mutatis mutandis*, to nonlinear maximum likelihood.

iterate yields

$$F(\beta_k + d_k) \approx F(\beta_k) + g'_k d_k + \tfrac{1}{2} d'_k H_k d_k, \tag{8.3}$$

where g_k and H_k are the gradient and Hessian, respectively, at the kth iteration. An associated quadratic function in d_k can be defined as

$$h(\lambda) = g'_k d_k + \tfrac{1}{2} d'_k H_k d_k. \tag{8.4}$$

A stationary point of the quadratic function, d_k, will satisfy

$$H_k d_k = -g_k. \tag{8.5}$$

When the direction d_k is a solution of the system of equations (8.5), d_k is the Newton direction. If, further, the step length is unity, the method is called *Newton's method*. If the step length is other than unity, the method is a modified Newton method. The Newton method is very powerful because it makes full use of the curvature information. However, it has three primary defects, two of which are remediable.

The first remediable problem is that the Newton step, $\lambda = 1$, is not always a good choice. One reason for choosing λ to be other than unity is because if $\lambda \equiv 1$, $\hat{\beta}^{k+1}$ is not guaranteed to be closer to a solution than $\hat{\beta}^k$. One way around this is to project a ray from $\hat{\beta}^k$ and then search along this ray for an acceptable value of λ; this is called a *line search*. In particular, solving

$$\min_{\lambda > 0} f(\hat{\beta}_k + \lambda d_k) \tag{8.6}$$

produces an *exact line search*. Because this can be computationally expensive, frequently an algorithm will find a value λ that roughly approximates the minimum; this is called an *inexact line search*. Proofs for some theorems on the convergences of various methods require that the line search be exact.

The second problem is that computing first derivatives for a method that only uses gradient information is much less onerous than also computing second derivatives for a method that explicitly uses Hessian information, too. Analytic derivatives are well known to be more reliable than their numerical counterparts. However, it is frequently the case that the user must calculate and then code these derivatives, a not insubstantial undertaking. When the user must calculate and code derivatives, the user often relies, instead, solely on numerical derivatives (Dennis 1984, p. 1766). Some packages offer automatic differentiation, in which a specialized subroutine calculates analytic derivatives, automatically, thus easing appreciably the burden on the user. See Nocedal and Wright (1999, Chap. 7) for a discussion of automatic derivatives. Automatic differentiation is not perfect, and on rare occasions the automatic derivatives are not numerically efficient. In such a case, it may be necessary to rewrite or otherwise simplify the expressions for the derivatives. Of course, it is also true that user-supplied analytic derivatives may need rewriting or simplification.

The third and irremediable problem with the Newton method is that for points far from the solution, the matrix H_k in (8.5) may not be positive definite. In such a case the direction does not lead toward the minimum. Therefore, other methods have been developed so that the direction matrix is always positive definite. One such class of methods is the class of quasi-Newton methods.

The quasi-Newton methods have a direction that is the solution of the following system of equations:

$$B_k d_k = -g_k, \tag{8.7}$$

where $B_{k+1} = B_k + U_k$, where U_k is an updating matrix. B_0 often is taken to be the identity matrix, in which case the first step of the quasi-Newton method is a steepest descent step. Different methods of computing the update matrix lead to different algorithms [e.g., Davidson–Fletcher–Powell (DFP) or Broyden–Fletcher–Goldfarb–Shannon (BFGS)]. Various modifications can be made to ensure that B_k is always positive definite. On each successive iteration, B_k acquires more information about the curvature of the function; thus, an approximate Hessian is computed, and users are spared the burden of programming an analytic Hessian. Both practical and theoretical considerations show that this approximate Hessian is generally quite good for the purpose of obtaining point estimates (Kelley 1999, Sec. 4); whether this approximate Hessian can be used for computing standard errors of the point estimates is another matter entirely.

Let β^* represent the vector that minimizes the sum-of-squared residuals, and let H^* be the Hessian at that point. There do exist theorems which prove that $B_k \to H^*$ when the number of iterations is greater than the number of parameters (see, e.g., Bazaara et al. 1993, p. 322, Th. 8.8.6). Perhaps on the basis of such theorems it is sometimes suggested that the approximate Hessian provides reliable standard errors (e.g., Bunday and Kiri 1987; Press et al. 2002, p. 398). However, the assumptions of such theorems are restrictive and quite difficult to verify in practice. Therefore, in practical situations it is not necessarily true that B_k resembles H_k (Gill et al. 1981, p. 120). In fact, the approximate Hessian should *not* be used as the basis for computing standard errors. To demonstrate this important point, we consider a pair of examples from Wooldridge (2000). For both examples the package used is TSP v4.5, which employs automatic differentiation and offers both the quasi-Newton method BFGS and a modified Newton–Raphson method.

The first example (Wooldridge 2000, p. 538, Table 17.1) estimates a probit model with "inlf" as the outcome variable [both examples use the dataset from Mroz (1987)]. The PROBIT command with options HITER = F and HCOV = FNB uses the BFGS method to compute point estimates and prints out standard errors using both the approximation to the Hessian and the Hessian itself, as well as the OPG (outer product of the gradient) estimator from the BHHH (Berndt et al. 1974) method for purposes of comparison. [Exactly the same point estimates are obtained when Newton's method is used.] The algorithm converged in 13 iterations. The results are presented in Table 8.1.

Table 8.1   Probit Results for Mroz Data (Outcome Variable "inlf")

Variable	Coef.	Approx. Hessian		Hessian		OPG	
		\multicolumn{6}{c}{Standard Errors Based on[a]:}					
C	0.270	0.511	(0.53)	0.509	(0.53)	0.513	(0.53)
NWIFEINC	−0.012	0.005	(−2.51)	0.005	(−2.48)	0.004	(−2.71)
EDUC	0.131	0.025	(5.20)	0.025	(5.18)	0.025	(5.26)
EXPER	0.123	0.019	(6.61)	0.019	(6.59)	0.019	(6.60)
EXPERSQ	−0.002	0.001	(−3.15)	0.001	(−3.15)	0.001	(−3.13)
AGE	−0.053	0.008	(−6.26)	0.008	(−6.24)	0.009	(−6.12)
KIDSLT6	−0.868	0.118	(−7.35)	0.119	(−7.33)	0.121	(−7.15)
KIDSGE6	0.036	0.045	(0.81)	0.043	(0.83)	0.042	(0.86)

[a] t-Statistics in parentheses.

Table 8.2   Tobit Results for Mroz Data (Dependent Variable "hours")

Variable	Coef.	Approx. Hessian		Hessian		OPG	
		\multicolumn{6}{c}{Standard Errors Based on[a]:}					
C	965.3	0.415	(2327.1)	446.4	(2.16)	449.3	(2.14)
NWIFEINC	−8.814	0.004	(−2101.2)	4.459	(−1.98)	4.416	(−1.99)
EDUC	80.65	0.020	(4073.7)	21.58	(3.74)	21.68	(3.72)
EXPER	131.6	0.016	(8261.5)	17.28	(7.61)	16.28	(8.10)
EXPERSQ	−1.864	0.001	(−3790.1)	0.538	(−3.47)	0.506	(−3.68)
AGE	−54.41	0.007	(−7978.7)	7.420	(−7.33)	7.810	(−6.97)
KIDSLT6	−894.0	0.105	(−8501.6)	111.9	(−7.99)	112.3	(−7.96)
KIDSGE6	−16.22	0.036	(−454.7)	38.64	(−0.42)	38.74	(−0.42)

[a] t-Statistics in parentheses.

As can be seen by examining the t-statistics, all three methods of computing the standard error return similar results. The same cannot be said for the second example, in which Tobit estimation is effected for the same explanatory variables and "hours" is the outcome variable (Wooldridge 2000 p. 544, Table 17.2). This time, the BFGS method converges in 15 iterations. Results are presented in Table 8.2.

Observe that the approximate Hessian standard errors are in substantial disagreement with the standard errors produced by the Hessian, the latter being about 1000 times larger than the former. In contrast to the approximate Hessian standard errors, the OPG standard errors, are in substantial agreement with the Hessian standard errors. It can be deduced that for the probit problem, the likelihood surface is very well behaved, as all three methods of computing the standard error are in substantial agreement. By contrast, the likelihood surface for the tobit problem is not so well behaved.

Thus far, all the algorithms considered have been for unconstrained optimization. These algorithms can be applied to nonlinear least squares, but it is often better to apply specialized nonlinear least squares algorithms. The reason for this is that in the case of nonlinear least squares, the gradient and the Hessian have specific forms that can be used to create more effective algorithms. In particular, for NLS the gradient and Hessian are given by

$$g(\hat{\beta}) = J(\hat{\beta})' f(\hat{\beta}) \tag{8.8}$$

$$H(\hat{\beta}) = J(\hat{\beta})' J(\hat{\beta}) + Q(\hat{\beta}), \quad Q(\hat{\beta}) = \sum_{i=1}^{n} f_i(\hat{\beta}) G_i(\hat{\beta}), \tag{8.9}$$

where $J(\hat{\beta})$ is the Jacobian matrix, $f(\hat{\beta})$ is the vector of residuals, and G_i is the ith contribution-to-the-Hessian matrix.

These specialized methods are based on the assumption that $Q(\hat{\beta})$ can be neglected (i.e., that the problem is a small-residual problem). By *small-residual problem* is meant that $\|f(\beta^*)\|$ is smaller than the largest eigenvalue of $J'(\beta^*)J(\beta^*)$. If, instead, they are of the same size, there is no advantage to using a specialized method.

One specialized method is the Gauss–Newton method, which uses $J'J$ as an approximation to the Hessian; this is based on the assumption that $Q(\hat{\beta})$ is negligible. In combination with a line search, it is called *damped Gauss–Newton*. For small residual problems, this method can produce very rapid convergence; in the most favorable case, it can exhibit quadratic convergence, even though it uses only first derivatives. Its nonlinear maximum likelihood analog is called the BHHH method. Another specialized method is the Levenberg–Marquardt method, which is an example of a *trust region algorithm*. The algorithms discussed so far compute a direction and then choose a step length. A trust region method first computes a step length and then determines the direction of the step. The nonlinear maximum likelihood analogue is the quadratic hill-climbing method of Goldfeld et al. (1966).

Finally, we note that mixed methods can be very effective. For example, use a quasi-Newton method, with its wider radius of convergence, until the iterate is within the domain of attraction for the Newton method, and then switch to the Newton method.

8.3   SOME NUMERICAL DETAILS

It is important to distinguish between an algorithm and its implementation. The former is a theoretical approach to a problem and leaves many practical details unanswered. The latter is how the approach is applied practically. Two different implementations of the same algorithm can produce markedly different results. For example, a damped quasi-Newton method only dictates that a line search be used; it does not specify how the line search is to be conducted. The Newton–Raphson method only dictates that second derivatives are to be used; it does not specify how the derivatives are to be calculated: by forward differences, central differences, or analytically.

Consider the forward difference approximation to the derivative of a univariate function: $f'(x) = (f(x + h) - f(x))/h + R(h)$, where $R(h)$ is the remainder. This contrasts sharply with the central difference approximation: $f'(x) = (f(x + h) - f(x - h))/2h + R(h)$. For the function $f(x) = x^3$, it is easy to show that the remainder for the forward difference is $3x^2 + 3xh + h^2$, whereas for the central difference it is only $3x^2 + h^2$. Generally, the method of central differences produces smaller remainders, and thus more accurate derivatives, than the method of forward differences. Of course, with analytic derivatives the remainder is zero.

Another crucial point that can lead to different results is that the iterative process defined by (8.1) needs some basis for deciding when to stop the iterations. Some common termination criteria are:

- $|SSR^{k+1} - SSR^k| < \varepsilon_s$; when the successive change in the sum of squared residuals is less than some small value.
- $\max_i[|\hat{\beta}_i^{k+1} - \hat{\beta}_i^k|] < \varepsilon_p$; when the largest successive change in some coefficient is less than some small value.
- $||g(\hat{\beta}^k)|| < \varepsilon_g$; when the magnitude of the gradient is less than some small value.
- $g'H^{-1}g < \varepsilon$; this criterion involves both the gradient and the Hessian.

The first three criteria are scale dependent; that is, they depend on the units of measurement. There do exist many other stopping rules, some of which are scale-independent versions of the first three criteria above. The important point is that termination criteria must not be confused with convergence criteria.

Convergence criteria are used to decide whether a candidate point is a minimum. For example, at a minimum, the gradient must be zero and the Hessian must be positive definite. A problem with many solvers is that they conflate convergence criteria and stopping rules (i.e., they treat stopping rules as if they were convergence criteria). It is obvious, however, that while the stopping rules listed above are necessary conditions for a minimum, none or even all of them together constitutes a sufficient condition. Consider a minimum that occurs in a flat region of the parameter space: Successive changes in the sum-of-squared residuals will be small at points that are far from the minimum. Similarly, parameters may not be changing much in such a region. In a flat region of the parameter space, the gradient may be very close to zero, but given the inherent inaccuracy of finite-precision computation, there may be no practical difference between a gradient that is "close to zero" and one that is numerically equivalent to zero.

Finally, the user should be aware that different algorithms can differ markedly in the speed with which they approach a solution, especially in the final iterations. Algorithms such as the (modified) Newton–Raphson that make full use of curvature information, converge very quickly. In the final iterations they exhibit *quadratic convergence*. At the other extreme, algorithms such as steepest descent exhibit *linear convergence*. Between these two lie the quasi-Newton methods, which exhibit *superlinear convergence*. To make these concepts concrete, define

Table 8.3   Comparison of Convergence Rates

Steepest Descent	BFGS	Newton
1.827e-04	1.70e-03	3.48e-02
1.826e-04	1.17e-03	1.44e-02
1.824e-04	1.34e-04	1.82e-04
1.823e-04	1.01e-06	1.17e-08

$h^k = \beta^k - \beta^*$, where β^* is a local minimum. The following sequences can be constructed:

- *Linear:* $||h^{k+1}||/||h^k|| \leq c$ $h^{k+1} = O(||h^k||)$
- *Superlinear:* $||h^{k+1}||/||h^k|| \to 0$ $h^{k+1} = o(||h^k||)$
- *Quadratic:* $||h^{k+1}||/||h^k||^2 \leq c$ $h^{k+1} = O(||h^k||^2)$

Nocedal and Wright (1999, p. 199) give an example for the final few iterations of a steepest descent, BFGS, and Newton algorithm, all applied to the same function. Their results are presented in Table 8.3. Observe that the Newton method exhibits quadratic convergence with the final few steps: e-02, e-04, and e-08. Conversely, steepest descent is obviously converging linearly, whereas the quasi-Newton method BFGS falls somewhere in between. These rates of convergence apply not only to the parameters but also to the value of the objective function (i.e., the sum-of-squared residuals or the log likelihood). In the latter case, simply define $h^k = \log L^k - \log L^*$, where $\log L^*$ is the value at the maximum.

Because quadratic convergence in the final iterations is commonly found in solutions obtained by the Newton method, if a user encounters only superlinear convergence in the final iterations of a Newton method, the user should be especially cautious about accepting the solution. Similarly, if a quasi-Newton method exhibits only linear convergence, the user should be skeptical of the solution.

8.4   WHAT CAN GO WRONG?

A good example is the Misra1a problem from the nonlinear suite of the StRD when it is given to the Microsoft Excel Solver. Not only is it a lower-difficulty problem it is a two-parameter problem which lends itself to graphical exposition. The equation is

$$y = \beta_1(1 - \exp(-\beta_2 x)) + \epsilon \qquad (8.10)$$

with the 14 observations given in Table 8.4.

The Excel Solver is used to minimize the sum-of-squared residuals, with Start I starting values of 500 for β_1 and 0.0001 for β_2. The Excel Solver

Table 8.4   Data for Misra1a Problem

Obs.	y	x	Obs.	y	x
1	10.070	77.60	8	44.820	378.40
2	14.730	114.90	9	50.760	434.80
3	17.940	141.10	10	55.050	477.30
4	23.930	190.80	11	61.010	536.80
5	29.610	239.90	12	66.400	593.10
6	35.180	289.00	13	75.470	689.10
7	40.020	332.80	14	81.780	760.00

offers various options. The default method of derivative calculation is forward differences, with an option for central differences. The default algorithm is an unspecified "Newton" method, with an option for an unspecified "Conjugate" method. On Excel97 the default convergence tolerance ("ct") is 0.001, although whether this refers to successive changes in the sum-of-squared residuals, coefficients, or some other criterion is unspecified. There is also an option for "automatic scaling," which presumably refers to recentering and rescaling the variables—this can sometimes have a meliorative effect on the ability of an algorithm to find a solution (Nocedal and Wright 1999, pp. 27, 94). Using Excel97, five different sets of options were invoked:

- **Options A:** ct = 0.001 (Excel97 default)
- **Options B:** ct = 0.001 and automatic scaling
- **Options C:** ct = 0.0001 and central derivatives
- **Options D:** ct = 0.0001 and central derivatives and automatic scaling
- **Options E:** ct = 0.00001 and central derivatives and automatic scaling

For each set of options, the Excel Solver reported that it had found a solution. (Excel97, Excel2000, and ExcelXP all produced the same answers.) These five solutions are given in Table 8.5, together with the certified values from NIST. The correct digits are underlined. For example, solutions A and B have no correct digits, while solutions C and D each have the first significant digit correct.[2] Solution E has four correct digits for each of the coefficients and five for the SSR. Additionally, for each of the six points the gradient of the objective function at that point is presented in brackets. These gradient values were produced in three independent ways: via the nonlinear least squares command in TSP, taking care to note that TSP scales its gradient by SSR/$(n-2)$; by implementing equation (8) in package R, which can produce $J(\hat{\beta})$ and $f(\hat{\beta})$; and programming from first principles using Mathematica. All three methods agreed to several digits. A contour plot showing the five solutions as well as the starting values (labeled "S") is given in Figure 8.1.

---

[2]The *significant digits* are those digits excluding leading zeros.

Table 8.5 "Solutions" for the Misra1a Problem Found by `Excel` Solver[a]

	$\hat{\beta}_1$	$\hat{\beta}_2$	SSR
NIST	238.94212918	0.00055015643181	0.12455138894
	[−1.5E-9]	[−0.00057]	
Solution A	454.12442033	0.00026757574438	16.725122137
(8 iterations)	[0.23002]	[420457.7]	
Solution B	552.84275702	0.00021685528323	23.150576131
(8 iterations)	[0.16068]	[454962.9]	
Solution C	244.64697774	0.00053527479056	0.16814681493
(31 iterations)	[−0.00744]	[−0.00011]	
Solution D	241.96737442	0.00054171455690	0.15384239922
(33 iterations)	[0.09291]	[37289.6]	
Solution E	238.93915212	0.00055016470282	0.12455140816
(37 iterations)	[−5.8E-5]	[−23.205]	

[a]Accurate digits underlined, component of gradient in brackets.

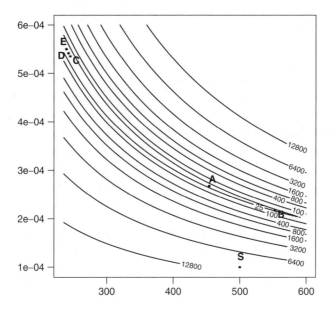

Fig. 8.1 SSR contours for Misra1A.

It is also useful to consider the trace of each "solution" as well as the trace of an accurate solution produced by the package S-Plus v6.2, which took 12 iterations to produce a solution with 9, 9, and 10 digits of accuracy for β_1, β_2, and SSR, respectively. Of import is the fact that the S-Plus solver employs a Gauss–Newton algorithm, and this is a small-residual problem. The final five sum-of-squared residuals, as well as the difference of each from its final value, are given in Table 8.6.

Table 8.6 Convergence of Solutions

A	B	C	D	E	S-Plus
SSR					
18.649	29.133	0.17698	0.8774	0.1538	5770.4200
18.640	28.966	0.17698	0.5327	0.1367	1242.3170
18.589	27.670	0.16900	0.3972	0.1257	1.1378420
18.310	27.026	0.16900	0.2171	0.1246	0.1245559
16.725	23.151	0.16815	0.1538	0.1245	0.1245514
Differences					
1.924	5.982	0.00883	0.7236	0.0293	5770.2954
1.915	5.815	0.00883	0.3789	0.0122	1242.1924
1.864	4.519	0.00085	0.2434	0.0012	1.0132906
1.585	3.875	0.00085	0.0633	0.0001	0.0000045
0.000	0.000	0.00000	0.0000	0.0000	0.0000000

Several interesting observations can be made about the Excel Solver solutions presented in Table 8.5. First, only solution E might be considered a correct solution, and even the second component of its gradient is far too large.[3] Solutions A and B have no accurate digits. Observe that the gradient of solution C appears to be zero, but examining the sum-of-squared residuals shows that the gradient obviously is "not zero enough" (i.e., 0.1681 is not nearly small enough). The gradient at solution E is not nearly zero, but it clearly has a smaller sum-of-squared residuals than that of solution C, so despite its larger gradient, may be preferred to solution C. This demonstrates the folly of merely examining gradients (and Hessians); examination of the trace can also be crucial.

The Excel Solver employs an unspecified Newton algorithm, with an unknown convergence rate. Rather than assume that this method is quadratically convergent, let us assume that it is superlinearly convergent. Examining Table 8.6, all the Excel solutions exhibit linear convergence, even solution E, for which $h_{k+1} \approx 0.1h_k$. In particular, examining the trace of solution C shows that the Solver is searching in a very flat region that can be characterized by at least two plateaus. Even though each component of the gradient appears to be zero, the trace does not exhibit the necessary convergence, so we do not believe point C to be a solution.

Figure 8.2a does not show sufficient detail, and some readers may think, especially given the gradient information provided in Table 8.6, that point C is a local minimum. It is not, as shown clearly in Figure 8.2b.

We have just analyzed five different solutions from one package. It is also possible to obtain five different solutions from five different packages, something Stokes (2003) accomplished when trying to replicate published probit results. It turned out that for the particular data used by Stokes, the maximum likelihood estimator did not exist. This did not stop several packages from reporting that

[3]The determination of "too large" is made with benefit of 20–20 hindsight (see Section 5.1).

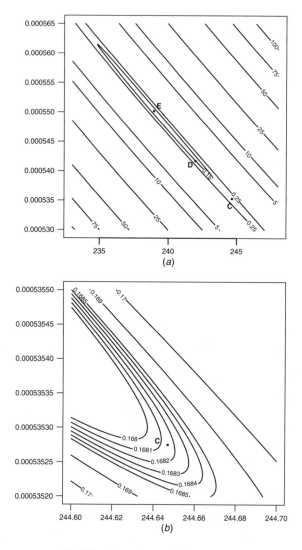

Fig. 8.2 Comparison of SSR Contours.

their algorithms had converged to a solution—a solution that did not exist! His paper is instructive reading.

8.5 FOUR STEPS

Stokes was not misled by his first solver because he did not accept its output uncritically. In fact, critically examining its output is what led him to use a second package. When this produced a different answer, he knew that something was definitely wrong. This was confirmed by the different answers from his third, fourth, and fifth packages. With his first package, Stokes varied the algorithm, the convergence tolerance, starting values, and so on. This is something that every user should do with every nonlinear estimation. Suppose that a user has done this and identified a possible solution. How might he verify the solution? McCullough and Vinod (2004) recommend four steps:

1. Examine the gradient. Is $||g|| \approx 0||$?
2. Inspect the sequence of function values. Does it exhibit the expected rate of convergence?
3. Analyze the Hessian. Is it positive semidefinite? Is it ill conditioned?
4. Profile the objective function. Is the function approximately quadratic?

Gill et al. (1981, p. 313) note that if the first three conditions hold, very probably a solution has been found, regardless of whether the program has declared convergence. The fourth step justifies the use of the usual t-statistics for coefficients reported by most packages. These t-statistics are Wald statistics and, as such, are predicated on the assumption that the objective function, in this case the sum-of-squares function, is approximately quadratic in the vicinity of the minimum. If the function is not approximately quadratic, the Wald statistic is invalid and other methods are more appropriate (e.g., likelihood ratio intervals); this topic is addressed in detail in Section 8.6. The easy way to determine whether the objective function is approximately quadratic is to profile the objective function. Each of these four steps is discussed in turn.

8.5.1 Step 1: Examine the Gradient

At the solution, each component of the gradient should be zero. This is usually measured by the magnitude of the gradient, squaring each component, summing them, and taking the square root of the sum. The gradient will be zero only to the order of the covariate(s). To see this, for the Misra1a problem, multiply the y and x vectors by 100 and 10, respectively. The correct solution changes from $\beta_1^* = 238.94212918$, $\beta_2^* = 0.00055015643181$, SSR $= 0.12455138894$ to $\beta_1^* = 23894.212918$, $\beta_2^* = 0.0000055015643181$, SSR $= 1245.5138894$, and the gradient changes from $[-1.5E-9, -0.000574]$ to $[1.5E-7, -57.4]$. The moral is that what constitutes a zero gradient depends on the scaling of the problem. See Gill et al. (1981, Secs. 7.5 and 8.7) for discussions of scaling. Of course, a package that does not permit the user to access the gradient is of little use here.

8.5.2 Step 2: Inspect the Trace

Various algorithms have different rates of convergence. By *rate of convergence* we mean the rapidity with which the function value approaches the extremum as the parameter estimates get close to the extremal estimates. As an example, let β^* be the vector that minimizes the sum of squares for a particular problem. If the algorithm in question is Newton–Raphson, which has quadratic convergence, then as $\hat{\beta}^k \to \beta^*$ the sum-of-squared residuals will exhibit quadratic convergence, as shown in Table 8.3.

Suppose, then, that Newton–Raphson is used and the program declares convergence. However, the trace exhibits only linear convergence in its last few iterations. Then it is doubtful that a true minimum has been found. This type of behavior can occur when, for example, the solver employs parameter convergence as a termination criterion and the current parameter estimate is in a very flat region of the parameter space. Then it makes sense that the estimated parameters are changing very little, and neither is the function value when the solver ceases iterating. Of course, a solver that does not permit the user to access the function value is of little use here.

8.5.3   Step 3: Analyze the Hessian

As in the case of minimizing a sum-of-squares function, the requirement for a multivariate minimum, is that the gradient is zero and the Hessian is positive definite (see Chapter 6 for additional details). The easiest way to check the Hessian is to do an eigensystem analysis and make sure all that the eigenvalues are positive. The user should be alert to the possibility that his package does not have accurate eigenroutines. If the developer of the package does not offer some positive demonstration that the package's matrix routines are accurate, the user should request proof.

In the case of a symmetric definite matrix (e.g., the covariance matrix), the ratio of the largest eigenvalue to the smallest eigenvalue is the *condition number*. If this number is high, the matrix is said to be *ill conditioned*. The consequences of this ill conditioning are threefold. First, it indicates that the putative solution is in a "flat" region of the parameter space, so that some parameter values can change by large amounts while the objective function changes hardly at all. This situation can make it difficult to locate and to verify the minimum of a NLS problem (or the maximum of a NML problem). Second, this ill conditioning leads to serious cumulated rounding error and a loss of accuracy in computed numbers. As a general rule, when solving a linear system, one digit of accuracy is lost for every power of 10 in the condition number (Judd 1998, p. 68). A PC has 16 digits. If the condition number of the Hessian is on the order of 10^9, the coefficients will be accurate to no more than seven digits. Third, McCullough and Vinod (2003) show that if the Hessian is ill conditioned, the quadratic approximation fails to hold in at least one direction. Thus, a finding in step 3 that the Hessian is ill conditioned implies automatically that Wald inference will be unreliable for at least one of the coefficients.

Of course, a package that does not permit the user to access the Hessian (perhaps because it cannot even compute the Hessian) is of little use here.

8.5.4   Step 4: Profile the Objective Function

The first three steps were concerned with obtaining reliable point estimates. Point estimates without some measure of variability are meaningless, so reliable standard errors are also of interest. The usual standard errors produced by non-linear routines, the t-statistics, are more formally known as *Wald statistics*. For their validity they depend on the objective function being quadratic in the vicinity of the solution (Box and Jenkins 1976). Therefore, it is also of interest to determine whether, in fact, the objective function actually is locally quadratic at the solution. To do this, profile methods are very useful.

The essence of profiling is simplicity itself. Consider a nonlinear problem with three parameters, α, β, and θ, and let $\hat{\alpha}^*$, $\hat{\beta}^*$, and $\hat{\theta}^*$ be the computed solution. The objective function ($\log L$ for MLE, SSR for NLS) has value f^* at the solution. For the sake of exposition, suppose that $\hat{\alpha}^* = 3$ with a standard error of $se(\alpha) = 0.5$ and that a profile of α is desired. For plus or minus some number of standard deviations, say four, choose several values of α (e.g., $\alpha_1 = 1.0$, $\alpha_2 = 1.5$, $\alpha_3 = 2.0, \dots, \alpha_7 = 4.0$, $\alpha_8 = 4.5$, $\alpha_9 = 5.0$). Fix $\alpha = \alpha_1$, reestimate the model allowing β and θ to vary, and obtain the value of the objective function at this new (constrained) extremum, f_1. Now fix $\alpha = \alpha_2$ and obtain f_2. The sequence of pairs $\{\alpha_i, f_i\}$ traces out the profile of $\hat{\alpha}$. If the profile is quadratic, Wald inference for that parameter is justified.

Visually, it is easier to discern deviations from linear shape than deviations from quadratic shape. The following transformation makes it easier to assess the validity of the quadratic approximation:

$$\tau(\alpha) = \text{sign}(\alpha - \hat{\alpha}^*)\sqrt{S(\alpha) - S(\hat{\alpha}^*)}/s, \tag{8.11}$$

where s is the standard error of the estimate.[4]

A plot of $\tau(\alpha_i)$ versus α_i will be a straight line if the quadratic approximation is valid. Now, let the studentized parameter be

$$\delta(\alpha_i) = \frac{\alpha_i - \hat{\alpha}^*}{se(\alpha_i)}. \tag{8.12}$$

A plot of $\tau(\alpha_i)$ versus $\delta(\alpha_i)$ will be a straight line with unit slope through the origin.

As a concrete example, consider profiling the parameter β_1 from the StRD Misra1a problem, the results of which are given in Table 8.7. Figure 8.3 shows the plot of SSR versus β_1 to be approximately quadratic and the plot of $\tau(\beta_1)$ versus β_1 to be a straight line. Many packages plot τ versus the parameter instead of τ versus δ. Usually, it is not worth the trouble to convert the former to the latter, although, on occasion, it may be necessary to achieve insight into the problem.

Profile methods are discussed at length in Bates and Watts (1988, Sec. 6) and in Venables and Ripley (1999, Sec. 8.5). Many statistical packages offer them (e.g., SAS and S-Plus). Many econometrics packages do not, although Gauss is an exception. What a user should do if he or she finds that Wald inference is unreliable is the subject of the next section.

---

[4]For NML, replace the right-hand side of (8.11) with $\text{sign}(\alpha - \hat{\alpha}^*)\sqrt{2(\log L^* - \log L)}$.

Table 8.7 Results of Profiling β_1

i	β_1	$\delta(\beta_1)$	SSR	$\tau(\beta_1)$
9	244.3561	2.0	0.1639193	1.947545
8	243.0026	1.5	0.1469733	1.469780
7	241.6491	1.0	0.1346423	0.986010
6	240.2956	0.5	0.1271061	0.496118
5	238.9421	0.0	0.1245514	0.000000
4	237.5886	−0.5	0.1271722	-0.502497
3	236.2351	−1.0	0.1351700	-1.011463
2	234.8816	−1.5	0.1487542	-1.527035
1	233.5280	−2.0	0.1681440	-2.049382

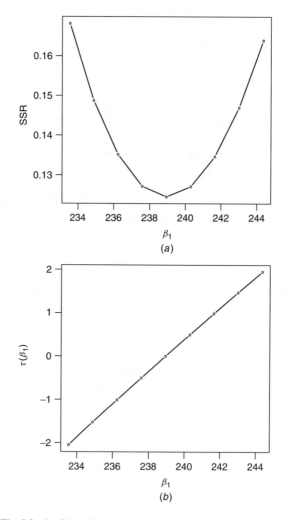

Fig. 8.3 Profiles of (a) β_1 versus SSR and (b) β_1 versus $\tau(\beta_1)$.

8.6 WALD VERSUS LIKELIHOOD INFERENCE

It is commonly thought that Wald inference and likelihood ratio (LR) inference
are equivalent (see, e.g., Rao 1973, p. 418). As noted earlier, the Wald interval
is valid only if the objective function is quadratic. However, the LR interval is
much more generally valid. In the scalar case, the 95% Wald interval is given by

$$\hat{\theta} \pm 1.96 \text{se}(\hat{\theta}), \tag{8.13}$$

and the LR interval is given by

$$\left\{ \theta, 2 \ln \frac{L(\hat{\theta})}{L(\theta)} \le 3.84 \right\}. \tag{8.14}$$

The Wald interval is exact if

$$\frac{\hat{\theta} - \theta}{\text{se}(\hat{\theta})} \sim N(0, 1), \tag{8.15}$$

whereas the LR interval is exact as long as there exists some transformation $g(\cdot)$
such that

$$\frac{g(\hat{\theta}) - g(\theta)}{\text{se}(g(\hat{\theta}))} \sim N(0, 1) \tag{8.16}$$

and it is not necessary that the specific function $g(\cdot)$ be known, just that it exists.
Consequently, when the Wald interval is valid, so is the LR interval, but not
conversely. Thus, Gallant's (1987, p. 147) advice is to "avoid the whole issue
as regards inference and simply use the likelihood ratio statistic in preference to
the Wald statistic."[5]

The assertion that the LR intervals are preferable to the Wald intervals merits
justification. First, a problem for which the profiles are nonlinear is needed. Such
a problem is the six-parameter Lanczos3 problem from the StRD, for which the
profiles were produced using the package R (Ihaka and Gentleman, 1996), with
the "profile" command from the MASS library of Venables and Ripley (1999).
These profiles are presented in Figure 8.4. None of the profiles is remotely linear,
so it is reasonable to expect that LR intervals will provide better coverage than
Wald intervals. To assess this claim, a Monte Carlo study is in order.

Using the NIST solution as true values and a random generator with mean
zero and standard deviation equal to the standard error of the estimate of the
NIST solution, 3999 experiments are run. For each run, 95% intervals of both
Wald and LR type are constructed. The LR intervals are computed using the
"confint.nls" command, which actually only approximates the likelihood ratio

---

[5]Also see Chapter 4 for alternatives to LR intervals that can be used to interpret results when the
likelihood function has multiple modes.

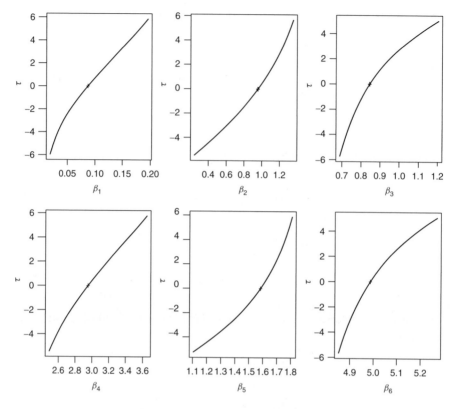

Fig. 8.4 Profiles for Lanczos3 parameters.

Table 8.8 Monte Carlo Results[a]

Errors	Test	b1	b2	b3	b4	b5	b6
Normal	Wald	91.725	92.125	92.650	92.300	92.550	92.875
	LR	94.200	94.125	94.275	94.175	94.200	94.275
$t(10)$	Wald	92.550	93.200	93.250	92.875	93.200	93.275
	LR	95.025	94.975	94.800	94.975	94.825	94.825
$\chi^2(3)$	Wald	92.700	93.300	93.275	93.500	93.075	93.450
	LR	95.275	95.250	95.550	95.425	95.625	95.475

[a]Coverage for 95% confidence intervals.

interval. The proportion of times that an interval contains the true parameter is its *coverage*. This setup is repeated for each of three types of error: normal, Student's t with 10 degrees of freedom (rescaled), and chi-square with 3 degrees of freedom (recentered and rescaled). The results are presented in Table 8.8. As can be seen, for each type of error and for each parameter, the LR interval provides substantially better coverage than the Wald interval.

8.7 CONCLUSIONS

Software packages frequently have *default options* for their nonlinear solvers. For example, the package may offer several algorithms for nonlinear least squares, but unless the user directs otherwise, the package will use Gauss–Newton. As another example, the convergence tolerance might be 1E-3; perhaps switching the tolerance to 1E-6 would improve the solution. It should be obvious that "solutions" produced by use of default options should not be accepted by the user until the solution has been verified by the user (see McCullough 1999b).

Although many would pretend otherwise, nonlinear estimation is far from automated, even with today's sophisticated software. There is more to obtaining trustworthy estimates than simply tricking a software package into declaring convergence. In fact, when the software package declares convergence, the researcher's job is just beginning—he has to verify the solution offered by the software. Software packages differ markedly not only in their accuracy but also in their ability to verify potential solutions. A desirable software package is one that makes it easy to guard against false convergence. Some relevant features are as follows:

- The user should be able to specify starting values.
- For nonlinear least squares, at least two algorithms should be offered: a modified Newton and a Gauss–Newton; a Levenberg–Marquardt makes a good third. The NL2SOL algorithm (Dennis et al. 1981) is highly regarded. For unconstrained optimization (i.e., for nonlinear maximum likelihood), at least two algorithms should be offered: a modified Newton and the BFGS. Again, the Bunch et al. (1993) algorithm is highly regarded.
- For nonlinear routines, the user should be able to fix one parameter and optimize over the rest of the parameters, in order to calculate a profile (all the better if the program has a "profile" command).
- The package should either offer LR statistics or enable the user to write such a routine.
- For routines that use numerical derivatives, the user should be able to supply analytic derivatives. Automatic differentiation is very nice to have when dealing with complicated functions.
- The user should be able to print out the gradient, the Hessian, and the function value at every iteration.

Casually perusing scholarly journals, and briefly scanning those articles that conduct nonlinear estimation, will convince the reader of two things. First, many researchers run their solvers with the default settings. This, of course, is a recipe for disaster, as was discovered by a team of statisticians working on a large pollution study (Revkin 2002). They simply accepted the solution from their solver, making no attempt whatsoever to verify it, and wound up publishing an incorrect solution. Second, even researchers who do not rely on default options practically never attempt to verify the solution. One can only wonder how many incorrect nonlinear results have been published.

Acknowledgments

Thanks to P. Spelluci and J. Nocedal for useful discussions, and to T. Harrison for comments.

CHAPTER 9

Spatial Regression Models

James P. LeSage

9.1 INTRODUCTION

With the progress of geographical information system (GIS) technology, large samples of socioeconomic demographic information based on spatial locations such as census tracts and zip-code areas have become available. For example, U.S. Bureau of the Census information at the tract level is now available covering a period of 40 years, adjusted to reflect the census 2000 tract boundaries. GIS software also allows survey information and other spatial observations that contain addresses or extended zip codes to be translated to map coordinates through geocoding. In this chapter we discuss computational issues associated with regression modeling of this type of information.

9.2 SAMPLE DATA ASSOCIATED WITH MAP LOCATIONS

9.2.1 Spatial Dependence

Spatial dependence in a collection of sample data implies that cross-sectional observations at location i depend on other observations at locations $j \neq i$. Formally, we might state that

$$y_i = f(y_j), \qquad i = 1, \ldots, n, \quad j \neq i. \tag{9.1}$$

The dependence typically involves several cross-sectional observations j that are located near i on a map. Spatial dependence can arise from theoretical as well as statistical considerations.

From a theoretical viewpoint, consumers in a neighborhood may emulate each other, leading to spatial dependence. Local governments might engage in competition that leads to local uniformity in taxes and services. Pollution can create systematic patterns over space, and clusters of consumers who travel to a more distant store to avoid a high-crime zone would also generate these patterns. Real estate appraisal practices that place reliance on nearby homes sold recently

Numerical Issues in Statistical Computing for the Social Scientist, by Micah Altman, Jeff Gill, and Michael P. McDonald
ISBN 0-471-23633-0 Copyright © 2004 John Wiley & Sons, Inc.

would be one example of a theoretical motivation for spatial dependence in the underlying data-generating process.

Spatial dependence can arise from unobservable latent variables that are spatially correlated. Consumer expenditures collected at spatial locations such as Census tracts exhibit spatial dependence, reflecting that local customs and practices influence consumption patterns in addition to economic factors such as income and education. For example, one observes spatial patterns of tobacco consumption that are not explained by income and education levels alone. It seems plausible that unobservable characteristics that are difficult to quantify such as the quality of life and locational amenities may also exhibit spatial dependence. To the extent that this type of influence is an important factor determining variation in the social–economic demographic variables we are modeling, spatial dependence is likely to arise.

9.2.2 Specifying Dependence Using Weight Matrices

There are several ways to quantify the structure of spatial dependence between observations, but a common specification relies on an $n \times n$ spatial weight matrix W with elements $W_{ij} > 0$ for observations $j = 1 \cdots n$ sufficiently close (as measured by some distance metric) to observation i.

Note that we require knowledge of the location associated with the observational units to determine the closeness of individual observations to other observations. Sample data that contain address labels could be used in conjunction with GIS or other dedicated software to measure the spatial proximity of observations. Assuming that address matching or other methods have been used to produce latitude–longitude coordinates for each observation in map space, we can rely on a Delaunay triangularization scheme to find neighboring observations. To illustrate this approach, Figure 9.1 shows a Delaunay triangularization centered on an observation located at point A. The space is partitioned into triangles such that there are no points in the interior of the circumscribed circle of any triangle. Neighbors could be specified using Delaunay contiguity defined as two points being a vertex of the same triangle. The neighboring observations to point A that could be used to construct a spatial weight matrix are B, C, E, and F.

One way to specify the spatial weight matrix W would be to set column elements associated with neighboring observations B, C, E, F equal to 1 in row A. This would reflect that these observations are neighbors to observation A. Typically, the weight matrix is standardized such that row sums equal unity, producing a row-stochastic weight matrix. The motivation for this will be clear in a moment.

An alternative approach would be to rely on neighbors ranked by distance from observation A. We can simply compute the distance from A to all other observations and rank these by size. In the case of Figure 9.1 we would have a nearest neighbor E; the nearest two neighbors E, and C; the nearest three neighbors E, C, D; and so on. Again, we could set elements $W_{Aj} = 1$ for observations j in row A to reflect any number of nearest neighbors to observation

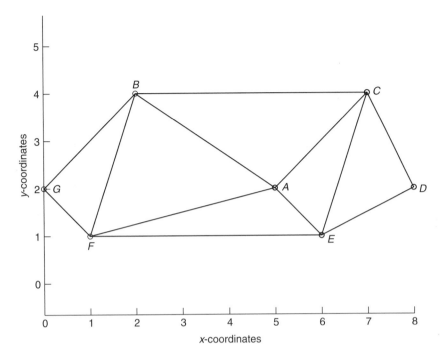

Fig. 9.1 Delaunay triangularization.

A, and transform to row-stochastic form. An important point is that computing the distance of each observation to all other observations is computationally intensive, whereas use of Delaunay triangle algorithms allow construction of a spatial weight matrix for a sample of 60,000 census tracts in about 2 seconds.

The spatial weight matrix can be used to specify the structure of spatial dependence $y_i = f(y_j)$ in our spatial regression model. To see the role of W in determining the spatial dependence structure of the model, consider the case where observations 2, 4, and 5 are neighbors to observation 1. This would lead to

$$y_1 = \rho \begin{bmatrix} 0 & \frac{1}{3} & 0 & \frac{1}{3} & \frac{1}{3} & 0 & 0 & \cdots & 0 \end{bmatrix} y + X_1 \beta + \varepsilon_1$$

$$= \rho \left[\frac{1}{3} y_2 + \frac{1}{3} y_4 + \frac{1}{3} y_5 \right] + X_1 \beta + \varepsilon_1, \tag{9.2}$$

where X_1 denotes the first row of a typical $n \times k$ matrix of explanatory variables, β is an associated $k \times 1$ parameter vector, and ε_1 is the first row of an $n \times 1$ disturbance vector, ε. We assume a normal, constant variance disturbance, $\varepsilon \sim \mathcal{N}(0, \sigma^2 I_n)$. The term created by multiplying the first row of W times the vector y is often called a *spatial lag*, and ρ represents a scalar parameter to be estimated along with β and σ^2. Of course, every observation in the n observation vector y is modeled as spatially dependent on nearby observations defined by nonzero values in the corresponding rows of the spatial weight matrix W. This approach to regression modeling of spatial dependence, known as the *spatial autoregressive* (SAR) *model*, can be written in the alternative forms

$$(I_n - \rho W)y = X\beta + \varepsilon$$

$$y = \rho W y + X\beta + \varepsilon \tag{9.3}$$

$$y = (I_n - \rho W)^{-1} X\beta + (I_n - \rho W)^{-1}\varepsilon,$$

where I_n denotes an identity matrix of order n. The second expression for the SAR model makes it clear why $W_{ii} = 0$, as this precludes an observation y_i from predicting itself directly. It also provides a rationale for the use of row-stochastic W, which makes each observation y_i a function of the spatial lag Wy, an explanatory variable representing an average of spatially neighboring values. We need not treat all neighboring relationships in an equal fashion; neighbors could be weighted by distance, length of adjoining property boundaries, or any number of other schemes that have been advocated in the literature on spatial regression relationships (see Bavaud 1998).

[It is also possible to estimate spatial autoregressive models with limited dependent variables. Methods for estimating logit, probit, and tobit variants of spatial autoregressive models are available in the collection of algorithms referenced at the end of the chapter and are discussed in LeSage (2000).]

We can extend the model in (9.3) to a *spatial Durbin model* (SDM) that allows for explanatory variables from neighboring observations, created by $W\tilde{X}$:

$$(I_n - \rho W)y = X\beta + W\tilde{X}\gamma + \varepsilon$$

$$y = \rho W y + X\beta + W\tilde{X}\gamma + \varepsilon \tag{9.4}$$

$$y = (I_n - \rho W)^{-1} X\beta + (I_n - \rho W)^{-1} W\tilde{X}\gamma + (I_n - \rho W)^{-1}\varepsilon,$$

where the matrix $\tilde{X}$ is equal to X with the constant term excluded. The $(k-1) \times 1$ parameter vector γ measures the marginal impact of the explanatory variables from neighboring observations on the outcome variable y. Multiplying $\tilde{X}$ by W produces spatial lags of the explanatory variables that reflect an average of neighboring observations' X-values.

Another model that has been used is the *spatial error model* (SEM):

$$y = X\beta + u$$

$$u = \rho W + \varepsilon \tag{9.5}$$

$$y = X\beta + (I_n - \rho W)^{-1}\varepsilon.$$

Here, the spatial dependence resides in the disturbance process u, as in the case of serial correlation in time series regression models.

9.2.3 Estimation Consequences of Spatial Dependence

In some applications the spatial structure of the dependence may be a subject of interest or provide a key insight. In other cases it may be a nuisance similar to serial correlation in time series models. In either case, inappropriate treatment of sample data with spatial dependence can lead to inefficient and/or biased and inconsistent estimates.

For the SAR and SDM models, least squares estimates for β are biased and inconsistent. To see this, note that we can define $Z = [Wy \quad X]$ and $\gamma = [\rho \quad \beta]'$ and consider the least squares estimate $\hat{\gamma} = (Z'Z)^{-1}Z'y$. We cannot show unbiasedness in the estimates γ because the matrix Z is not fixed in repeated sampling. The presence of neighboring values Wy in the matrix Z creates a situation similar to that found in simultaneous equation models from econometrics, where least squares estimates are biased and inconsistent.

For the case of the SEM model, least squares estimates for β are unbiased but inefficient, and the associated least squares measures of dispersion are biased, leading to bias in inferences drawn using conventional t-statistics from least squares.

By way of summary, there are numerous examples where social scientists use least squares to examine election returns, performance on standardized tests by school district or schools, voting behavior of congressional representatives, household behavior by census tract, and so on. These samples of cross-sectional spatially distributed observations are highly likely to exhibit spatial dependence, creating at best biased inferences regarding the importance of various explanatory variables and at worst biased and inconsistent estimates.

In the next section we discuss maximum likelihood estimation of the SAR, SDM, and SEM models that are capable of producing unbiased estimates in the face of spatial dependence. One approach to testing for the problem of spatial dependence would be to compare least squares estimates to maximum likelihood estimates from SAR, SDM, or SEM models, which nest least squares as a special case. If the parameter ρ measuring spatial dependence is significantly different from zero, least squares estimates will probably deviate greatly from the spatial regression models.

9.3 MAXIMUM LIKELIHOOD ESTIMATION OF SPATIAL MODELS

Maximum likelihood estimation of the SAR, SDM, and SEM models described here involves maximizing the log-likelihood function (concentrated for β and σ^2) with respect to the parameter ρ. In the case of row-stochastic spatial weight matrices W, the parameter ρ is restricted to the interval $(-1/\lambda_{\min}, 1/\lambda_{\max})$, where $\lambda_{\min} < 0$ and $\lambda_{\max} > 0$ represent the minimum and maximum eigenvalues of the spatial weight matrix (see, e.g., Sun et al., 1999, Lemma 2).

For the case of the SAR model, we have

$$
\begin{aligned}
\ln L &= C + \ln|I_n - \rho W| - (n/2)\ln(e'e) \\
e &= e_o - \rho e_d \\
e_o &= y - X\beta_o \\
e_d &= Wy - X\beta_d
\end{aligned}
\tag{9.6}
$$

$$\beta_o = (X'X)^{-1}X'y$$

$$\beta_d = (X'X)^{-1}X'Wy,$$

where C represents a constant not involving the parameters. This same approach can be applied to the SDM model by simply defining $X = [X \quad W\tilde{X}]$ in (9.6).

The SEM model has a concentrated log likelihood taking the form

$$\ln L = C + \ln|I_n - \rho W| - (n/2)\ln(e'e)$$

$$\tilde{X} = X - \rho W X$$

$$\tilde{y} = y - \rho W y \tag{9.7}$$

$$\beta^\star = (\tilde{X}'\tilde{X})^{-1}\tilde{X}\tilde{y}$$

$$e = \tilde{y} - \tilde{X}\beta^\star.$$

The computationally troublesome aspect of this is the need to compute the log determinant of the $n \times n$ matrix $(I_n - \rho W)$. Operation counts for computing this determinant grow with the cube of n for dense matrices. While W is a $n \times n$ matrix, in typical problems W will be sparse. For a spatial weight matrix constructed using Delaunay triangles between n points in two dimensions, the average number of neighbors for each observation will equal 6, so the matrix will have $6n$ nonzeros out of n^2 possible elements, leading to $6/n$ as the proportion of nonzeros.

9.3.1 Sparse Matrix Algorithms

A sparse matrix is one that contains a large proportion of zeros. As a concrete example, consider the spatial weight matrix for a sample of 3107 U.S. counties used in Pace and Barry (1997). This matrix is sparse since the largest number of neighbors to any county is eight and the average number of neighbors is four. To understand how sparse matrix algorithms conserve on storage space and computer memory, consider that we only need record the nonzero elements along with an indication of their row and column position. This requires a 1×3 vector for each nonzero element, consisting of a row index, a column index, and the element value. Since nonzero elements represent a small fraction of the total $3107 \times 3107 = 9,653,449$ elements in the weight matrix, we save on computer memory. For our example of the 3107 U.S. counties, only 12,429 nonzero elements were found in the weight matrix, representing a very small fraction (about 0.4%) of the total elements. Storing the matrix in sparse form requires only three times 12,429 elements, or more that 250 times less computer memory than would be needed to store 9,653,449 elements.

In addition to storage savings, sparse matrices result in lower operation counts as well, speeding computations. In the case of nonsparse (dense) matrices, matrix multiplication and common matrix decompositions such as the Cholesky require $O(n^3)$ operations, whereas for sparse W these operation counts can fall as low as $O(n_{\neq 0})$, where $n_{\neq 0}$ denotes the number of nonzero elements. As an example, suppose that we store the sparse matrix information in an $nz \times 3$ matrix **sparseM**, where nz represents the number of nonzero elements in the matrix. The c-language code loop needed to carry out the matrix-vector multiplication Wy needed to form the spatial lag vector in our SAR model is shown below.

```
// do W*y matrix times vector multiplication
for(i=1; i<nz; i++){
    ii  =  (int) sparseM[i][1];
    jj  =  (int) sparseM[i][2];
    val =        sparseM[i][3];
    Wy[ii] = Wy[ii] + val*y[jj];
}
```

Here the vectors **Wy**, **y**, and the matrix **sparseM** all begin at 1 rather than the traditional zero-based position used by the c-language. Note that for each of the nz nonzero elements in W, we require (1) indexing into the matrix **sparseM**, (2) a scalar multiplication operation, and (3) an accumulation, bringing the operation count to nz. For the case of our 3107×3107 matrix W, this would be 12,429 operations, which is in stark contrast to dense matrix multiplication that would require $3107^3 = 2.9993e + 010$ operations.

As a concrete example of the savings, we consider computing the largest and smallest eigenvalues, $\lambda_{\min}$ and $\lambda_{\max}$, needed to set the feasible range on ρ in the SAR, SDM, and SEM estimation problems. This required 45 minutes in the case of the 3107×3107 matrix using nonsparse algorithms, and less than 1 second using sparse matrix algorithms.

Pace and Barry (1997) suggested using direct sparse matrix algorithms such as the Cholesky or LU decompositions to compute the log determinant over a grid of values for the parameter ρ restricted to the interval 0, 1. They rule out negative values for ρ, arguing that these are of little practical interest in many cases. This accelerates the computations slightly because the time required to compute eigenvalues is eliminated.

9.3.2 Vectorization of the Optimization Problem

This still leaves the problem of maximizing the log likelihood over the grid of values for ρ, which can be solved without optimization algorithms using a vector evaluation of the SAR or SDM log-likelihood functions. Using a grid of q values of ρ in the interval $[0, 1)$, we can write the log-likelihood function as a vector in ρ:

$$
\begin{pmatrix} \ln L(\beta, \rho_1) \\ \ln L(\beta, \rho_2) \\ \vdots \\ \ln L(\beta, \rho_q) \end{pmatrix} \propto \begin{pmatrix} \ln |I_n - \rho_1 W| \\ \ln |I_n - \rho_2 W| \\ \vdots \\ \ln |I_n - \rho_q W| \end{pmatrix} - (n/2) \begin{pmatrix} \ln (\phi(\rho_1)) \\ \ln (\phi(\rho_2)) \\ \vdots \\ \ln (\phi(\rho_q)) \end{pmatrix}, \quad (9.8)
$$

where $\phi(\rho_i) = e_o' e_o - 2\rho_i e_d' e_o + \rho_i^2 e_d' e_d$. [For the SDM model, we replace X with $[X \; W\tilde{X}]$ in (9.6).] Given the vector of log-likelihood function values, one can simply find the associated value of ρ where the vector obtains a maximum. A finer grid can be constructed around this optimizing value to increase the numerical precision of the resulting estimate.

Note that the SEM model cannot be vectorized and must be solved using more conventional optimization methods such as a simplex algorithm. Since the optimization problem is univariate, involving only the parameter ρ, this does not present any problems. Further, the grid of values for the log determinant over the feasible range for ρ can be used to speed evaluation of the log-likelihood function during optimization with respect to ρ. This is accomplished by storing values for the log determinant in a table and using table lookup when evaluating the likelihood function rather than computing the log determinant for each value of ρ used in the optimization search.

The computationally intense part of this approach is still calculating the log determinant, which takes 18 seconds for the 3107 U.S. county sample and 746 seconds for a sample of 60,611 observations representing all 1990 census tracts in the continental United States. These times were based on a grid of 100 values from $\rho = 0$ to 1 using sparse matrix algorithms in MATLAB version 6.1 on a 1200-megahertz Athalon computer. Note that if the optimum ρ occurs on the boundary (i.e., 0), this indicates the need to consider negative values of ρ.

A final aspect of the estimation problem is determination of measures of dispersion for the estimates that can be used for inference. An asymptotic variance matrix based on the Fisher information matrix shown below for the parameters $\theta = (\rho, \beta, \sigma^2)$ can be used to provide measures of dispersion for the estimates of ρ, β, and σ^2:

$$[I(\theta)]^{-1} = -E \left[\frac{\partial^2 L}{\partial \theta \, \partial \theta'} \right]^{-1}. \tag{9.9}$$

For problems involving a small number of observations, we can use analytical expressions for the theoretical information matrix presented in Anselin (1988). This approach is computationally impossible when dealing with large problems involving thousands of observations. The expressions used to calculate terms in the information matrix involve operating on very large matrices that would take a great deal of computer memory and computing time. In these cases we can evaluate the numerical Hessian matrix using the maximum likelihood estimates of ρ, β, and σ^2 and our sparse matrix representation of the likelihood. Given the ability to evaluate the likelihood function rapidly, numerical methods can be used to compute approximations to the gradients shown in (9.9).

9.3.3 Trade-offs between Speed and Numerical Accuracy

There are trade-offs between numerical accuracy and speed in a number of the operations required to solve the estimation problem. First, one can ignore the eigenvalue calculations needed to set bounds on the feasible interval for ρ, or set an iterative convergence criterion for the algorithm used to compute these eigenvalues that will increase speed at the cost of accuracy. For row-stochastic weight matrices, the maximum for ρ takes on a theoretical value of unity, and a minimum value for ρ of 0 or -1 would be reasonable in many applications. Positive spatial dependence reflected by $\rho > 0$ can be interpreted as spatial clustering of values for y at similar locations in geographic or map space. Larger values near 1 reflect increased clustering. Negative spatial dependence represented by $\rho < 0$ would represent a situation where dissimilar numerical values tend to cluster in geographic space, whereas $\rho = 0$ indicates a situation where numerical values are arranged in a haphazard fashion on the map. Unlike conventional correlation coefficients, spatial correlation or dependence measured by ρ can, theoretically, take on values less than -1. However, spatial clustering of dissimilar numerical values reflected by negative values of ρ tends to be of less interest in applied problems than does spatial clustering of similar values. An implication of this is that values of ρ less than -1 are rarely encountered in applied practice.

Viewing ρ as a spatial correlation coefficient suggests that excessive accuracy in terms of decimal digits may not be necessary. This suggests another place where speed improvements can be made at the cost of accuracy. Changing the grid over ρ values from increments of 0.01 to a coarse grid based on 0.1 increment reduces the 746-second time needed for a 60,611-observation sample to 160 seconds. Spline interpolation of the log determinant values for a very fine grid based on a 0.0001 increment can be computed in 0.03 second, and the generally smooth nature of the log determinant makes these very accurate. Following this line of reasoning, numerous methods have been proposed in the literature that replace direct calculation of the log determinant with approximations or estimates. For example, Barry and Pace (1999) propose a Monte Carlo log determinant estimator, Smirnov and Anselin (2001) provide an approach based on a Taylor series expansion, and Pace and LeSage (2003) suggest using a Chebyshev expansion to replace the Taylor expansion, demonstrating improved accuracy. LeSage and Pace (2001) report experimental results indicating that the Barry and Pace (1999) Monte Carlo estimator achieves robustness that is rather remarkable. As an illustration of the speed improvements associated with these approaches, the Monte Carlo log determinant estimator required only 39 seconds to compute the log determinant for the 60,611-observation census tract sample, which compares favorably with 746 seconds, or 160 seconds in the case of the coarse grid over ρ for this sample.

A final place where accuracy and speed trade-offs exist is in the level of precision used to compute the numerical Hessian required to construct measures of inference such as asymptotic t-statistics. Small changes in the tolerance used to compute numerical differences from 1e-05 to 1e-08 can increase the time required by 50%. Use of a log determinant approximation computed and stored for a grid of values can be used with table lookup to speed the numerous log-likelihood function evaluations required to compute the numerical Hessian.

Ultimately, questions regarding these trade-offs between speed and accuracy depend on the specific problem at hand. For exploratory work, where changes are made to the model specification, the sample size of observations used, or the spatial weight matrix structure, one might rely on fast methods that trade off accuracy in the estimates. A switch to more accurate methods would typically be used to produce final estimation results that will be reported.

9.3.4 Applied Illustrations

We illustrate the speed versus accuracy trade-off as well as the importance of proper handling of data where spatial dependence is present. A sample of 10,418 census blocks in the state of Ohio were used to establish a regression relationship between 1998 expenditures on tobacco, measured as the log budget share of all expenditures and explanatory variables from the 1990 census. There may be spatially local phenomena that contribute to smoking behavior that are not easily measured by census variables. If so, we would expect to find a statistically significant value for the parameter ρ, indicating the presence of spatial dependence. This would also suggest that least squares estimates of the relationship between the explanatory variables and expenditures on tobacco would result in different inferences than those from a spatial autoregressive model.

The explanatory variables used were a constant term; four variables on education [measuring (1) the proportion of population with less than 8 years' education, (2) the proportion with less than a high school education, (3) the proportion with a college degree, and (4) those with a graduate or professional degree]; the proportion of the census block population that was female; the proportion of population that was born in the state; and the proportion living in the same house at the time of the 1990 census and in 1985. The outcome and explanatory variables were logged to convert the proportions to normally distributed magnitudes and to facilitate interpretation of the estimated coefficients. Given the double-log transform, we can interpret the estimated parameters as elasticities. For example, a coefficient of -1 on college graduates would indicate that a 10% increase in college graduates is associated with a 10% lower budget share for expenditures on tobacco.

Least squares estimates are reported in Table 9.1 alongside those from an SAR model. The first point to note is that the estimate for ρ was equal to 0.4684, with a large t-statistic, indicating the presence of spatial dependence in the pattern of expenditures on tobacco. Comparing the R-squared statistic of 0.6244 from the least squares model to 0.7237 for the SAR model, we find that about 10% of the variation in tobacco expenditures is accounted for by the spatially lagged explanatory variable.

The least squares coefficient estimates appear to be biased toward overstating the sensitivity of tobacco expenditures to all variables except "females," where the two models produce similar estimates. Since least squares ignores spatial dependence, it attributes spatial variation in tobacco expenditures to the explanatory variables, leading to estimates that overstate their importance. Another difference

Table 9.1 Model of Tobacco Expenditures at the Census Block Level

Variable	OLS	t-Statistic	SAR	t-Statistic
Constant	−4.7301	−247.36	−2.5178	−134.30
0–8 years' education	0.8950	27.93	0.6118	22.60
<12 years' education	0.9083	32.26	0.5582	23.75
College degree	−1.0111	−29.41	−0.5991	−20.15
Graduate or professional degree	−1.1115	−25.86	−0.9059	−24.53
Females	0.3536	10.89	0.3547	12.92
Born in state	−0.0536	−2.61	−0.0158	−0.90
Same house in 1985 and 1990	−0.6374	−47.76	−0.4877	−42.53
ρ	—	—	0.4703	135.00
σ^2	0.0170	—	0.0125	—
R^2	0.6244	—	0.7239	—

Table 9.2 Comparison of Exact and Approximate Estimates

Variable	SAR Exact	t-Statistic	SAR Approx.	t-Statistic
Constant	−2.5178	−134.30	−2.5339	−134.18
0-8 years' education	0.6118	22.60	0.6138	22.67
<12 years' education	0.5582	23.75	0.5608	23.86
College degree	−0.5991	−20.15	−0.6021	−20.24
Graduate or professional degree	−0.9059	−24.53	−0.9074	−24.57
Females	0.3547	12.92	0.3547	12.92
Born in state	−0.0158	−0.90	−0.0160	−0.92
Same house in 1985 and 1990	−0.4877	−42.53	−0.4888	−42.61
ρ	0.4703	135.00	0.4669	133.76
σ^2	0.0125	—	0.0125	—
R^2	0.7239	—	0.7237	—

between the least squares and SAR estimates involves the importance of persons 'born in the state,' which is negative and significantly different from zero in the OLS model, but not significant in the SAR model.

Table 9.2 illustrates the difference between estimates based on the Barry and Pace (1999) Monte Carlo estimator for the log determinant and exact solution of the estimation problem using sparse matrix algorithms. The time required for the exact solution was 426 seconds, whereas the approximate solution took 9 seconds. The estimates as well as t-statistics are identical to two decimal places in nearly all cases, and we would draw identical inferences from both sets of estimates.

9.4 BAYESIAN SPATIAL REGRESSION MODELS

One might suppose that application of Bayesian estimation methods to SAR, SDM, and SEM spatial regression models where the number of observations is very large would result in estimates nearly identical to those from maximum likelihood methods. This is a typical result when prior information is dominated by a large amount of sample information. Bayesian methods can, however, be used to relax the assumption of normally distributed constant-variance disturbances made by maximum likelihood methods, resulting in heteroscedastic Bayesian variants of the SAR, SDM, and SEM models. In these models, the prior information exerts an impact, even in very large samples. The true benefits from applying a Bayesian methodology to spatial problems arise when we extend the conventional model to relax the assumption of normally distributed disturbances with constant variance.

9.4.1 Bayesian Heteroscedastic Spatial Models

We introduce a more general version of the SAR, SDM, and SEM models that allows for nonconstant variance across space, as well as outliers. When dealing with spatial datasets one can encounter what have become known as *enclave effects*, where a particular region does not follow the same relationship as the majority of spatial observations. As an example, suppose that we were examining expenditures on alcohol at the census block level similar to our model of tobacco expenditures. There may be places where liquor control laws prohibit sales of alcohol, creating aberrant observations or outliers that are not explained by census variables such as education or income. This will lead to fat-tailed errors that are not normally distributed, but more likely to follow a Student's-t distribution.

This extended version of the SAR, SDM, and SEM models involves introduction of nonconstant variance to accommodate spatial heterogeneity and outliers that arise in applied practice. Here we can follow LeSage (1997, 2000) and introduce a set of variance scalars $(v_1, v_2, \ldots, v_n)$, as unknown parameters that need to be estimated. This allows us to assume that $\varepsilon \sim N(0, \sigma^2 V)$, where $V = \text{diag}(v_1, v_2, \ldots, v_n)$. The prior distribution for the v_i terms takes the form of an independent $\chi^2(r)/r$ distribution. The χ^2 distribution is a single-parameter distribution, where this parameter is labeled r. This allows estimation of the additional n parameters v_i in the model by adding the single parameter r to our estimation procedure.

This type of prior was used in Geweke (1993) to model heteroscedasticity and outliers in the context of linear regression. The specifics regarding the prior assigned to the v_i terms can be justified by considering that the prior mean is equal to unity and the variance of the prior is $2/r$. This implies that as r becomes very large, the terms v_i will all approach unity, resulting in $V = I_n$, the traditional assumption of constant variance across space. On the other hand, small values of r lead to a skewed distribution that permits large values of v_i that deviate greatly from the prior mean of unity. The role of these large v_i values is to accommodate outliers or observations containing large variances by downweighting these observations. Note that $\varepsilon \sim N(0, \sigma^2 V)$ with V diagonal implies a generalized least squares (GLS) correction to the vector y and explanatory variables matrix X. The GLS correction involves dividing through by $\sqrt{v_i}$, which leads to large

v_i values functioning to downweight these observations. Even in large samples, this prior will exert an impact on the estimation outcome.

A formal statement of the Bayesian heteroscedastic SAR model is

$$y = \rho W y + X\beta + \varepsilon$$

$$\varepsilon \sim N(0, \sigma^2 V) \quad V = \text{diag}(v_1, \dots, v_n)$$

$$\pi(\beta) \sim N(c, T) \tag{9.10}$$

$$\pi(r/v_i) \sim \text{IID}\chi^2(r)$$

$$\pi(1/\sigma^2) \sim \Gamma(d, v),$$

where we have added a normal-gamma conjugate prior for β and σ, a diffuse prior for ρ, and the chi-squared prior for the terms in V. The prior distributions are indicated using π.

With very large samples involving upward of 10,000 observations, the normal-gamma priors for β, σ should exert relatively little influence. Setting c to zero and T to a very large number results in a diffuse prior for β. Diffuse settings for σ are $d = 0$ and $v = 0$. For completeness, we develop the results for the case of a normal-gamma prior on β and σ.

In contrast to the case of the priors on β and σ, assigning an informative prior to the parameter ρ associated with spatial dependence would exert an impact on the estimation outcomes even in large samples. This is due to the important role played by spatial dependence in these models. In typical applications where the magnitude and significance of ρ is a subject of interest, a diffuse prior should be used. It is possible, however, to rely on an informative prior for this parameter.

9.4.2 Estimation of Bayesian Spatial Models

An unfortunate situation arises with this extension in that the addition of the chi-squared prior greatly complicates the posterior distribution. Assume for the moment, diffuse priors for β and σ. A key insight is that if we knew V, this problem would look like a GLS version of the previous constant-variance maximum likelihood problem. That is, conditional on V, we would arrive at expressions similar to those in our earlier model, where the y and X are transformed by dividing through by: $\sqrt{\text{diag}(V)}$. We rely on a Markov Chain Monte Carlo (MCMC) estimation method that exploits this fact.

MCMC is based on the idea that a large sample from the posterior distribution of our parameters can be used in place of an analytical solution when this is difficult or impossible (see Chapter 5). We designate the posterior using $p(\theta|D)$, where θ represents the parameters and D the sample data. If the samples from $p(\theta|D)$ were large enough, we could approximate the form of the posterior density using kernel density estimators or histograms, eliminating the need to know the precise analytical form of this complicated density. Simple statistics can be used to construct means and variances based on the sample from the posterior.

The parameters β, V, and σ in the heteroscedastic SAR model can be estimated by drawing sequentially from the conditional distributions of these parameters, a process known as *Gibbs sampling* because of its origins in image analysis (Geman and Geman 1984). It is also labeled *alternating conditional sampling*, which seems a more accurate description. Gelfand and Smith (1990) demonstrate that sampling from the sequence of complete conditional distributions for all parameters in the model produces a set of estimates that converge in the limit to the true (joint) posterior distribution of the parameters. That is, despite the use of conditional distributions in our sampling scheme, a large sample of the draws can be used to produce valid posterior inferences regarding the joint posterior mean and moments of the parameters.

9.4.3 Conditional Distributions for the SAR Model

To implement this estimation method, we need to determine the conditional distributions for each parameter in our Bayesian heteroscedastic SAR model. The conditional distribution for β follows from the insight that given V, we can rely on standard Bayesian GLS regression results to show that

$$p(\beta|\rho, \sigma, V) \sim N(\bar{b}, \sigma^2 B)$$

$$\bar{\beta} = (X'V^{-1}X + \sigma^2 T^{-1})^{-1}(X'V^{-1}(I_n - \rho W)y + \sigma^2 T^{-1}c) \quad (9.11)$$

$$B = \sigma^2(X'V^{-1}X + \sigma^2 T^{-1})^{-1}.$$

We see that the conditional for β is a multinormal distribution, from which it is easy to sample a vector β.

Given the other parameters, the conditional distribution for σ takes the form (see Gelman et al. 1995)

$$p(\sigma^2|\beta, \rho, V) \propto (\sigma^2)^{-(n/2+d+1)} \exp\left[-e'V^{-1}e + \frac{2v}{2\sigma^2}\right] \quad (9.12)$$

$$e = (I_n - \rho W)y - X\beta, \quad (9.13)$$

which is proportional to an inverse gamma distribution with parameters $(n/2)+d$ and $e'V^{-1}e+2v$. Again, this would be an easy distribution from which to sample a scalar value for σ.

Geweke (1993) shows that the conditional distribution of V given the other parameters is proportional to a chi-square density with $r+1$ degrees of freedom. Specifically, we can express the conditional posterior of each v_i as

$$p\left(\frac{e_i^2 + r}{v_i}\bigg|\beta, \rho, \sigma^2, v_{-i}\right) \sim \chi^2(r+1), \quad (9.14)$$

where $v_{-i} = (v_1, \ldots, v_{i-1}, v_{i+1}, \ldots, v_n)$ for each i and e is as defined in (9.13). Again, this represents a known distribution from which it is easy to construct a sequence of scalar draws.

Finally, the conditional posterior distribution of ρ takes the form

$$
\begin{aligned}
p(\rho|\beta, \sigma, V) &\propto |I_n - \rho W|(s^2(\rho))^{-(n-k)/2} \\
s^2(\rho) &= ((I_n - \rho W)y - Xb(\rho))'V^{-1}((I_n - \rho W)y \\
&\quad - Xb(\rho))/(n - k) \\
b(\rho) &= (X'X)^{-1}X'(I_n - \rho W)y.
\end{aligned}
\tag{9.15}
$$

A problem arises here in that this distribution is not one for which established algorithms exist to produce random draws. Given the computational power of today's computers, one can produce a draw from the conditional distribution for ρ in these models using univariate numerical integration on each pass through the sampler. A few years back, this would have been unthinkable, as numerical integration represented one of the more computationally demanding tasks.

To carry out numerical integration efficiently, we use logs to transform the conditional posterior in (9.15), and the Barry and Pace (1999) Monte Carlo estimator for the log determinant term, along with the vectorized expression for $s(\rho)^2 = \phi(\rho_i) = e'_o e_o - 2\rho_i e'_d e_o + \rho_i^2 e'_d e_d$. This produces a simple numerical integration problem that can be solved rapidly using Simpson's rule. We arrive at the entire conditional distribution using this numerical integration approach and then produce a draw from this distribution using inversion.

To see how inversion works, consider that we could arrange the distribution of ρ values in a vector sorted from low to high. There would be a large number of similar (or identical) ρ values in the vector near the mode of the distribution and only a few similar values at the beginning and end of the vector representing the two tails of the distribution. This is because the vector reflects the frequency of ρ values in the conditional distribution. Suppose that the vector of ρ values representing the conditional distribution contained 1000 values. A uniform draw from the set of integers between 1 and 1000 can be used to index into the vector representing the conditional distribution to select a ρ value. Values selected in this fashion reflect draws by inversion from the conditional distribution of ρ. Since values near the modal value will represent a majority of the values in the vector, these will most often be the resulting value drawn for ρ, as they should be.

Keep in mind that on the next pass through the MCMC sampler, we need to integrate the conditional posterior again. This is because the distribution is conditional on the changing values for the other parameters (v_i, β, σ) in the model. New values for the parameters produce an altered expression for s^2 in the conditional distribution for ρ, as well as a new value for the log determinant. Nonetheless, given that we have computed the log determinant over a grid of ρ values and stored these values for table lookup during sampling, this part of the integration problem can be carried out rapidly.

It should be noted that an alternative approach to sampling from the conditional distribution using a Metropolis–Hastings algorithm is described in LeSage (2000), where a normal or Student's t is recommended as the proposal distribution.

9.4.4 MCMC Sampler

By way of summary, an MCMC estimation scheme involves starting with arbitrary initial values for the parameters, which we denote β^0, σ^0, V^0, and ρ^0. We then sample sequentially from the following set of conditional distributions for the parameters in our model (see Chapter 5 for details).

1. $p(\beta|\sigma^0, V^0, \rho^0)$, which is a multinormal distribution with mean and variance defined in (9.11). This updated value for the parameter vector β we label β^1.

2. $p(\sigma|\beta^1, V^0, \rho^0)$, which is chi-square distributed with $n + 2d$ degrees of freedom, as shown in (9.13). Note that we rely on the updated value of the parameter vector $\beta = \beta^1$ when evaluating this conditional density. We label the updated parameter $\sigma = \sigma^1$ and note that we will continue to employ the updated values of previously sampled parameters when evaluating the next conditional densities in the sequence.

3. $p(v_i|\beta^1, \sigma^1, v_{-i}, \rho^0)$, which can be obtained from the chi-square distribution shown in (9.14). Note that this draw can be accomplished as a vector, providing greater speed.

4. $p(\rho|\beta^1, \sigma^1, V^1)$, which we sample using numerical integration and inversion. It is possible to constrain ρ to an interval such as $(0,1)$ using rejection sampling. This simply means that we reject values of ρ outside this interval. Note also that it is easy to implement a normal or some alternative prior distribution for this parameter.

We now return to step 1, employing the updated parameter values in place of the initial values β^0, σ^0, V^0, and ρ^0. On each pass through the sequence we collect the parameter draws which are used to construct a joint posterior distribution for the parameters in our model. As already noted, Gelfand and Smith (1990) demonstrate that MCMC sampling from the sequence of complete conditional distributions for all parameters in the model produces a set of estimates that converge in the limit to the true (joint) posterior distribution of the parameters. Another point to note is that the parameter draws can be used to test hypotheses regarding any function of interest involving the parameters.

9.4.5 Illustration of the Bayesian Model

To illustrate the Bayesian model, we use the same Ohio census block-level data sample but use the log budget share of expenditures on alcohol as the outcome variable. A set of 2500 draws were made from the conditional posterior distributions of the model, with the first 500 draws discarded to allow the sampling scheme to settle into a steady state. Tests for convergence of the sampler were based on longer runs, involving up to 10,000 draws, that produced estimates identical to two decimal places to those from 2500 draws. The hyperparameter r, representing our prior concerning heteroscedasticity, was set to a small value of 7, reflecting a prior belief in nonconstant variance for this model. Diffuse priors were used for all other parameters in the model.

It happens that census blocks in urban entertainment areas exhibit large magnitudes for the v_i estimates, indicating that census demographics associated with these blocks do not account for the budget share of alcohol expenditures. Of course, expenditures need not take place in the same location where people reside, so use of census block sociodemographic characteristics is unlikely to model adequately the spatial pattern of these expenditures.

Maximum likelihood estimates for the parameters of the SAR model are presented in Table 9.3 along with those from the Bayesian heteroscedastic spatial SAR model. The first point to note is that the spatial correlation parameter ρ is much higher for the robust Bayesian model that accommodates nonconstant variance. Aberrant observations will distort the spatial pattern in the outcome variable, making it more difficult to estimate the extent of spatial dependence in the data.

The estimates indicate that low levels of education are associated with lower expenditures for alcohol, since the coefficients on education of less than eight years and less than high school are both negative and significant. In contrast, college graduates and professionals are positively associated with expenditures on alcohol. Women exert a negative influence on alcohol consumption, as might be expected. A difference between the maximum likelihood and Bayesian estimates is that a higher proportion of population 'born in the state' is positive and significant in the case of maximum likelihood and negative and insignificant for the Bayesian model. Another difference can be found in the estimate for 'females,' which is higher in the Bayesian model.

Table 9.3 Estimates for the Spatial Alcohol Expenditures Model

Variable	SAR ML	t-Statistic	SAR Bayesian	t-Statistic
Constant	−3.1861	−4591.83	−2.6146	−65.07
0-8 years' education	−0.1210	−15.31	−0.1215	−19.07
<12 years' education	−0.1447	−20.81	−0.1219	−23.47
College degree	0.1324	15.69	0.1280	20.57
Graduate or professional degree	0.1372	12.98	0.1166	15.43
Females	−0.0707	−8.85	−0.0989	−14.35
Born in state	0.0093	1.85	−0.0029	−0.80
Same house in 1985 and 1990	0.1038	31.68	0.1063	44.49
ρ	0.3279	548.01	0.4453	52.62
σ^2	0.0010	—	0.0003	—
R^2	0.4976	—	0.5001	—

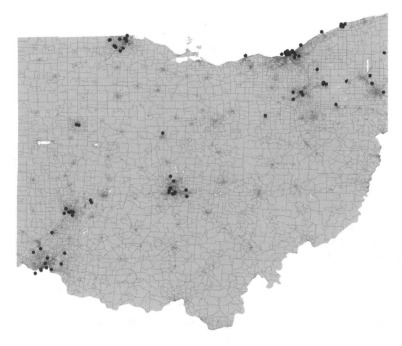

Fig. 9.2   Map of large v_i estimates.

A final point is that the lower coefficient estimate for ρ in the maximum likelihood results leads to a slight overstatement of the influence of education for those with less than 12 years as well as college and graduate/professionals.

A map of Ohio with the census blocks where the estimates for v_i exceeded a value of 4 is shown in Figure 9.2. There were 87 values where this was true, reflecting a variance four times that assumed by the homoscedastic maximum likelihood model. Although this represents a small fraction of the 10,418 census blocks used as observations, these 87 values exhibit a distinct spatial pattern. Figure 9.2 indicates that these aberrant observations exist primarily in the eight large metropolitan areas of Ohio: Akron, Canton, Cincinnati, Cleveland, Columbus, Dayton, Toledo, and Youngstown–Warren. This should not be surprising in that entertainment venues in these urban areas reflect places where alcohol expenditures are high. The practical impact of this is that the Bayesian estimates place less weight on observations from these urban areas. Large v_i estimates lead to downweighting the associated observations, as noted earlier.

9.5   CONCLUSIONS

The use of sparse matrix algorithms along with estimates or approximations to the troublesome log determinant of an $n \times n$ by n matrix that appears in the likelihood allow large spatial estimation problems to be carried out in a reasonable amount of time. Because Bayesian models also rely on the likelihood, many of the computational advances described here for maximum likelihood estimation can also be applied to Bayesian estimation. In this connection it should be noted that it took only 342 seconds to produce 2500 draws for the problem involving 10,514 census block observations. This time is based on using the Barry and Pace (1999) Monte Carlo estimator for the log determinant as well as univariate numerical integration of the conditional distribution for ρ on every pass through the sampler. This timing result is based on using MATLAB version 6.1 on a 1200-megahertz Athalon processor. Use of c-language functions that interface with MATLAB reduced this time to 228 seconds.

Another issue discussed was the trade-off between computational speed and accuracy. One can analyze the impact of: (1) using approximations, (2) changing tolerances, and (3) placing reasonable restrictions or bounds on parameter values during optimization. Flexible software algorithms should allow users to change these aspects of the program so that exploratory work might be carried out using faster but less accurate versions of the algorithm. Public-domain algorithms for use with MATLAB that take this approach are available at <http://www. spatial-econometrics.com>, along with c-language functions that can be used in conjunction with MATLAB or as stand-alone functions.

CHAPTER 10

Convergence Problems in Logistic Regression

Paul Allison

10.1 INTRODUCTION

Anyone with much practical experience using logistic regression will have occasionally encountered problems with convergence. Such problems are usually both puzzling and exasperating. Most researchers haven't a clue as to why certain models and certain data sets lead to convergence difficulties. For those who do understand the causes of the problem, it is often unclear whether and how the problem can be fixed.

In this chapter we explain why numerical algorithms for maximum likelihood estimation of the logistic regression model sometimes fail to converge and consider a number possible solutions. Along the way, we look at the performance of several popular computing packages when they encounter convergence problems of varying kinds.

10.2 OVERVIEW OF LOGISTIC MAXIMUM LIKELIHOOD ESTIMATION

We begin with a review of the logistic regression model and maximum likelihood estimation of its parameters. For a sample of n cases ($i = 1, \ldots, n$) there are data on a dichotomous outcome variable y_i (with values of 1 and 0) and a vector of explanatory variables $\mathbf{x}_i$ (including a 1 for the intercept term). The logistic regression model states that

$$\Pr(y_i = 1 | \mathbf{x}_i) = \frac{1}{1 + \exp(-\boldsymbol{\beta}\mathbf{x}_i)}, \tag{10.1}$$

where $\boldsymbol{\beta}$ is a vector of coefficients. Equivalently, the model may be written in logit form:

$$\ln\left[\frac{\Pr(y_i = 1 | \mathbf{x}_i)}{\Pr(y_i = 0 | \mathbf{x}_i)}\right] = \boldsymbol{\beta}\mathbf{x}_i. \tag{10.2}$$

Numerical Issues in Statistical Computing for the Social Scientist, by Micah Altman, Jeff Gill, and Michael P. McDonald
ISBN 0-471-23633-0 Copyright © 2004 John Wiley & Sons, Inc.

Assuming that the n cases are independent, the log-likelihood function for this model is

$$\ell(\boldsymbol{\beta}) = \sum_i \boldsymbol{\beta} \mathbf{x}_i \, y_i - \sum_i \ln[1 + \exp(\boldsymbol{\beta} \mathbf{x}_i)]. \tag{10.3}$$

The goal of maximum likelihood estimation is to find a set of values for $\boldsymbol{\beta}$ that maximize this function. One well-known approach to maximizing a function such as this is to differentiate it with respect to $\boldsymbol{\beta}$, set the derivative equal to 0, and then solve the resulting set of equations. The first derivative of the log likelihood is

$$\frac{\partial \ell(\boldsymbol{\beta})}{\partial \boldsymbol{\beta}} = \sum_i \mathbf{x}_i \, y_i - \sum_i \mathbf{x}_i \, \hat{y}_i, \tag{10.4}$$

where $\hat{y}_i$ is the predicted value of y_i:

$$\hat{y}_i = \frac{1}{1 + \exp(-\boldsymbol{\beta} \mathbf{x}_i)}. \tag{10.5}$$

The next step is to set the derivative equal to 0 and solve for $\boldsymbol{\beta}$:

$$\sum_i \mathbf{x}_i \, y_i - \sum_i \mathbf{x}_i \, \hat{y}_i = 0. \tag{10.6}$$

Because $\boldsymbol{\beta}$ is a vector, (10.6) is actually a set of equations, one for each of the parameters to be estimated. These equations are identical to the "normal" equations for least squares linear regression, except that by (10.5), $\hat{y}_i$ is a non-linear function of the $\mathbf{x}_i$'s rather than a linear function.

For some models and data (e.g., "saturated" models), the equations in (10.6) can be solved explicitly for the ML estimator $\hat{\boldsymbol{\beta}}$. For example, suppose that there is a single dichotomous x variable such that the data can be arrayed in a 2×2 table, with observed cell frequencies f_{11}, f_{12}, f_{21}, and f_{22}. Then the ML estimator of the coefficient of x is given by the logarithm of the *cross-product* ratio:

$$\hat{\beta} = \ln \frac{f_{11} f_{22}}{f_{12} f_{21}}. \tag{10.7}$$

For most data and models, however, the equations in (10.6) have no explicit solution. In such cases, the equations must be solved by numerical methods, of which there are many. The most popular numerical method is the Newton–Raphson algorithm. Let $\mathbf{U}(\boldsymbol{\beta})$ be the vector of first derivatives of the log likelihood with respect to $\boldsymbol{\beta}$, and let $\mathbf{I}(\boldsymbol{\beta})$ be the matrix of second derivatives. That is,

$$\mathbf{U}(\boldsymbol{\beta}) = \frac{\partial \ell(\boldsymbol{\beta})}{\partial \boldsymbol{\beta}} = \sum_i \mathbf{x}_i \, y_i - \sum_i \mathbf{x}_i \, \hat{y}_i$$

$$\mathbf{I}(\boldsymbol{\beta}) = \frac{\partial^2 \ell(\boldsymbol{\beta})}{\partial \boldsymbol{\beta} \partial \boldsymbol{\beta}'} = \sum_i \mathbf{x}_i \mathbf{x}_i' \, \hat{y}_i (1 - \hat{y}_i). \tag{10.8}$$

The vector of first derivatives $\mathbf{U}(\boldsymbol{\beta})$ is called the *gradient*, while the matrix of second derivatives $\mathbf{I}(\boldsymbol{\beta})$ is called the *Hessian*. The Newton–Raphson algorithm is then

$$\boldsymbol{\beta}_{j+1} = \boldsymbol{\beta}_j - \mathbf{I}^{-1}(\boldsymbol{\beta}_j)\mathbf{U}(\boldsymbol{\beta}_j), \tag{10.9}$$

where $\mathbf{I}^{-1}$ is the inverse of $\mathbf{I}$. Chapter 6 showed what can go wrong with this process as well as some remedies.

To operationalize this algorithm, a set of starting values $\boldsymbol{\beta}_0$ is required. Choice of starting values is usually not critical for solving the problem above; usually, setting $\boldsymbol{\beta}_0 = \mathbf{0}$ works fine. (Note, however, that when Newton–Raphson and similar algorithms are used to solve other types of problems, starting values may be quite important; see Chapters 4 and 8.) The starting values are substituted into the right-hand side of (10.9), which yields the result for the first iteration, $\boldsymbol{\beta}_1$, These values are then substituted back into the right-hand side, the first and second derivatives are recomputed, and the result is $\boldsymbol{\beta}_2$. The process is repeated until the maximum change in each parameter estimate from one iteration to the next is less than some criterion, at which point we say that the algorithm has converged. Once we have the results of the final iteration, $\hat{\boldsymbol{\beta}}$, a by-product of the Newton–Raphson algorithm is an estimate of the covariance matrix of the coefficients, which is just $-\mathbf{I}^{-1}(\hat{\boldsymbol{\beta}})$. Estimates of the standard errors of the coefficients are obtained by taking the square roots of the main diagonal elements of this matrix.

10.3 WHAT CAN GO WRONG?

A problem that often occurs in trying to maximize a function is that the function may have local maxima, that is, points that are larger than any nearby point but not as large as some more distant point. In those cases, setting the first derivative equal to zero will yield equations that have more than one solution. If the starting values for the Newton–Raphson algorithm are close to a local maximum, the algorithm will probably iterate to that point rather than the global maximum. Fortunately, problems with multiple maxima cannot occur with logistic regression because the log likelihood is globally concave, meaning that the function can have at most one maximum (Amemiya 1985).

Unfortunately, there are many situations in which the likelihood function has *no* maximum, in which case we say that the maximum likelihood estimate does not exist. Consider the set of data on 10 observations in Table 10.1.

For these data it can be shown that the ML estimate of the intercept is zero. Figure 10.1 shows a graph of the log likelihood as a function of the slope "beta". It is apparent that although the log likelihood is bounded above by zero, it does not reach a maximum as beta increases. We can make the log likelihood as close to zero as we choose by making beta sufficiently large. Hence, there is no maximum likelihood estimate.

This is an example of a problem known as *complete separation* (Albert and Anderson 1984), which occurs whenever there exists some vector of coefficients

**Table 10.1 Data Exhibiting
Complete Separation**

x	y	x	y
−5	0	1	1
−4	0	2	1
−3	0	3	1
−2	0	2	1
−1	0	5	1

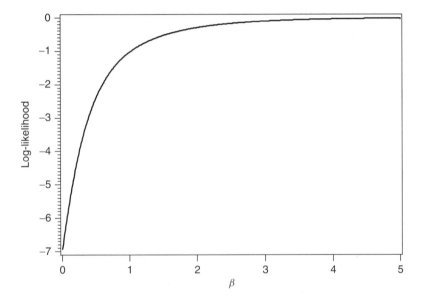

Fig. 10.1 Log-likelihood as a function of the slope under complete separation.

b such that $y_i = 1$ whenever $\mathbf{b}x_i > 0$ and $y_i = 0$ whenever $\mathbf{b}x_i < 0$. In other words, complete separation occurs whenever a linear function of **x** can generate perfect predictions of y. For our hypothetical dataset, a simple linear function that satisfies this property is $0 + 1(x)$. That is, when x is greater than 0, $y = 1$, and when x is less than 0, $y = 0$.

A related problem is known as *quasi-complete separation*. This occurs when there exists some coefficient vector **b** such that $\mathbf{b}x_i \geq 0$ whenever $y_i = 1$ and $\mathbf{b}x_i \leq 0$ whenever $y_i = 0$, and when equality holds for at least one case in each category of the outcome variable. Table 10.2 displays a dataset that satisfies this condition. What distinguishes this dataset from the preceding one is that there are two additional observations, each with x values of 0 but having different values of y.

The log-likelihood function for these data, shown in Figure 10.2, is similar in shape to that in Figure 10.1. However, the asymptote for the curve is not 0 but a

Table 10.2   Data Exhibiting Quasi-Complete Separation

x	y	x	y
−5	0	1	1
−4	0	2	1
−3	0	3	1
−2	0	2	1
−1	0	5	1
0	0	0	1

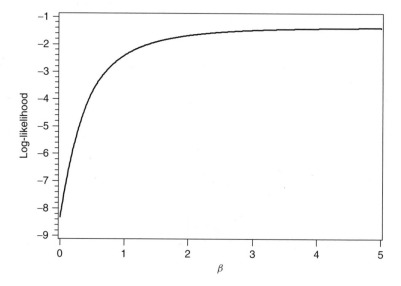

Fig. 10.2   Log-likelihood as a function of the slope: quasi-complete separation.

number that is approximately −1.39. In general, the log-likelihood function for quasi-complete separation will not approach zero, but some number lower than that. In any case, the curve has no maximum, so again, the maximum likelihood estimate does not exist.

Of the two conditions, complete and quasi-complete separation, the latter is far more common. It most often occurs when an explanatory variable x is a dummy variable, and for one value of x, either every case has the event $y = 1$ or every case has the event $y = 0$. Consider the following 2×2 table:

		y	
		1	0
x	1	5	0
	0	15	10

If we form the linear function $c = 0 + (1)x$, we have $c \geq 0$ when $y = 1$ and $c \leq 0$ when $y = 0$. Further, for all the cases in the second row, $c = 0$ for both values of y. So the conditions of quasi-complete separation are satisfied.

To get some intuitive sense of why this leads to nonexistence of the maximum likelihood estimator, consider equation (10.7), which gives the maximum likelihood estimator of the slope coefficient for a 2×2 table. For our quasi-complete table, that would be

$$\hat{\beta} = \ln \left[\frac{5 \times 10}{15 \times 0} \right]. \tag{10.10}$$

But this is undefined because there is a zero in the denominator. The same problem would occur if there were a zero in the numerator because the logarithm of zero is also undefined. If the table is altered to read

		y	
		1	0
x	1	5	0
	0	0	10

there is *complete* separation, with zeros in both the numerator and the denominator.

So the general principle is evident: Whenever there is a zero in any cell of a 2×2 table, the maximum likelihood estimate of the logistic slope coefficient does not exist. This principle also extends to multiple explanatory variables: *For any dichotomous explanatory variable in a logistic regression, if there is a zero in the 2×2 table formed by that variable and the outcome variable, the ML estimate for the regression coefficient will not exist.*

This is by far the most common cause of convergence failure in logistic regression. Obviously, it is more likely to occur when the sample size is small. Even in large samples, it will frequently occur when there are extreme splits on the frequency distribution of either the outcome or explanatory variables. Consider, for example, a logistic regression predicting whether a person had some rare disease whose overall prevalence is less than 1 in 1000. Suppose further that the explanatory variables include a set of seven dummy variables representing different age categories. Even if the sample contained 20,000 cases, it would hardly be surprising if no one had the disease for at least one of the age categories, leading to quasi-complete separation.

10.4 BEHAVIOR OF THE NEWTON–RAPHSON ALGORITHM UNDER SEPARATION

We just saw that when there are explicit formulas for the maximum likelihood estimate and there is either complete or quasi-complete separation, the occurrence of zeros in the formulas prevents computation. What happens when the Newton–Raphson algorithm is applied to data exhibiting either kind of separation? That depends on the particular implementation of the algorithm. The classic behavior is this: At each iteration, the parameter estimate for the variable (or variables) with separation gets larger in magnitude. Iterations continue until the fixed iteration limit is exceeded. At whatever limit is reached, the parameter estimate is large and the estimated standard error is extremely large. If separation is complete, the log likelihood will be reported as zero.

10.4.1 Specific Implementations

What actually happens depends greatly on how the algorithm is implemented. To determine how available software handles complete and quasi-complete separation, I tried estimating logistic regression models for the datasets in Tables 10.1 and 10.2 using several popular statistical packages. For some packages (SAS, Stata), more than one command or procedure was evaluated. Keep in mind that these tests were run in September 2002 using software versions that were available to me at the time. Results are summarized in Table 10.3. What follows is a detailed discussion of each of the headings in this table.

10.4.2 Warning Messages

Ideally, the program should detect the separation and issue a clear warning message to the user. In their classic paper on separation in logistic regression, Albert and Anderson (1984) proposed one "empirical" method that has been

Table 10.3 Performance of Packages under Complete and Quasi-Complete Separation

	Warning Messages[a]		False Convergence		Report Estimates		LR Statistics	
	Comp.	Quasi.	Comp.	Quasi.	Comp.	Quasi.	Comp.	Quasi.
GLIM			*	*	*	*		
JMP	A	A	*	*	*	*	*	*
LIMDEP			*	*	*	*		
Minitab					*	*		
R	A	A	*	*	*	*		
SAS GENMOD		A	*		*	*	*	*
SAS LOGISTIC	C	C			*	*		
SAS CATMOD	A	A	*	*	*	*		
SPSS	C			*		*		
Stata LOGIT	C	C				*		
Stata MLOGIT			*	*	*	*		
Systat	C			*	*	*		

[a]C, clear warning; A, ambiguous warning.

implemented in PROC LOGISTIC in SAS software (SAS Institute 1999). It has the following steps:

1. If the convergence criterion is satisfied within eight iterations, conclude that there is no problem.
2. For all iterations after the eighth, compute the probability of the observed response predicted for each observation, which is given by

$$\hat{y}_i = \frac{1}{1 + \exp[(2y_i - 1)\hat{\boldsymbol{\beta}}\mathbf{x}_i]}.$$

    If the probability predicted is 1 for all observations, conclude that there is complete separation and stop the iterations.
3. If the probability of the observed response is large (≥ 0.95) for some observations (but not all), the examine estimated standard errors for that iteration. If they exceed some criterion, conclude that there is quasi-complete separation and stop the iteration.

The check for complete separation is very reliable, but the check for quasi-complete separation is less so. For more reliable checks of quasi-complete separation, methods based on linear programming algorithms have been proposed by Albert and Anderson (1984) and Santner and Duffy (1986).

When the dataset in Table 10.1 was used with PROC LOGISTIC, it printed the message

    Complete separation of data points detected.

WARNING: The maximum likelihood estimate does not
         exist.
WARNING: The LOGISTIC procedure continues in spite of
         the above warning. Results shown are based
         on the last maximum likelihood iteration.
         Validity of the model fit is questionable.

For quasi-complete separation, the message was

    Quasicomplete separation of data points detected.

WARNING: The maximum likelihood estimate may not exist.
WARNING: The LOGISTIC procedure continues in spite of
         the above warning. Results shown are based
         on the last maximum likelihood iteration.
         Validity of the model fit is questionable.

Although PROC LOGISTIC came close to the ideal, most other software left much to be desired with regard to detection and warnings. SPSS did a good job for complete separation, presenting the following warning:

```
Estimation terminated at iteration number 21 because a
perfect fit is detected. This solution is not unique.
Warning \# 18582 Covariance matrix cannot be computed.
Remaining statistics will be omitted.
```

But SPSS gave no warning message for quasi-complete separation. For complete separation, the LOGIT command in STATA gave the message

```
outcome = x>-1 predicts data perfectly
```

For quasi-complete separation, the message was

```
outcome = x>0 predicts data perfectly except for x==0
subsample: x dropped and 10 obs not used
```

For complete separation, Systat produced the message

```
Failure to fit model or perfect fit
```

but said nothing in the case of quasi-complete separation.

Several other programs produced warning messages that were ambiguous or cryptic. SAS CATMOD marked certain coefficient estimates with # and said that they were "regarded to be infinite." JMP identified some coefficient estimates as "unstable." R said that some of the fitted probabilities were numerically 0 or 1. For quasi-complete separation, the GENMOD procedure in SAS reported that the negative of the Hessian matrix was not positive definite. Finally, several programs gave no warning messages whatsoever (GLIM, LIMDEP, Stata MLOGIT, and Minitab).

10.4.3 False Convergence

Strictly speaking, the Newton–Raphson algorithm should not converge under either complete or quasi-complete separation. Nevertheless, the only program that exhibited this classic behavior was Minitab. No matter how high I raised the maximum number of iterations, this program would not converge. With the exception of SAS LOGISTIC and Stata LOGIT (both of which stopped the iterations once separation had been detected), the remaining programs all reported that the convergence criterion had been met. In some cases (GLIM, R, SPSS, and Systat) it was necessary to increase the maximum iterations beyond the default to achieve this apparent convergence.

Lacking information on the convergence criterion for most of these programs, I do not have a definite explanation for the false convergence. One possibility is that the convergence criterion is based on the log likelihood rather than the parameter estimates. Some logistic regression programs determine convergence by examining the change in the log likelihood from one iteration to the next. If that change is less than some criterion, convergence is declared and the iterations cease. Unfortunately, as seen in Figures 10.1 and 10.2, with complete or quasi-complete separation the log-likelihood may change imperceptibly from one iteration to the next, even as the parameter estimate is rapidly increasing in magnitude. So to avoid the false appearance of convergence, it is essential that convergence be evaluated by looking at changes in the parameter estimate across iterations rather than changes in the log-likelihood.

The explanation is a bit more complicated with the SAS GENMOD procedure, however. SAS documentation is quite explicit that the criterion is based on parameter estimates, not on the log-likelihood (SAS Institute 1999). But GEN-MOD uses a *modified* Newton–Raphson algorithm that performs a line search at each step to stabilize its convergence properties. In most cases this improves the rate of convergence. However, if the likelihood cannot be improved along the current Newton step, the algorithm returns the current parameter values as the updated values and therefore determines that the convergence criterion has been met (G. Johnston 2002 personal communication).

Whatever the explanation, the combination of apparent convergence and lack of clear warning messages in many programs means that some users are likely to be misled about the viability of their parameter estimates.

10.4.4   Reporting of Parameter Estimates and Standard Errors

Some programs do a good job of detecting and warning about complete separation but then fail to report any parameter estimates or standard errors (SPSS, STATA LOGIT). This might seem sensible since nonconvergent estimates are essentially worthless as parameter estimates. However, they may still serve a useful diagnostic purpose in determining which variables have complete or quasi-complete separation.

10.4.5   Likelihood Ratio Statistics

Some programs (SAS GENMOD, JMP) report optional likelihood-ratio chi-square tests for each coefficient in the model. Unlike Wald chi-squares, which are essentially useless under complete or quasi-complete separation, the likelihood ratio test is still a valid test of the null hypothesis that a coefficient is equal to zero. Thus, even if a certain parameter cannot be estimated, it may still be possible to judge whether or not it is significantly different from zero.

10.5   DIAGNOSIS OF SEPARATION PROBLEMS

We are now in a position to make some recommendations about how the statistical analyst should approach the detection of problems of complete or quasi-complete separation. If you are using software that gives clear diagnostic messages (SAS LOGISTIC, STATA LOGIT), one-half of the battle is won. But there is still a need to determine which variables are causing the problem and to get a better sense of the nature of the problem.

The second step (or the first step with programs that do not give good warning messages) is to carefully examine the estimated coefficients and their standard

errors. Variables with nonexistent coefficients will invariably have large parameter estimates, typically greater than 5.0, and huge standard errors, producing
Wald chi-square statistics that are near zero. If any of these variables is a
dummy (indicator) variable, the next step is to construct the 2×2 table for
each dummy variable with the outcome variable. A frequency of zero in any
single cell of the table implies quasi-complete separation. Less commonly, if
there are two diagonally opposed zeros in the table, the condition is complete
separation.

Once you have determined which variables are causing separation problems, it
is time to consider possible solutions. The potential solutions are somewhat different for complete and quasi-complete separation, so I will treat them separately.
I begin with the more common problem of quasi-complete separation.

10.6   SOLUTIONS FOR QUASI-COMPLETE SEPARATION

10.6.1   Deletion of Problem Variables

In practice, the most widely used method for dealing with quasi-complete separation is simply to delete from the model any variables whose coefficients did
not converge. *I do not recommend this method.* If a variable has quasi-complete
separation with the outcome variable, it is reasonable to suppose that the variable
has a strong (albeit noninfinite) effect on the outcome variable. Deleting variables
with strong effects will certainly obscure the effects of those variables and is also
likely to bias the coefficients for other variables in the model.

10.6.2   Combining Categories

As noted earlier, the most common cause of quasi-complete separation is a
dummy predictor variable such that for one level of the variable, either every
observation has the event or no observation has the event. For those cases in
which the problem variable is one of a set of variables representing a single
categorical variable, the problem can often be solved easily by combining categories. For example, suppose that marital status has five categories: never married,
currently married, divorced, separated, and widowed. This variable could be
represented by four dummy variables, with currently married as the reference
category. Suppose further that the sample contains 50 persons who are divorced
but only 10 who are separated. If the outcome variable is 1 for employed and 0
for unemployed, it's quite plausible that all 10 of the separated persons would be
employed, leading to quasi-complete separation. A natural and simple solution is
to combine the divorced and separated categories, turning two dummy variables
into a single dummy variable.

Similar problems often arise when a quantitative variable such as age is divided
into a set of categories with dummy variables for all but one of the categories.
Although this can be a useful device for representing nonlinear effects, it can
easily lead to quasi-complete separation if the number of categories is large and

the number of cases within some categories is small. The solution is to use a smaller number of categories or perhaps to revert to the original quantitative representation of the variable.

If the dummy variable represents an irreducible dichotomy such as sex, this solution is clearly not feasible. However, there is another simple method that often provides a very satisfactory solution.

10.6.3 Do Nothing and Report Likelihood Ratio Chi-Squares

Just because maximum likelihood estimates do not exist for some coefficients because of quasi-complete separation, that does not mean that they do not exist for other variables in the logistic regression model. In fact, if one leaves the offending variables in the model, the coefficients, standard errors, and test statistics for the remaining variables are still valid maximum likelihood estimates. So one attractive strategy is just to leave the problem variables in the model. The coefficients for those variables could be reported as $+\infty$ or $-\infty$. The standard errors and Wald statistics for the problem variables will certainly be incorrect, but as noted above, likelihood ratio tests for the null hypothesis that the coefficient is zero are still valid. If these statistics are not available as options in the computer program, they can be obtained easily by fitting the model with and without each problem variable, then taking twice the positive difference in the log-likelihoods.

If the problem variable is a dummy variable, the estimates obtained for the *other* variables have a special interpretation. They are the ML estimates for the subsample of cases that fall into the category of the dummy variable, in which observations differ on the outcome variable. For example, suppose that the outcome variable is whether or not a person smokes cigars. A dummy variable for sex is included in the model, but none of the women smoke cigars, producing quasi-complete separation. If sex is left in the model, the coefficients for the remaining variables (e.g., age, income, education) represent the effects of those variables among men only. (This can easily be verified by actually running the model for men only.) The advantage of doing it in the full sample with sex as a covariate is that one also gets a test of the sex effect (using the likelihood ratio chi-square) while controlling for the other predictor variables.

10.6.4 Exact Inference

As mentioned previously, problems of separation are most likely to occur in small samples and/or when there is an extreme split on the outcome variable. Of course, even without separation problems, maximum likelihood estimates may not have good properties in small samples. One possible solution is to abandon maximum likelihood entirely and do exact logistic regression. This method was originally proposed by Cox (1970) but was not computationally feasible until the advent of the LogXact program and, more recently, the introduction of exact methods to the LOGISTIC procedure in SAS.

Exact logistic regression is designed to produce exact p-values for the null hypothesis that each predictor variable has a coefficient of zero, conditional on all the other predictors. These p-values, based on permutations of the data rather than on large-sample chi-square approximations, are essentially unaffected by complete or quasi-complete separation. The coefficient estimates reported with this method are usually conditional maximum likelihood estimates, and these may not be correct when there is separation. In that event, both `LogXact` and `PROC LOGISTIC` report median unbiased estimates for the problem coefficients. If the true value is β, a median unbiased estimator β_u has the property

$$\Pr(\beta_u \leq \beta) \geq \tfrac{1}{2}, \qquad \Pr(\beta_u \geq \beta) \geq \tfrac{1}{2}, \tag{10.11}$$

Hirji et al. (1989) demonstrated that the median unbiased estimator is generally more accurate than the maximum likelihood estimator for small sample sizes.

I used `PROC LOGISTIC` to do exact estimation for the data in Tables 10.1 and 10.2. For the completely separated data in Table 10.1, the p-value for the coefficient of x was 0.0079. The median unbiased estimate was 0.7007. For the quasi-completely separated data in Table 10.2, the p-value was 0.0043 with a median unbiased estimate 0.9878.

Despite the attractiveness of exact logistic regression, it is essential to emphasize that it is computationally feasible only for quite small samples. For example, I recently tried to estimate a model for 150 cases with a 2:1 split on the outcome variable and five explanatory variables. The standard version of `LogXact` was unable to handle the problem. An experimental version of `LogXact` using a Markov chain Monte Carlo method took three months of computation to produce the p-value to three decimal places for just one of the five explanatory variables.

10.6.5 Bayesian Estimation

In those situations where none of the preceding solutions is appropriate, a natural approach is to do Bayesian estimation with a prior distribution on the regression coefficients (Hsu and Leonard 1997; Kahn and Raftery 1996). In principle, this should be accomplished easily with widely available software such as `WinBUGS`. However, my very limited experience with this approach suggests that the results obtained are extremely sensitive to the choice of prior distribution.

10.6.6 Penalized Maximum Likelihood Estimation

As this chapter was nearing the final stages of the editorial process, I learned of a very promising new method for dealing with separation. Firth (1993) proposed the use of penalized maximum likelihood estimation to reduce bias in logistic regression in small samples. Heinze and Schemper (2002) have shown that this method always yields finite estimates of parameters under complete or quasi-complete separation. Their simulation results indicate that these estimates have relatively little bias, even under extreme conditions. In fact, the bias is appreciably less than that found for median unbiased estimates associated with exact logistic regression. Unlike exact logistic regression, penalized maximum likelihood is computationally feasible even for large samples.

Firth's procedure replaces the gradient vector $U(\beta)$ in equation (10.6) with

$$U(\beta*) = \sum_i \mathbf{x}_i y_i - \sum_i \mathbf{x}_i \hat{y}_i - \sum_i h_i \mathbf{x}_i (0.5 - \hat{y}_i),$$

where h_i is the ith diagonal element of the "hat" matrix and $\mathbf{W} = \text{diag}\{\hat{y}_i(1-\hat{y}_i)\}$. Once this replacement is made, the Newton–Raphson algorithm of equation (10.9) can proceed in the usual way. Standard errors are also calculated as usual by taking the square roots of the diagonal elements of $-\mathbf{I}(\beta*)^{-1}$. However, Heinze and Schemper (2002) point out that Wald tests based on the standard errors for variables causing separation can be highly inaccurate (as with conventional maximum likelihood). Instead, they recommend chi-square tests based on differences between penalized log likelihoods.

The penalized maximum likelihood method has been implemented in a macro for the SAS system (Heinze 1999) and a library for S-PLUS (Ploner 2001). Using the SAS macro, I estimated models for the data in Tables 10.1 and 10.2. For the data with complete separation (Table 10.1), the parameter estimate was 0.936, with a p-value of 0.0032 (based on a penalized likelihood ratio chi-square). For the data with quasi-complete separation (Table 10.2), the parameter estimate was 0.853 with a p-value of 0.0037. These estimates and p-values are similar to those reported in Section 10.6.4 using the exact method with median unbiased estimation.

10.7 SOLUTIONS FOR COMPLETE SEPARATION

It is fortunate that complete separation is less common than quasi-complete separation, because when it occurs, it is considerably more difficult to deal with. For example, it is not feasible to leave the problem variable in the model because that makes it impossible to get maximum likelihood estimates for any other variables; nor will combining categories for dummy variables solve the problem. Exact logistic regression may be useful with small samples and a single predictor variable causing complete separation. But it is not computationally feasible for larger samples and cannot produce coefficient estimates for any additional predictor variables. That is because the permutation distribution is degenerate when one conditions on a variable, causing complete separation. Bayesian estimation may be a feasible solution, but it does require an informative prior distribution on the problem parameters, and the results may be sensitive to the choice of that distribution.

The one approach that seems to work well in cases of complete separation is the penalized maximum likelihood method described in Section 10.6. As in the case of quasi-complete separation, this method produces finite estimates that

Table 10.4   Ordered Data Exhibiting Complete Separation

x	1	2	3	4	5	6	7	8	9	10	11	12
y	1	1	1	1	2	2	2	2	3	3	3	3

have good sampling properties, with only a modest increase in the computational burden over conventional maximum likelihood. Furthermore, unlike exact logistic regression, it yields estimates for all predictor variables, not just the one with complete separation.

10.8   EXTENSIONS

In this chapter we have focused entirely on problems of nonconvergence with binary logistic regression. But it is important to stress that complete and quasi-complete separation also lead to the nonexistence of maximum likelihood estimates under other "link" functions for binary outcome variables, including the probit model and the complementary log-log model. For the most part, software treatment of data with separation is the same with these link functions as with the logit link. The possible solutions that I described for the logistic model should also work for these alternative link functions, with one exception: The computation of exact p-values is available only for the logit link function.

Data separation can also occur for the unordered multinomial logit model; in fact, complete and quasi-complete separation were first defined in this more general setting (Albert and Anderson 1984). Separation problems can also occur for the ordered (cumulative) logit model, although to my knowledge, separation has not been defined rigorously for this model. Table 10.4 displays data with complete separation for a three-valued, ordered outcome variable.

These data could be modified to produce quasi-complete separation by adding a new observation with $x = 5$ and $y = 1$. PROC LOGISTIC in SAS correctly identifies both of these conditions and issues the same warning messages that we saw earlier.

CHAPTER 11

Recommendations for Replication and Accurate Analysis

11.1 GENERAL RECOMMENDATIONS FOR REPLICATION

The furthering of science depends on reproducible results regardless of the specific field. Without the ability to reproduce research, science is reduced to a matter of faith, faith that what is published is indeed the truth. Without the ability to reproduce research, we cannot "stand on the shoulders of giants" to further expand our knowledge of the world. We have noted throughout this book where research results may depend on the computational algorithms used in the analysis. Here, we discuss, as a way to summarize our findings, the implications for reproducing research.

A widely accepted component of publication is that research findings should be replicable and that references to data and analysis must be specific enough to support replication (Clubb and Austin 1985; Feigenbaum and Levy 1993; King 1997). Many journals in social science specifically require that sufficient detail be provided in publications to replicate published results.[1] Additionally, several government organizations, such as the National Science Foundation, require that data resulting from funded projects be made available to other users (Hildreth and Aborn 1985; Sieber 1991). Recently, in recognition of the role of social science analyses in public policy, and despite some protest in the academic community, the Freedom of Information Act was expanded to include research data produced using public funding (National Research Council 2002). Although a debate continues over whether data sharing should be mandated, all sides agree that some

---

[1]Information gathered from the "Contributors" section of the 1997 edition of many journals show that the exact terms of the requirement to share data vary in principle and in practice. *Science, Nature, American Economic Review, Journal of the American Statistical Association, Journal of the American Medical Association, Social Science Quarterly, Lancet, American Political Science Review*, and the various journals published by the American Psychological Association explicitly document in submission guidelines that a condition of publication is the provision of important data on which the article is based. The *Uniform Requirements for Manuscripts Submitted to Biomedical Journals* has, arguably, the weakest policy, stating simply that the editor may recommend that important data be made available.

Numerical Issues in Statistical Computing for the Social Scientist, by Micah Altman, Jeff Gill, and Michael P. McDonald
ISBN 0-471-23633-0   Copyright © 2004 John Wiley & Sons, Inc.

form of replaceability is essential, and at a minimum, a researcher should be able to reproduce published results. What is not widely realized is that even the simplest verifications of previous results are surprisingly difficult.

In research for this book, we reproduced over a half-dozen previously published research articles whose data are derived from public sources. We expected reproduction to be simple and straightforward, but we found that it is surprisingly difficult and were not always successful, even in consultation with the original authors. This is disquieting, especially in light of how infrequently replications are performed and the disincentives for seasoned scholars to perform them (Feigenbaum and Levy 1993).

Some troubles with replication lie in the publication process. Dewald et al. (1986) attempted to replicate a complete year's run of articles from the *Journal of Money, Credit and Banking* and found that inadvertent errors in published empirical articles are commonplace. We, too, have found examples of publication errors that prevented faithful replication. In one case we discovered that an entire table of results was the victim of a publication error (Altman and McDonald 2003). Undoubtedly there are errors in this book, so we should be careful to call the kettle black. But acceptance of a manuscript is not the end of the publication process; researchers should pay careful attention to proofing their publications to reduce the occurrence of publication errors.

Issues of statistical computing presented in this book suggest that even without publication errors, replication may fail. In Chapter 2 we discuss how differences in the procedures for computing even simple statistical quantities, such as the standard deviation, may lead to wildly different results. In Chapter 3 we demonstrate that as a consequence of inaccurate statistical computation, statistical packages may produce results that differ from the truth for descriptive statistics, analysis of variance, linear regression, and nonlinear models. In Chapters 5, 7, and 8 we show how complex statistical models are particularly sensitive to the computational procedures used to estimate them. In this chapter we discuss the concepts we have covered and place them in the context of reproduction of research.

11.1.1 Reproduction, Replication, and Verification

There are various ways of "reproducing" previous research. For clarity we adopt the nomenclature of the Committee on National Statistics (quoted by Herrnson 1995; p. 452): A researcher attempts to *replicate* a finding when she performs an identical analysis of the data used in the original article but taken from the original data sources. Replication may be a time-intensive process, so a researcher might instead *verify* published results by performing the same analysis on data provided by the original researcher. Once a finding has been reproduced, a researcher might wish to perform a *secondary analysis* of the original data by reanalyzing it using a different model or additional variables. To this we add *independent reproduction*, the process of performing the same analysis on theoretically similar data.

Table 11.1   Tasks Involved in Reproducing Research

	Recreating Data from Original Sources	Inputting Data into a Software Program	Analyzing Data Using Original Methodology
Replication	Yes	Yes	Yes
Verification	—	Yes	Yes
Secondary analysis	—	Yes	—
Independent reproduction	—	—	Yes

Table 11.1 presents a typology of three fundamental tasks that are performed in reproducing quantitative analysis: recreating these original data, loading these original data into a software package for analysis, and analyzing data using the original methodology. Surprisingly, reproduction of research is at risk for each of the three tasks.

11.1.2   Recreating Data

Data are not immutable. Sometimes data are revised when scoring errors, or errors in methodology are uncovered. Sometimes, but not always, data versions are documented. For example, a careful reading of the National Election Survey (NES) Cumulative Data File codebook reveals that the data file has undergone several revisions. Two major version changes stand out: interviews were dropped from the 1990 and 1992 panel studies because it is believed that an interviewer faked his 1990 interviews, and the 1996 NES was reweighted after its release because of an error discovered in the sample design for that year (Miller and the American National Election Studies 1999).

Data sets are stored in repositories or archives, such as the Inter-University Consortium for Political and Social Research (ICPSR) and are often available for electronic download. ICPSR maintains a publication-related archive where authors may store and make publicly available verification datasets. Data archives are invaluable for the preservation of research data and the replication of research. A *canonical data source* is the original, root data source. Many of the data sets available at ICPSR are canonical; others are redistributions of data available from other canonical sources, such as the U.S. Bureau of the Census. Ideally, a data repository should not only make available the most recent version of a canonical data source, but should further provide access to all previous versions. Unfortunately, typically this is not the current practice; ICPSR, for example, provides no access to previous versions of data.

The unavailability of previous versions of data may affect the ability of researchers to replicate research. During the course of our replications of Nagler (1994), presented in Chapter 3, we discovered that data contained in the 1984 Current Population Survey Voter Supplement File, published by the Bureau of the Census, has undergone undocumented version changes that made it impossible for even the original author to replicate the published research (Altman and

McDonald 2003). Fortunately, the author kept a private archive of these data, which made it possible to verify the results.[2]

The complexity of statistical data can create challenges for replication. Data structure, analysis units, aggregations, relationships among variables, and conditions of the measurement process can make appropriate use and reuse of data difficult (see David 1991; Robbin and Frost-Kumpf 1997). Use of data documentation standards such as the *Data Documentation Initiative Specification*, and systems that support them, such as the *Virtual Data Center* (Altman et al. 2001), can aid in later use and replication. (The Virtual Data Center is freely available from <http://thedata.org> and provides all the functionality necessary to create and maintain an on-line data archive.)

To guarantee verification and replication of research, researchers should document data, carefully using a documentation standard. We recommend that researchers archive a verification database either on their own or through a service such as ICPSR's publication-related archive. This step may be enough for verification but may not be sufficient for replication. Researchers should document carefully all the steps necessary to recreate their data. We have found that providing computer programs to construct a dataset is particularly helpful in verification and replication. We also recommend that when constructing a verification dataset, researchers include any unique case identifiers, such as respondent identification numbers, so that the cases of a verification file and a replication file may be compared.

11.1.3 Inputting Data

Surprisingly, inputting data into a statistical software package is not necessarily a straightforward process. Measurement error occurs when the precision level of a statistical software package is exceeded. Observations may be silently truncated when the storage capacity of observations and variables is exceeded. It is possible that users will identify variable locations and formats incorrectly. Any of these errors may prevent the reproduction of research.

As we explain in Chapter 2, computer representations of numbers do not necessarily match pencil-and-paper representations. Truncation and rounding at the precision level of the computer software may introduce small measurement errors; but for most ordinary analysis, these small errors are practically invisible to the researcher and are of not great concern. For more complex analysis, reading a double-precision variable into a program as a single-precision variables, or worse, as an integer, may prevent reproduction of research. Where possible, researchers creating a verification dataset should document precision levels of variables carefully, and researchers reproducing research should note precision levels carefully when inputting data into their statistical software program.

Statistical software programs may also truncate observations or variables (rows or columns) of large datasets. Programs either have hard limitations on the number

---

[2]Ironically, the substantive results of the research were overturned when estimating a model using the verification dataset and the most recent version of the software used in the analysis. We exactly replicated the published results closely but not using the most of these data and of the program.

of records and columns they can handle or are simply limited by system resources. In Chapter 3 we read test data into statistical software packages and noted where programs encountered difficulty in reproducing data internally. We found that although storage limitations are usually documented, when the limitations are reached statistical programs will sometimes truncate columns or rows silently without notifying the user. A missing variable may be obvious, but a missing observation may go undetected, which could, in turn, affect reproduction of research.

Common programming errors occur when incorrectly identifying where a column for a variable begins and ends, using the wrong delimiter when reading in delimited data, or specifying the wrong data format when reading formatted data. Care should be taken in programming, especially when relying on a statistical software program's "wizard" to guess at variable locations and formats. Sometimes misspecified data locations and formats will cause a statistical program to produce error messages. Sometimes summary statistics will reveal errors. But this is not always the case.

We recommend that when inputting a dataset into a statistical software program, researchers at least run simple diagnostics, such as a generating summary statistics, to verify that data have been read in properly. Supplying means, frequencies of values, and the number of observations of variables in a verification dataset will aid those who reproduce research to verify that data have indeed been read in properly when they create diagnostic summary statistics. Such methods of verification are weak, however. Thus we recommend that when researchers produce data to archive or share with others, they use the data-level CRC methods described in Chapter 3 to ensure the integrity of the data with respect to data input.

11.1.4 Analyzing Data

Once data are loaded into a statistics program, a researcher is ready to attempt to reproduce research. There are two aspects of recreating a model, specifying and implementing a model. Both have pitfalls that may be avoided with some prudent foresight.

11.1.4.1 Specifying a Model

The first step of analysis is to interpret correctly the model used by the researcher. For familiar models such as linear regression, perhaps all that is needed is to list the components of the model, in this case the explanatory and outcome variables. For more complex models, such as those employing maximum likelihood, a researcher attempting to reproduce research must know the functional form of the model and exactly how the model was applied to the data (see Chapter 3).

When researchers create datasets intended to support verification, they often neglect to document the answers to such questions as: How was the model specified in the relevant statistical programming language? Which variables from the data were entered into the model, and how were these data coded or recoded? What are the rules for selecting observations for analysis? Archiving the programming code used to run the analysis is a positive step toward reproducibility, but this may not be enough. Statistical software programs may differ in their syntax and in subtle but important details, such as the precedence of operators or the exact specification of parameters to statistical functions. These idiosyncrasies may raise barriers against researchers who use software different from that of the original author. Even if the same software is used in the verification, the syntax of the statistical programming language may change over time. Mechanisms to allow a statistical software program to declare a legacy syntax and operate under it are rare.[3]

We recommend that researchers who create a verification dataset provide the programs used to generate the research. But as this may not be enough, we recommend that researchers explicitly write out the model being estimated, using parentheses and brackets where necessary to identify precedence clearly. Since syntax for variable transformation and case selection may be idiosyncratic to the version of the statistical software program, we recommend that researchers document these steps of their analysis.

11.1.4.2 *Implementation*

There are often many ways to solve the same problem, and there is no guarantee that algorithms to solve a particular problem are the same across statistical software packages. Nor do programs that implement the same algorithms necessarily implement them in an identical manner. Methods for computing statistical distributions, generating random numbers, and finding the optima of nonlinear functions may vary across statistical packages and programming languages, and as a consequence may produce varied results (see Chapter 3).

Many packages offer users a variety of options, and these options and their defaults will not generally be consistent across statistical packages. Options may govern the choice of algorithm, implementation, or aspects of both. Even familiar procedures such as regression may offer implementation-specific options; for example, the *regress* command in Stata (v 6.0) has four options for calculating the standard error of the coefficients. Maximum likelihood algorithms have even more options. There are over 400 different combinations of algorithmic options and over a dozen continuous parameters that can be used to control the behavior of the MLE solver in Gauss (v 4.0). Since these options can affect results, any option used in the estimation, even the default, should be noted in the documentation accompany the replication dataset.

Optimization search algorithms typically simply climb "uphill" until a maximum is found. They are prone to stop when they find a local optimum of

---

[3]Stata is a notable exception. The Stata "version" command allows the programmer to declare that the program should be run with a particular version of the Stata language syntax *and* implementation. For example, although Stata upgraded its random number generators in v6.0, the old PRNG algorithms were preserved and are still used when a program explicitly declares itself to use v5.0 syntax. The programming language PERL supports a more common and limited version of version declaration. A PERL program may require a particular version of the language to run. However, in this case, the designers of PERL assumed that the requirements were always minimum requirements and that any version at newer than the stated requirement is also accepted.

multiple-peaked likelihood functions or even fail to converge altogether (see Chapters 2, 4, and 8). The reported local optimum may depend on where the search algorithm started its uphill climb. Researchers should provide the vector of starting coefficients used find their published solution, even if they use the default, because defaults may not be consistent across software packages.

Monte Carlo analysis has its own set of consideration for reproduction of research. There are many types of pseudorandom number generators (PRNGs) that are known to differ in their performance (see Chapter 2 for details), and some are known to work rather poorly. Poor choices of generators can, as shown in Chapter 5, limit the accuracy of the simulation, or worse, yield biased results. Further, because PRNGs are deterministic functions, the output can be recreated exactly by using the same seed value. Thus, we recommend that the documentation accompanying Monte Carlo analysis use a high-quality PRNG and report the generator and initial seed values used.

11.2 RECOMMENDATIONS FOR PRODUCING VERIFIABLE RESULTS

Some guidelines are needed to aid researchers in creating replication datasets and accompanying documentation. Our experiences with testing the StRD benchmarks, which are essentially a series of replications, and our experiences with replicating two published sources, provide some lessons that researchers can use to create their own replication datasets. We summarize them in a checklist of replication guidelines:

1. The model being estimated should be written out explicitly using standard mathematical notation and using parentheses and brackets where necessary to identify precedence clearly.
2. The original data that produced the results should be made available. If the data are drawn from other sources, such as data found at the ICPSR archive, these sources should be noted.
3. All variables constructed from nonarchived data should be accompanied by a list of rules used to create the variables.
4. If possible, each observation in the replication dataset should include a unique identifier. If a unique identifier is available in a dataset used to construct the replication dataset, the unique identifier from the original dataset should be provided.
5. The precision of variables used in the analysis should be noted. The choice between single- and double-precision storage can affect results because floating point numbers do not have a true binary representation.
6. Any transformations made to the data should be documented. This includes selection criteria used to subset the data, any recalculation of variables, the construction of any new variables, and aggregation.

7. Descriptive statistics should be made available to help researchers ascertain the integrity of the data as read into their own programs. We recommend that a standard of calculating and reporting cyclical redundancy checks on the internal data matrix be applied to ensure the integrity of the data.

8. The statistical software and operating system and hardware platform used to analyze the data should be noted, including the version number and the version numbers of any accompanying libraries used in the estimation.

9. Note for maximum likelihood estimation (and similar problems): (a) the selection of algorithms (b) the values for all the control parameters used (even if these are the default values in the current version), and (c) starting values for the coefficient vector.

10. Where random numbers are used in the analysis, (a) note which algorithm was used to generate the random sequence, and (b) provide the seed value used.

11. Report the method for calculating the standard errors (or other measures of uncertainty), including the method for calculating the variance–covariance matrix if this was used in calculation of the standard errors.

11.3 GENERAL RECOMMENDATIONS FOR IMPROVING THE NUMERIC ACCURACY OF ANALYSIS

The recommendations above are a guide to reproducing research, but what can researchers do to ensure that their results are accurate before publication? For commonly used statistical analysis, choosing a statistically accurate software package and applying common sense in interpreting the results and standard diagnostics is probably enough. In other research, especially that involving complex maximum likelihood estimation or simulation, it may be necessary to repeat certain portions with different techniques and to track down all the implications of particular methods. Here we provide recommendations to improve numerical accuracy. We start with the steps that we believe to be most generally applicable, and finish with steps geared toward numerically intensive models:

1. *Trust, but verify.* Most packages correctly solve routine statistical analyses, such as univariate statistics, linear regression, and many simple nonlinear regression models. Choose software that is regularly updated, uses well-documented algorithms, and buttresses itself with standard tests of accuracy and reliability. A number of packages now publish the results of numerical benchmarks in their documentation, and benchmarks of accuracy are becoming a more regular part of statistical software reviews. Be wary of software packages that are not tested regularly for accuracy or that do not perform well.

2. *Check for substantive and statistical plausibility.* Examine results to ensure that they make sense and fit these data. Check for evidence of misspecification, outliers, or other signs of measurement or coding errors. These precautions have

two benefits. First, by checking for substantive and statistical plausibility, one is less likely to be misled if numerical accuracy affects the model. Second, if estimation is susceptible to numerical inaccuracy, cleaning data and considering respecification of the model may eliminate numerical inaccuracy and improve the overall analysis.

3. *Test complex or problematic models for accuracy.* In problematic cases where a model gives implausible results or where complex models or simulations are used, especially for the first time, researchers should test explicitly for numerical accuracy. Ideally, when implementing a new model, a researcher should create a comprehensive set of tests whose answers are known. Sometimes these tests may be generated through Monte Carlo simulation by solving the problem in closed form or by using multiple-precision arithmetic in programs such as `Maple` or `Mathematica`.

4. *Apply sensitivity tests to complex models.* Indications of numerical inaccuracy for a model may be found in the sensitivity of results to data perturbations, option variation, and variations in software and platforms used in estimation.

5. *Choose an optimization algorithm suited to your estimation problem.* Computational problems are often exacerbated by attempting to force a recalcitrant model into a limited statistical software application. Seek out documentation of the algorithms supplied by your statistical package. If your problem exceeds the capabilities of the software, seek a more appropriate algorithm rather than modifying your model. (See Chapter 4 for a discussion of optimization algorithms.)

6. *Incorporate all sources of uncertainty in your estimates.* The most common computational techniques for generating point estimates and confidence regions assume a single mode of likelihood that is approximately locally quadratic. Nonquadratic and/or multiple modes violate these assumptions and can cause reported confidence intervals to be much too narrow. We strongly recommend that rather than simply reporting standard errors, one use bootstrapping and/or highest-posterior-density regions to provide full information about the distribution of the posterior. Moreover, computational implementation selection is analogous to model selection, although generally a milder problem. Although every effort should be made to find an accurate solution, multiple plausible computational strategies will sometimes continue to yield different results. In this case one should summarize the uncertainty that stems from the continued existence of these competing computational solutions. Bayesian model averaging and Leamer bounds, both discussed in Chapter 4, are two techniques for doing just that.

11.4 RECOMMENDATIONS FOR PARTICULAR STATISTICAL MODELS

11.4.1 Nonlinear Least Squares and Maximum Likelihood

Even with infinite precision, nonlinear optimization algorithms may fail to find an optimum (see Chapter 2). The likelihood function for common statistical models may have multiple optima (Chapters 4 and 8) or no optimum at all (Chapter 10), either as a consequence of the functional form of the equation to be solved or as a consequence of numerical issues.

There are a number of methods to find an optimum of a nonlinear problem, and as we have seen repeatedly, two implementations of even the same algorithms may find different solutions to the same problem. One should choose a statistical package that allows one to vary optimization algorithms, supply analytical derivatives, and otherwise control the nonlinear and maximum likelihood solver, such as modifying convergence criteria and starting values. We recommend using analytic derivatives where possible, and where not, using central differences rather than forward differences for numerical derivatives, as central differences are generally more accurate (see Chapter 8).

How can we one be confident that a local optimum is the global optimum, or even if it is an optimum at all? In Chapter 4 we provide tests that estimate the probability that a local optimum is the global optimum, given the number of local optima identified. In Chapter 8 we recommend four steps that may be taken to diagnose if an optimum identified by a solver is a false convergence point (i.e., not a true optimum):

1. Verify that the gradient is indeed zero while keeping in mind that this may be a function of the scaling of the data.
2. Examine the trace or note how the solver converged. If the solver converged much too slowly relative to the expected algorithmic rate of convergence, it is likely that the function is in a flat region of the objective function and may have stopped short of the true solution.
3. Check the Hessian for positive definiteness. (These flat regions may also pose problems to algorithms that invert the Hessian, a topic we describe in the next section; checking the Hessian for positive definiteness will diagnose this problem also.)
4. Profile the objective function, as this information provides the shape of the function around the estimated solution.

If multiple optima are identified, and these modes compete as the global optimum, or the shape of the likelihood function is nonsymmetric around the optimum, we recommend in Chapter 4 a method of summarizing the highest posterior density (HPD) to describe the uncertainty of the estimates.

11.4.2 Robust Hessian Inversion

Every practitioner has experienced the frustration of an estimation being aborted because the Hessian was not invertible (sometimes whether they knew the cause or not). Noninvertible Hessians may signal limitations in data or in numerical methods, and it is advisable to be on one's guard for both. The alternatives have not generally been clear, and the standard recommendation of respecification is often unsatisfactory on theoretical grounds.

Since the Hessian can be noninvertible for both specification reasons and data reasons, it is important to have several alternatives for dealing with this problem. Some strategies we have recommended include:

- Consider a more accurate implementation of the likelihood by identifying actual or potentially degenerate regions of the sample space (see Chapters 2 and 3).
- Use more accurate libraries for statistical distributions used in the likelihood. These include some of the standard references and those given in the SIAM libraries (again, see Chapters 2 and 3).
- Specify the analytical derivative such that the gradient function is part of the software's estimation process (this is discussed in Chapter 8).
- Change the optimization algorithm (across a set of algorithms suited to the problem) and observe whether or not it makes a difference. This may be a sign of numerical instability or it may be a sign of problems with the specified parametric form (see Chapters 9 and 10).

If these methods aren't feasible or don't work, which often happens, we provide an innovative new library for doing generalized inverses. This computes an invertible approximation to the Hessian that is guaranteed to be invertible. This can produce estimates when everything else fails and the results are informative. The standard cautions still apply to interpreting the results, and you should follow examine diagnostics carefully and use sensitivity tests as described in Chapters 4 and 8. We also strongly recommend that in addition to reporting standard errors you describe all relevant portions of the posterior distribution of parameters using HPDs (Chapter 4).

11.4.3 MCMC Estimation

In some ways Markov chain Monte Carlo estimation is distinct as an estimation process because it uses a nonindependent sequence of user-generated values according to some transition kernel. The two primary reliability concerns, which are entirely unique to MCMC work in this context, are mixing and convergence. Convergence is the state whereby the chain has reached its stationary distribution such that all future values are samples from the target distribution. Therefore, until the Markov chain has converged, the generated values are not useful for estimation purposes. Mixing refers to the rate with which the chain moves through the support of the posterior, which we care about after convergence so as to establish that the empirical quantities used for inference are adequately representative of the actual posterior. These are two very serious concerns, and a lack of appropriate attention by the researcher is almost guaranteed to produce nonsensical results.

In other ways, Markov chain Monte Carlo estimation depends on numerical concerns that affect other, more familiar estimation processes. The quality of the underlying pseudorandom number generator is critical, as demonstrated in Chapter 5. Chapters 2 and 3 discuss how to select generators, such as the

Mersenne twister, that have extremely long periods and good distributional properties, and how to generate *truly random* seeds automatically for these generators
using physical sources of randomness.

It is important to run the chain for a reasonably long period of time, just as
in standard Monte Carlo simulation. Also, as we have said many times over the
preceding pages, it is important to evaluate and consider carefully default values
provided by the software. These include the starting values, the random number
seed, and various tuning parameters, such as the form of the candidate-generating
distribution in the Metropolis–Hastings algorithm and its descendants.

In this volume we have clearly focused on the second set of problems: Those
that affect MCMC work also affect other numerical estimation methods. This is
primarily because the problems of mixing and convergence are very well studied
in the MCMC literature. This work can be segmented into the following typology:

- **Analytical determination.** There is a lengthy and at times frustrated literature that seeks to use statistical and general probabilistic theory to bound
  convergence times of chains built on specific transition kernels. The goal is
  to determine in advance the total variation distance to the target distribution
  with some specified tolerance. Fortunately, all standard MCMC algorithms
  converge at a rate related to how close the second-largest eigenvalue of the
  transition matrix (discrete state spaces) or kernel density (continuous state
  spaces) is to 1. So sometimes this is a very productive enterprise; see Amit
  (1991, 1996), Frigessi *et al.* (1993), Rosenthal (1993, 1995a,b), Frieze, et al.
  (1994), Ingrassia (1994), Meyn and Tweedie (1994b), Roberts and Polson
  (1994), and Roberts and Smith (1994).

- **Empirical diagnostics.** The most common way to worry about convergence is to use one of many diagnostics that evaluate some property of
  an observed Markov chain and make inferential claim about a lack of
  convergence. That is, these diagnostics typically look for evidence of nonconvergence rather than convergence, in much the same way that hypothesis
  testing makes assertions about a lack of statistical reliability. The well-
  known variants are given in Gill (2002), various essays in Gilks et al.
  (1996), Gamerman (1997), and Carlin and Louis (2000).

- **Perfect sampling.** The revolutionary approach of Propp and Wilson (1996)
  is still emerging as a feasible general method for setting up MCMC estimation. The key idea is to use "coupling from the past" to produce a sample
  from the exact stationary distribution. Currently, this idea works well only
  for discrete state spaces, but there is a lot of working going on at this time
  to generalize the principle. See the recent works by Häggström (2000),
  Häggström and Nelander (1999), Hobert et al. (1999), Kendall and Moller
  (2000), Casella et al. (2001), and Corcoran and Tweedie (2002).

- **Mixture-improving algorithms.** Mixing problems can be discouraging.
  High correlation *between* the parameters of a chain tends to give slow
  convergence, whereas high correlation within a single-parameter (autocorrelation) chain leads to slow mixing and possibly individual *nonconvergence*

to the limiting distribution because the chain will tend to explore less space in finite time. Chains that mix poorly take longer to converge, and once they have converged take longer to fully describe the stationary distribution. Some solutions involve adding instrumental variables such as the slice sampler described in Chapter 5. Useful references include Asmussen et al. (1992), Gilks and Roberts (1996), and Brooks and Roberts (1999).

This is a very active research area with a steady stream of new developments.

The second set of problems are the focus of Chapter 5. Our primary concern there was the effect of necessarily periodic pseudorandom number generation on ergodicity. We provide evidence for mild concern in models with many parameters that are believed to require a very large number of iterations. Perhaps of greater everyday concern is the possibility of absorbing states that result from inaccuracies in the parametric form of the posterior distribution. In cases with relatively complex mathematical forms, it seems warranted to spend the time to determine analytically whether or not a Markov chain can wander into such regions. Naturally, this task becomes more complex with higher dimensions.

Perhaps our strongest concern here is with the poor performance of default pseudorandom number generators in common implementations of C, C++, and FORTRAN libraries. We recommend that researchers never use the standard furnished random number procedures, but instead, expend the relatively little effort required to download, test, and use a more sophisticated and reliable alternative.

We noted as well that problems from the pseudo-randomness of generated values do not affect MCMC procedures alone. In fact, regular Monte Carlo simulation for calculation of volumes and statistics are also sensitive to poor performance of the random number generating mechanism, albeit in a different manner.

11.4.4 Logistic Regression

Logistic regression (or *logit*) is used with great frequency in sociology and other branches of social science. These models are almost always estimated through MLE. Despite its ubiquity, many practitioners are unaware that MLE approaches to logit models need not have a solution: There are many cases in which the likelihood function for the logit model has no optimum. For example, for any dichotomous explanatory variable in a logistic regression, if there is a zero in the 2×2 table formed by that variable and the outcome variable, the ML estimate for the regression coefficient will not exist. (This can occur when there are extreme splits on the frequency distribution of either the outcome or explanatory variables.)

Crude stopping rules, still used by some statistics packages, can cause convergence to be falsely reported in this case, although the estimates returned are, in fact, meaningless. Even within the same statistics package, one routine for computing logit may give appropriate warnings while another reports false convergence. In Chapter 10 we describe how to detect false convergence in logit and how to deal with this (e.g., by combining categories, using computing logistic regression exactly, or using penalized maximum likelihood estimation).

11.4.5   Spatial Regression

With the progress of geographical information system (GIS) technology, large samples of socioeconomic demographic information based on spatial locations such as census tracts and zip-code areas has become available. In Chapter 9 we discuss computational issues associated with regression modeling combined demographic and spatial data of information.

Both MLE and Bayesian estimation methods can be applied to spatial regression models. Bayesian methods are preferable because they can be used to relax the assumption of normally distributed constant variance disturbances made by maximum likelihood methods.

Estimating these models is greatly aided by the use of sparse matrix algorithms. There are, however, trade-offs between computational speed and accuracy, comprising use of approximations, tolerance values, and bounding parameter values during optimization. Methods and software are discussed in Chapter 9.

11.5   WHERE DO WE GO FROM HERE?

We have focused substantially on the negative consequences of ignoring numerical issues in statistical computing. When things go amiss, it is sometimes obvious: Hessians do not invert, Markov chains do not converge, or maximum likelihood software gives ridiculous results. In these cases, if one knew nothing about numerical methods, it would be necessary either to change the model (which is not preferred, because this involves retheorizing the substantive problem, and one usually has a specific theory to test) or to get more data (which is not preferred, because it is often expensive and time consuming, or often even impossible). An understanding of numerical methods provides another, better option—fix the real problem, or at least gain insights into it.

One hundred years ago, researchers could never have imagined the advent of computing technology or the revolution it would bring to the field of statistics. One hundred years hence, we imagine that researchers looking back at the methods discussed in this book will find them "quaint." But even a century ago, researchers such as Duhem (1906) were concerned with a timeless question: What are the limits of the instruments we use to measure and form inferences about the world we study?

Knowledge of the limitations of our research tools helps us avoid making incorrect inferences. However, often this knowledge is inaccessible to social scientists because it is specific to statisticians or computer scientists. We hope that this book bridges the gap not only among these disciplines, but also among the many disciplines within the social sciences that commonly use statistical software.

Bibliography

Aarts, E. H. L. and Lenstra, J. K. (eds.) (1997). *Local Search in Combinatoric Optimization*. New York: John Wiley & Sons.

Achen, C. H. (1996). Implicit Substantive Assumptions Underlying the Generalized Event Count Estimator. *Political Analysis* **6**, 155–74.

Achen, C. H. (2003). Toward a New Political Methodology: Microfoundations and ART. *Annual Review of Political Science*, **5**, 423–450.

Acton, F. S. (1970). *Numerical Methods That Work.* New York: Harper & Row.

Acton, F. S. (1996). *Real Computing Made Real.* Princeton, NJ: Princeton University Press.

Ades, A. and Di Tella, R. (1997). National Champions and Corruption: Some Unpleasant Interventionist Arithmetic. *Economic Journal* **107**, 1023–42.

Adolph, C. and King, G. (2003). Analyzing Second-Stage Ecological Regressions: Comment on Herron and Shotts. *Political Analysis* **11**, 65–76.

Akhand, H. A. (1998). Central Bank Independence and Growth: A Sensitivity Analysis. *Canadian Journal of Economics* **31**, 303–17.

Albert, A. (1973). The Gauss–Markov Theorem for Regression Models with Possible Singular Covariances. *SIAM Journal on Applied Mathematics* **24**, 182–87.

Albert, A. and Anderson, J. A. (1984). On the Existence of Maximum Likelihood Estimates in Logistic Regression Models. *Biometrika* **71**, 1–10.

Altman, M. (1998). Modeling the Effect of Mandatory District Compactness on Partisan Gerrymanders. *Political Geography* **17**, 989–1012.

Altman, M. and McDonald, M. P. (2001). Choosing Reliable Statistical Software. *PS: Political Science and Politics* **XXXIV**, 681–87.

Altman, M. and McDonald, M. P. (2003). Replication with Attention to Numerical Accuracy. *Political Analysis* **11**, 302–307.

Altman, M., Andreev, L., Diggory, M., King, G., Sone, A., Verba, S., Kiskis, D. L., and Krot, M. (2001). A Digital Library for the Dissemination and Replication of Quantitative Social Science Research. *Social Science Computer Review* **19**(4), 458–71.

Numerical Issues in Statistical Computing for the Social Scientist, by Micah Altman, Jeff Gill, and Michael P. McDonald
ISBN 0-471-23633-0   Copyright © 2004 John Wiley & Sons, Inc.

Amato, T. W. (1996). On Difference Equations, Probability Models and the "Generalized Event Count" Estimator. *Political Analysis* **6**, 175–212.

Amemiya, T. (1981). Qualitative Response Models: A Survey. *Journal of Economic Literature* **19**, 1483–536.

Amemiya, T. (1985). *Advanced Econometrics*. Cambridge, MA: Harvard University Press.

Amemiya, T. (1994). *Introduction to Statistics and Econometrics*. Cambridge, MA: Harvard University Press.

Amit, Y. (1991). On Rates of Convergence of Stochastic Relaxation for Gaussian and Non-Gaussian Distributions. *Journal of Multivariate Analysis* **38**, 82–99.

Amit, Y. (1996). Convergence Properties of the Gibbs Sampler for Perturbations of Gaussians. *Annals of Statistics* **24**, 122–40.

Anderson, J. E. and Louis, T. A. (1996). Generating Pseudo-random Variables from Mixture Models by Exemplary Sampling. *Journal of Statistical Computation and Simulation* **54**, 45–53.

Andrews, D. F. (1974). A Robust Method for Multiple Linear Regression. *Technometrics* **16**, 523–31.

Andrews, D. F., Bickel, P. J., Hampel, F. R., Huber, P. J., Rogers, W. H., and Tukey, J. W. (1972). *Robust Estimates of Location*. Princeton, NJ: Princeton University Press.

Anselin, L. (1988). *Spatial Econometrics: Methods and Models*. Dordrecht, The Netherlands: Kluwer Academic Publishers.

ANSI (American National Standards Institute and Computer and Business Equipment Manufacturers Association) (1992). *American National Standard for Programming Language, FORTRAN, Extended*. ANSI X3.198-1992: ISO/IEC 1539: 1991(E). New York: American National Standards Institute.

Anton, H. and Rorres, C. (2000). *Elementary Linear Algebra*, 8th ed. New York: John Wiley & Sons.

Asmussen, S., Glynn, P. W., and Thorisson, H. (1992). Stationary Detection in the Initial Transient Problem. *ACM Transactions on Modeling and Computer Simulation* **2**, 130–57.

Aptech Systems, Inc. (1999). *Maximum Likelihood (Application Module) Software Manual*. Maple Valley, WA: Aptech Systems, Inc.

Athreya, K. B. and Ney, P. (1978). A New Approach to the Limit Theory of Recurrent Markov Chains. *Transactions of the American Mathematical Society* **245**, 493–501.

Athreya, K. B., Doss, H., and Sethuraman, J. (1996). On the Convergence of the Markov Chain Simulation Method. *Annals of Statistics* **24**, 69–100.

Atkinson, A. C. (1980). Tests of Pseudo-random Numbers. *Applied Statistics* **29**, 164–71.

Axler, S. J. (1997). *Linear Algebra Done Right*. New York: Springer-Verlag.

Bailey, D. H. (1993). Algorithm 719: Multiprecision Translation and Execution of FORTRAN Programs. *ACM Transactions on Mathematical Software* **19**, 288–319.

Bailey, D. H., Krasny, R. and Pelz, R. (1933). Multiple Precision, Multiple Vortex Sheet Roll-Up Computation. In *Proceedings of the Sixth SIAM Conference on Parallel Processing for Scientific Computing*, Vol. 1. Philadelphia, PA: SIAM Press. pp. 52–6.

Barnett, V. (1973). *Comparative Statistical Inference*. New York: John Wiley & Sons.

Barnett, V. and Lewis, T. (1978). *Outliers in Statistical Data*. New York: John Wiley & Sons.

Barry, R. and Pace, R. K. (1999). A Monte Carlo Estimator of the Log Determinant of Large Sparse Matrices. *Linear Algebra and Its Applications* **289**, 41–54.

Bartels, L. M. (1997). Specification Uncertainty and Model Averaging. *American Journal of Political Science* **41**, 641–74.

Bates, D. M. and Watts, D. G. (1988). *Nonlinear Regression Analysis and Its Applications*. New York: John Wiley & Sons.

Bauens, W. and Richards, J. F. (1985). A 1-1 Poly-*t* Random Variable Generator with Application to Monte Carlo Integration. *Journal of Econometrics* **29**, 19–46.

Bavaud, F. (1998). Models for Spatial Weights: A Systematic Look. *Geographical Analysis* **30**, 153–71.

Bazaraa, M., Sherali, H., and Shetty, C. (1993). *Nonlinear Programming: Theory and Algorithms, 2nd Edition*. New York: Wiley & Sons.

Beaton, A. E., Rubin, D. B., and Barone, J. L. (1976). The Acceptability of Regression Solutions: Another Look at Computational Accuracy. *Journal of the American Statistical Association* **71**, 158–68.

Beck, N., King, G., and Zeng, L. (2000). Improving Quantitative Studies of International Conflict: A Conjecture. *American Political Science Review* **94**, 21–36.

Belsley, D. A., Kuh, E., and Welsch, R. E. (1980). *Regression Diagnostics*. New York: John Wiley & Sons.

Berger, J. O. (1976). Admissible Minimax Estimation of a Multivariate Normal Mean with Arbitrary Quadratic Loss. *Annals of Statistics* **4**, 223–26.

Berger, J. O. (1984). The Robust Bayesian Viewpoint. In *Robustness of Bayesian Analysis*, J. B. Kadane (ed.). Amsterdam: North-Holland, pp. 63–144.

Berger, J. O. (1985). *Statistical Decision Theory and Bayesian Analysis*, 2nd ed. New York: Springer-Verlag.

Berger, J. O. (1990). Robust Bayesian Analysis: Sensitivity to the Prior. *Journal of Statistical Planning and Inference* **25**, 303–28.

Berger, J. and Berliner, L. M. (1986). Robust Bayes and Empirical Bayes Analysis with ϵ-Contaminated Priors. *Annals of Statistics* **14**, 461–86.

Berger, J. O. and O'Hagan, A. (1988). Ranges of Posterior Probabilities for Unimodal Priors with Specified Quantiles. In *Bayesian Statistics 3*, J. M. Bernardo, M. H. DeGroot, D. V. Lindley, and A. F. M. Smith (eds.). Oxford: Oxford University Press, pp. 45–65.

Bernardo, J. M. and Smith, A. F. M. (1994). *Bayesian Theory*. New York: John Wiley & Sons.

Berndt, E., Hall, R., Hall, B. and Hausman, J. (1974). Estimation and Inference in Nonlinear Structural Models. *Annals of Economic and Social Measurement* **3/4**, 653–65.

Bickel, P. J. and Freedman, D. A. (1981). Some Asymptotic Theory for the Bootstrap. *Annals of Statistics* **9**, 1196–1217.

Billingsley, P. (1995). *Probability and Measure*, 3rd ed. New York: John Wiley & Sons.

Binstock, A. and Rex, J. (1995). *Practical Algorithms for Programmers*. Boston, MA: Addison-Wesely.

Birnbaum, A. (1962). On the Foundations of Statistical Inference. *Journal of the American Statistical Association* **57**, 269–306.

Blalock, H. (1979). *Social Statistics*. New York: McGraw-Hill.

Bleau, B. L. (1994). *Forgotten Calculus: A Refresher Course with Applications to Economics and Business*. New York: Barrons Educational Series.

Blum, L., Blum, M., and Shub, M. (1986). A Simple Unpredictable Pseudo-random Number Generator. *SIAM Journal on Computing* **15**, 364–83.

Boisvert, R. F., Howe, S. E., and Kahaner, D. K. (1985). GAMS: A Framework for the Management of Scientific Software. *ACM Transactions on Mathematical Software* **11**, 313–55.

Bose, S. (1994a). Bayesian Robustness with Mixture Classes of Priors. *Annals of Statistics* **22**, 652–67.

Bose, S. (1994b). Bayesian Robustness with More Than One Class of Contaminations. *Journal of Statistical and Inference* **40**, 177–87.

Box, G. E. P. and Jenkins, G. (1976). *Time Series Analysis: Forecasting and Control*. San Francisco: Holden-Day.

Box, G. E. P. and Müller, M. E. (1958). A Note on Generation of Normal Deviates. *Annals of Mathematical Statistics* **28**, 610–11.

Box, G. E. P. and Tiao, G. C. (1973). *Bayesian Inference in Statistical Analysis*. New York: John Wiley & Sons.

Box, G. E. P., Hunter, W. G., and Hunter, J. S. (1978). *Statistics for Experiments: An Introduction to Design, Data Analysis, and Model Building*. New York: John Wiley & Sons.

Brémaud, P. (1999). *Markov Chains: Gibbs Fields, Monte Carlo Simulation, and Queues*. New York: Springer-Verlag.

Brent, R. P. (1978). ALGORITHM 524MP, a Fortran Multiple Precision Arithmetic Package. *ACM Transactions on Mathematical Software*.

Brockett, P. L., Cooper, W. W., Golden, L. L., and Xia, X. (1997). A Case Study in Applying Neural Networks to Predicting Insolvency for Property and Casualty Insurers. *Journal of the Operational Research Society* **48**, 1153–62.

Brooks, S. P. and Morgan, J. T. (1995). Optimisation Using Simulated Annealing. *The Statistician* **44**, 241–57.

Brooks, S. P. and Roberts, G. O. (1999). Convergence Assessment Techniques for Markov Chain Monte Carlo. *Statistics and Computing* **8**, 319–35.

Brown, P. J. (1982). Multivariate Calibration. *Journal of the Royal Statistical Society, Series B* **44**, 287–321.

Brown, P. and Gould, J. (1987). Experimental Study of People Creating Spreadsheets. *ACM Transactions on Office Information Systems* **3**, 258–72.

Brown, P. J. and Oman, S. D. (1991). Double Points in Nonlinear Calibration. *Biometrika* **78**, 33–43.

Brown, B. W., Gutierrez, D., Lovato, J., Spears, M., and Venier, J. (1998). DSTATTAB v 1.1. (computer program). Houston: University of Texas.

Bunch, D. S., Gay, D. M., and Welsch, R. E. (1993). Algorithm 717: Subroutines for Maximum Likelihood and Quasi-likelihood Estimation of Parameters in Nonlinear Regression Models. *ACM Transactions on Mathematical Software* **19**, 109–30.

Bunday, B. D. and Kiri, V. A. (1987). Maximum Likelihood Estimation: Practical Merits of Variable Metric Optimization Methods. *Statistician* **36**, 349–55.

Buonaccorsi, J. P. (1996). Measurement Error in the Response in the General Linear Model. *Journal of the American Statistical Association* **91**, 633–42.

Burden, B. and Kimball, D. C. (1998). A New Approach to the Study of Ticket Splitting. *American Political Science Review* **92**, 533–44.

Burr, I. (1942). Cumulative Frequency Functions. *Annals of Mathematical Statistics* **13**, 215–32.

Butcher, J. C. (1961). A Partition Test for Pseudo-random Numbers. *Mathematics of Computation* **15**, 198–99.

Cameron, A. C. and Johansson, P. (1997). Count Data Regression Using Series Expansions: With Applications. *Journal of Applied Econometrics* **12**, 203–23.

Cameron, A. and Trivedi, P. (1986). Econometric Models Based on Count Data: Comparisons and Applications of Some Estimators and Tests. *Journal of Applied Econometrics* **1**, 29–54.

Cameron, A. and Trivedi, P. (1990). Regression Based Tests for Overdispersion in the Poisson Model. *Journal of Econometrics* **46**, 347–64.

Campbell, S. L. and Meyer, C. D., Jr. (1979). *Generalized Inverses of Linear Transformations.* New York: Dover Publications.

Cao, G. and West, M. (1996). Practical Bayesian Inference Using Mixtures of Mixtures. *Biometrics* **52**, 1334–41.

Carley, K. M. and Svoboda, D. M. (1996). Modeling Organizational Adaptation as a Simulated Annealing Process. *Sociological Methods and Research* **25**, 138–48.

Carlin, B. P. and Louis, T. A. (2000). *Bayes and Empirical Bayes Methods for Data Analysis*, 2nd ed. New York: Chapman & Hall.

Carreau, M. (2000). Panel Blames NASA Chiefs in Mars Mess; Design Flaw in Polar Lander Cited. *Houston Chronicle*, March 29, Section A, Page 1.

Carroll, R. J., Ruppert, D., and Stefanski, L. A. (1995). *Measurement Error in Nonlinear Models.* London: Chapman & Hall.

Carroll, R. J., Maca, J. D., and Ruppert, D. (1999). Nonparametric Regression in the Presence of Measurement Error. *Biometrika* **86**, 541–44.

Casella, G. (1980). Minimax Ridge Regression Estimation. *Annals of Statistics* **8**, 1036–56.

Casella, G. (1985). Condition Numbers and Minimax Ridge Regression Estimators. *Journal of the American Statistical Association* **80**, 753–58.

Casella, G., Lavine, M., and Robert, C. P. (2001). Explaining the Perfect Sampler. *American Statistician* **55**, 299–305.

Chambers, J. M. (1973). Fitting Nonlinear Models: Numerical Techniques. *Biometrika* **60**, 1–13.

Chan, K. S. (1989). A Note on the Geometric Ergodicity of a Markov Chain. *Advances in Applied Probability* **21**, 702–3.

Chan, K. S. (1993). Asymptotic Behavior of the Gibbs Sampler. *Journal of the American Statistical Association* **88**, 320–26.

Chan, T. F., Golub, G. H., and LeVeque, R. J. (1983). Algorithms for Computing Sample Variance: Analysis and Recommendations. *Statistician* **37**, 242–47.

Chen, M–H., and Schmeiser, B. (1993). Performance of the Gibbs, Hit–and–Run, and Metropolis Samplers. *Journal of Computational and Graphical Statistics* **2**, 251–72.

Chen, M–H., Shao, Q–M., and Ibrahim, J. G. (2000). *Monte Carlo Methods* in *Bayesian Computation*. New York: Springer-Verlag.

Cheng, C., and Van Ness, J. W. (1999). *Statistical Regression with Measurement Error*. London: Edward Arnold.

Chib, S. and Greenberg, E. (1995). Understanding the Metropolis–Hastings Algorithm. *American Statistician.* **49**, 327–35.

Clarke, E. M., Wing, J. M., Alur, R., Cleaveland, R., Dill, D., Emerson, A., Garland, S., German, S., Guttag, J., Hall, A., Henzinger, T., Holzmann, G., Jones, C., Kurshan, R., Leveson, N., McMillan, K., Moore, J., Peled, D., Pnueli, A., Rushby, J., Shankar, N., Sifakis, J., Sistla, P., Steffen, B., Wolper, P., Woodcock, J., and Zave, P. (1996). Formal Methods: State of the Art and Future Directions. *ACM Computing Surveys* **28**, 626–43.

Clubb, J. M. and Austin, E. W. (1985). Sharing Research Data in the Social Sciences. In *Sharing Research Data*, S. E. Fienberg, M. E. Martin, and M. L. Straf (eds.). Washington, DC: National Academy of Sciences.

Collings, B. J. (1987). Compound Random Number Generators. *Journal of the American Statistical Association* **82**, 525–27.

Congdon, P. (2001). *Bayesian Statistical Modeling*. New York: John Wiley & Sons.

Cook, R. D. (1986). Assessment of Local Influence. *Journal of the Royal Statistical Society* **48**, 133–69.

Cook, R. D. and Weisberg, S. (1982). *Residuals and Influence in Regression*. New York: Chapman & Hall.

Cooper, B. E. (1972). Computing Aspects of Data Management. *Applied Statistics* **21**, 65–75.

Corcoran, J. N. and Tweedie, R. L. (2002). Perfect Sampling from Independent Metropolis–Hastings Chains. *Journal of Statistical Planning and Inference* **104**, 297–314.

Coveyou, R. R. (1960). Serial Correlation in the Generation of Pseudo-random Numbers. *Journal of the Association for Computing Machinery* **7**, 72–74.

Coveyou, R. R. (1970). Random Numbers Fall Mainly in the Planes (review). *ACM Computing Reviews*, 225.

Coveyou, R. R. and MacPherson, R. D. (1967). Fourier Analysis of Uniform Random Number Generators. *Journal of the Association for Computing Machinery* **14**, 100–119.

Cox, D. R. (1970). *Analysis of Binary Data*. London: Chapman & Hall.

Cox, D. R. and Hinkley, D. V. (1974). *Theoretical Statistics.* New York: Chapman & Hall.

Cragg, J. G. (1994). Making Good Inferences from Bad Data. *Canadian Journal of Economics* **28**, 776–99.

Creeth, R. (1985). Micro-computer Spreadsheets: Their Uses and Abuses. *Journal of Accountancy* **159**, 90–93.

Cuevas, A. and Sanz, P. (1988). On Differentiability Properties of Bayes Operators. In *Bayesian Statistics 3*, J. M. Bernardo, M. H. DeGroot, D. V. Lindley, and A. F. M. Smith (eds.). Oxford: Oxford University Press, pp. 569–77.

Dahl, O. J., Dijkstra, E. W., and Hoare, C. A. R. (1972). *Structured Programming.* San Diego, CA: Academic Press.

Datta, S. and McCormick, W. P. (1995). Bootstrap Inference for a First-Order Autoregression with Positive Innovations. *Journal of American Statistical Association* **90**, 1289–300.

David, M. H. (1991). The Science of Data Sharing: Documentation. In Sieber, J. E. (Ed.) *Sharing Social Science Data.* Newbury Park, CA: Sage Publications.

Davidson, R. and MacKinnon, J. G. (1984). Convenient Specification Tests for Logit and Probit Models. *Journal of Econometrics* **25**, 241–62.

Davidson, R. and MacKinnon, J. G. (1993). *Estimation and Inference in Econometrics.* Oxford: Oxford University Press.

Davison, A. C. and Hinkley, D. V. (1988). Saddle Point Approximations in Resampling Methods. *Biometrika* **75**, 417–31.

Davison, A. C. and Hinkley, D. V. (1997). *Bootstrap Methods and Their Application.* Cambridge: Cambridge University Press.

Davison, A. C., Hinkley, D. V., and Schechtman, E. (1986a). Efficient Bootstrapping Simulations. *Biometrika* **73**, 555–66.

Davison, A. C., Hinkley, D. V., and Worton, B. J. (1986b). Bootstrap Likelihood Methods. *Biometrika* **79**, 113–30.

Day, N. E. (1969). Estimating the Components of a Mixture of Normal Distributions. *Biometrika* **56**, 463–74.

de Haan, L. (1981). Estimation of the Minimum of a Function Using Order Statistics. *Journal of the American Statistical Association* **76**, 467–69.

Delampady, M. and Dey, D. K. (1994). Bayesian Robustness for Multiparameter Problems. *Journal of Statistical Planning and Inference* **50**, 375–82.

Den Haan, W. J. and Marcet, A. (1994). Accuracy in Simulations. *Review of Economic Studies* **61**, 3–18.

Dennis, J. E., Jr. (1984). A User's Guide to Nonlinear Optimization Algorithms. *Proceedings of the IEEE* **72**, 1765–76.

Dennis, J. E., Jr. and Schnabel, R. B. (1982). *Numerical Methods for Unconstrained Optimization and Nonlinear Equations.* Philadelphia: SIAM Press.

Dennis, J. E., Jr., Gay, D. M., and Welsch, R. E. (1981). ALGORITHM 573: NL2SOL–An Adaptive Nonlinear Least-Squares Algorithm. *ACM Transactions on Mathematical Software* **7**, 369–83.

Derigs, U. (1985). Using Confidence Limits for the Global Optimum in Combinatorial Optimization. *Operations Research* **22**, 1024–49.

DeShon, R. P. (1998). A Cautionary Note on Measurement Error Correlations in Structural Equation Models. *Psychological Methods* **3**, 412–23.

Dewald, W. G., Thursby, J. G., and Anderson, R. G. (1986). Replication in Empirical Economics: The *Journal of Money, Credit and Banking* Project. *American Economic Review* **76**, 587–603.

Dey, D. K. and Micheas, A. (2000). Ranges of Posterior Expected Losses and ϵ-Robust Actions. In *Robust Bayesian Analysis*, D. R. Insua and F. Ruggeri (eds.). New York: Springer-Verlag, pp. 71–88.

Dhrymes, P. J. (1994). *Topics in Advanced Econometrics*, Volume II: *Linear and Nonlinear Simultaneous Equations*. New York: Springer-Verlag.

DiCiccio, T. J. and Tibshirani, R. J. (1987). Bootstrap Confidence Intervals and Bootstrap Approximations. *Journal of the American Statistical Association* **82**, 163–70.

Dick, N. P. and Bowden, D. C. (1973). Maximum Likelihood Estimation for Mixtures of Two Normal Distributions. *Biometrics* **29**, 781–90.

Diekhoff, G. M. (1996). *Basic Statistics for the Social and Behavioral Sciences*. Upper Saddle River, NJ: Prentice Hall.

Dieter, U. (1975). Statistical Interdependence of Pseudo-random Numbers Generated by the Linear Congruential Method. In *Applications of Number Theory to Numerical Analysis*, S. K. Zaremba (ed.). San Diego, CA: Academic Press, pp. 287–318.

Doeblin, W. (1940). Éléments d'une théorie générale des chaînes simples constantes de Markoff. *Annales Scientifiques de l'Ecole Normale Superieure* **57**, 61–111.

Dongarra, J. J. and Walker, D. W. (1995). Software Libraries for Linear Algebra Computations on High Performance Computers. *SIAM Review* **47**, 151–80.

Dongarra, J. J., Du Croz, J., Duff, S., and Hammarling, S. (1990). Algorithm 679: A Set of Level 3 Basic Linear Algebra Subprograms. *ACM Transactions on Mathematical Software* **16**, 18–28.

Doob, J. L. (1990). *Stochastic Processes*. New York: John Wiley & Sons.

Dorsey, R. E., and Mayer, W. J. (1995). Genetic Algorithms for Estimation Problems with Multiple Optima Non-differentiability, and Other Irregular Features. *Journal of Business and Economic Statistics* **13**, 53–66.

Down, D., Meyn, S. P., and Tweedie, R. L. (1995). Exponential and Uniform Ergodicity of Markov Processes. *Annals of Probability* **23**, 1671–91.

Downham, D. Y. (1970). The Runs Up and Test. *Applied Statistics* **19**, 190–92.

Downham, D. Y. and Roberts, F. D. K. (1967). Multiplicative Congruential Pseudo-random Number Generators. *Computer Journal* **10**, 74–77.

Downing, D. D. (1996). *Calculus the Easy Way*, 3rd ed. New York: Barrons Educational Series.

Dreze, J. H. and Richard, J. F. (1983). Bayesian Analysis of Simultaneous Equation Systems. In *Handbook of Econometrics*, Z. Griliches and M. Intriligator (eds.). Amsterdam: North-Holland, pp. 369–77.

Drezner, Z. and Wesolowsky, G. O. (1989). On the Computation of the Bivariate Normal Integral. *Journal of Statistical Computation and Simulation* **35**, 101–17.

Dudewicz, E. J. (1975). Random Numbers: The Need, the History, the Generators. In *A Modern Course on Statistical Distributions in Scientific Work*, Volume 2, G. P. Patil, S. Kotz, and J. K. Ord (eds.). Boston: D. Reidel, pp. 25–36.

Dudewicz, E. J. (1976). Speed and Quality of Random Numbers for Simulation. *Journal of Quality Technology* **8**, 171–78.

Duhem, P. (1906). *The Aim and Structure of Physical Theory* (1991 translation, with commentary). Princeton, NJ: Princeton University Press.

Duncan, O. D. and Davis, B. (1953). An Alternative to Ecological Correlation. *American Sociological Review* **18**, 665–66.

Durbin, J. (1954). Errors in Variables. *International Statistical Review* **22**, 23–32.

Durbin, R. and Wilshaw, D. (1987). An Analogue Approach to the Traveling Salesman Problem Using an Elastic Net Method. *Nature* **326**, 689–91.

Eastlake, D., Crocker, S., and Schiller, J. (1994). *Randomness Recommendations for Security*. IETF RFC 1750.

Efron, B. (1979). Bootstrap Methods: Another Look at the Jackknife. *Annals of Statistics* **7**, 1–26.

Efron, B. (1982). *The Jackknife, the Bootstrap, and Other Resampling Plans*. Philadelphia, PA: Society for Industrial and Applied Mathematics.

Efron, B. and Morris, C. (1972). Limiting the Risk of Bayes and Empirical Bayes Estimators. Part II: The Empirical Bayes Case. *Journal of the American Statistical Association* **67**, 130–39.

Efron, B. and Tibshirani, R. J. (1993). *An Introduction to the Bootstrap*. New York: Chapman & Hall/CRC.

Eiben, A. E., Aarts, E. H. L., and Van Hee, K. M. (1991). Global Convergence of Genetic Algorithms: An Infinite Markov Chain Analysis. In *Proceedings of the First International Conference on Parallel Problem Solving from Nature* **4**, 12. Berlin: Springer-Verlag.

Eichenaver, J. and Lehn, J. (1986). A Nonlinear Congruential Pseudorandom Number Generator. *Statistische Hefte* **27**, 315–26.

Emerson, J. D. and Hoaglin, D. C. (1983). Resistant Lines for y versus x. In *Understanding Robust and Exploratory Data Analysis*, D. C. Hoaglin, F. Mosteller, and J. Tukey (eds.). New York: John Wiley & Sons, pp. 129–65.

Engle, R. F. (1984). Wald, Likelihood Ratio, and Lagrange Multiplier Tests in Econometrics. In *Handbook of Econometrics 2*, Z. Griliches and M. Intriligator (eds.). Amsterdam: North-Holland, pp. 775–826.

Entacher, K. (1998). Bad Subsequences of Well-Known Linear Congruential Pseudorandom Number Generators. *ACM Transactions on Modeling and Computer Simulation* **8**, 61–70.

Escobar, M. D. and West, M. (1995). Bayesian Density Estimation and Inference Using Mixtures. *Journal of the American Statistical Association* **90**, 577–88.

Everett, M. G. and Borgatti, S. P. (1999). The Centrality of Groups and Classes. *Journal of Mathematical Sociology* **23**, 181–201.

Fahrmeir, L. and Tutz, G. (2001). *Multivariate Statistical Modelling Based on Generalized Linear Models*, 2nd ed. New York: Springer-Verlag.

Falk, M. (1999). A Simple Approach to the Generation of Uniformly Distributed Random Variables with Prescribed Correlations. *Communications in Statistics, Part B, Simulation and Computation* **28**, 785–91.

Fedorov, V. V. (1974). Regression Problems with Controllable Variables Subject to Error. *Biometrika* **61**, 49–56.

Feigenbaum, S. and Levy, D. M. (1993). The Market for (Ir)Reproducible Econometrics. *Social Epistemology* **7**, 215–32.

Feldstein, M. S. (1974), Social Security, Induced Retirement and Capital Accumulation. *Journal of Political Economy* **82**, 905–26.

Ferguson, T. S. (1983). Bayesian Density Estimation by Mixtures of Normal Distributions. In *Recent Advances in Statistics*, H. Rizvi and J. Rustagi (eds.). New York: Academic Press, pp. 287–302.

Ferree, K. (1999). Iterative Approaches to $R \times C$ Ecological Inference Problems: Where They Can Go Wrong. Presented at *Summer Methods Conference*, July 1999, College Station, TX. <http://www.polmeth.ufl.edu/papers/99/ferre99.pdf>.

Ferrenberg, A. (1992). Monte Carlo Simulations: Hidden Errors from "Good" Random Number Generators. *Physical Review Letters* **69**, 3382–84.

Ferretti, N. and Romo, J. (1996). Unit Root Bootstrap Tests for AR(1) Models. *Biometrika* **83**, 849–60.

Fiacco, A. V. and McCormick, G. P. (1968). *Nonlinear Programming: Sequential Unconstrained Minimization Techniques*. New York: John Wiley & Sons.

Finch, S. J., Mendell, N. R., and Thode, H. C., Jr. (1989). Probabilistic Measures of Adequacy of a Numerical Search for a Global Maximum. *Journal of the American Statistical Association* **84**, 1020–23.

FIPS (1994). *Security Requirements for Cryptographic Modules*. Federal Information Processing Standards Publication 140–1. Springfield, VA: U.S. Department of Commerce/National Technical Information Service.

Firth, D. (1993). Bias Reduction of Maximum Likelihood Estimates. *Biometrika* **80**, 27–38.

Fishman, G. S. (1996). *Monte Carlo: Concepts, Algorithms, and Applications*. New York: Springer-Verlag.

Fishman, G. S. and Moore, L. R. (1982). A Statistical Evaluation of Multiplicative Congruential Random Number Generators with Modulus $2^{31} - 1$. *Journal of the American Statistical Association* **77**, 129–36.

Fletcher, R. (1987). *Practical Methods of Optimization*, 2nd ed. New York: John Wiley & Sons.

Fogel, L. J., Owens, A. J. and Walsh, M. J. (1967). *Artificial Intelligence Through Simulated Evolution*. New York: John Wiley & Sons.

Ford, J. A. and Moghrabi, I. A. (1994). Multi-step Quasi-Newton Methods for Optimization. *Journal of Computational and Applied Mathematics* **50**, 305–23.

Fox, J. (1997). *Applied Regression Analysis, Linear Models, and Related Methods.* Thousand Oaks, CA: Sage Publications.

Freedman, D. A. (1981). Bootstrapping Regression Models. *Annals of Statistics* **9**, 1218–28.

Freedman, D., Pisani, R., and Purves, R. (1997). *Statistics*, 3rd ed. New York: W.W. Norton.

Freedman, D. A., Klein, S. P., Ostland, M., and Roberts, M. R. (1998). Review of *A Solution to the Ecological Inference Problem*, by Gary King. *Journal of the American Statistical Association* **93**, 1518–22.

Freedman, D. A., Ostland, M., Roberts, M. R., and Klein, S. P. (1999). Response to King's Comment. *Journal of the American Statistical Association* **94**, 355–57.

Frieze, A., Kannan, R., and Polson, N. G. (1994). Sampling from Log-Concave Distributions. *Annals of Applied Probability* **4**, 812–37.

Frigessi, A., Hwang, C.-R., Sheu, S. J., and Di Stefano, P. (1993). Convergence Rates of the Gibbs Sampler, the Metropolis Algorithm, and Other Single-Site Updating Dynamics. *Journal of the Royal Statistical Society, Series B* **55**, 205–20.

Fryer, J. G. and Robertson, C. A. (1972). A Comparison of Some Methods for Estimating Mixed Normal Distributions. *Biometrika* **59**, 639–48.

Fuller, W. A. (1987). *Measurement Error Models.* New York: John Wiley & Sons.

Fuller, W. A. (1990). Prediction of True Values for Measurement Error Model. In *Statistical Analysis of Measurement Error Models and Applications*, P. Brown and W. Fuller (eds.). *Contemporary Mathematics* **112**, 41–57. Providence, RI: American Mathematical Society.

Fylstra, D., Lasdon, L., Watson, J., and Waren, A. (1998). Design and Use of the Microsoft Excel Solver. *Interfaces* **28**, 29–55.

Galassi, M., Davies, J., Theiler, J., Gough, B., Jungman, G., Booth, M., and Rossi, F. (ed.) (2003). *Gnu Scientific Library Reference Manual*, 2nd ed. Bristol, United Kingdom: Network Theory Ltd.

Gallant, A. R. (1987). *Nonlinear Statistical Models.* New York: John Wiley & Sons.

Gamerman, D. (1997). *Markov Chain Monte Carlo.* New York: Chapman & Hall.

Gan, L. and Jiang, J. (1999). A Test for Global Maximum. *Journal of the American Statistical Association* **94**, 847–54.

Ganesalingam, S. and McLachlan, G. J. (1981). Some Efficiency Results for the Estimation of the Mixing Proportion in a Mixture of Two Normal Distributions. *Biometrics* **37**, 23–33.

Garey, M. R. and Johnson, D. S. (1979). *Computers and Intractability: A Guide to the Theory of NP-Completeness.* San Francisco: W.H. Freeman.

Gelfand, A. E. and Dey, D. K. (1991). On Bayesian Robustness of Contaminated Classes of Priors. *Statistical Decisions* **9**, 63–80.

Gelfand, A. E. and Smith, A. F. M. (1990). Sampling-Based Approaches to Calculating Marginal Densities. *Journal of the American Statistical Association* **85**, 398–409.

Gelfand, A. E., Mallick, B. K., and Polasek, W. (1997). Broken Biological Size Relationships: A Truncated Semiparametric Regression Approach with Measurement Error. *Journal of the American Statistical Association* **92**, 836–45.

Gelman, A. (1992). Iterative and Non-iterative Simulation Algorithms. *Computing Science and Statistics* **24**, 433–38.

Gelman, A., Carlin, J. B., Stern, H. S., and Rubin, D. B. (1995). *Bayesian Data Analysis.* New York: Chapman & Hall.

Geman, S. and Geman, D. (1984). Stochastic Relaxation, Gibbs Distributions, and the Bayesian Restoration of Images. *IEEE Transactions on Pattern Analysis and Machine Intelligence* **6**, 721–41.

Gentle, J. E. (1990). Computer Implementation of Random Number Generators. *Journal of Computational and Applied Mathematics* **31**, 119–25.

Gentle, J. E. (1998). *Random Number Generation and Monte Carlo Methods.* New York: Springer-Verlag.

Gentle, J. E. (2002). *Elements of Statistical Computing.* New York: Springer-Verlag.

Geweke, J. (1989). Bayesian Inference in Econometric Models Using Monte Carlo Integration. *Econometrica* **57**, 1317–39.

Geweke, J. (1993). Bayesian Treatment of the Independent Student *t* Linear Model. *Journal of Applied Econometrics* **8**, 19–40.

Geyer, C. J. (1991). Constrained Maximum Likelihood Exemplified by Isotonic Convex Logistic Regression. *Journal of the American Statistical Association* **86**, 717–24.

Geyer, C. J. and Thompson, E. A. (1992). Constrained Monte Carlo Maximum Likelihood for Dependent Data. *Journal of the Royal Statistical Society, Series B* **54**, 657–99.

Gilks, W. R. and Roberts, G. O. (1996). Strategies for Improving MCMC. In *Markov Chain Monte Carlo in Practice*, W. R. Gilks, S. Richardson, and D. J. Spiegelhalter (eds.). New York: Chapman & Hall, pp. 89–114.

Gilks, W. R., Richardson, S., and Spiegelhalter, D. J. (1996). *Markov Chain Monte Carlo in Practice.* New York: Chapman & Hall, pp. 131–44.

Gill, J. (2000). *Generalized Linear Models: A Unified Approach.* Thousand Oaks, CA: Sage Publications.

Gill, J. (2002). *Bayesian Methods: A Social and Behavioral Sciences Approach.* New York: Chapman & Hall.

Gill, J. and King, G. (2000). *Alternatives to Model Respecification in Nonlinear Estimation.* Technical Report. Gainesville, FL: Department of Political Science, University of Florida.

Gill, P. E. and Murray, W. (1974). Newton-Type Methods for Unconstrained and Linearly Constrained Optimization. *Mathematical Programming* **7**, 311–50.

Gill, P. E., Golub, G. H., Murray, W., and Sanders, M. A. (1974). Methods for Modifying Matrix Factorizations. *Mathematics of Computation* **28**, 505–35.

Gill, P. E., Murray, W., and Wright, M. H. (1981). *Practical Optimization.* London: Academic Press.

Glanz, J. (2002). Studies Suggest Unknown Form of Matter Exists. *New York Times*, July 31, Section A, Page 12, Column 1.

Gleser, L. J. (1992). The Importance of Assessing Measurement Reliability in Multivariate Regression. *Journal of the American Statistical Association* **87**, 696–707.

Goffe, W. L., Ferrier, G. D., and Rogers, J. (1992). Simulated Annealing: An Initial Application in Econometrics. *Computer Science in Economics & Management* **5**, 133–46.

Goffe, W. L., Ferrier, G. D., and Rogers, J. (1994). Global Optimization of Statistical Functions with Simulated Annealing. *Journal of Econometrics* **60**, 65–99.

Goldberg, A. (1989). *Smalltalk-80: The Language.* Reading, MA: Addison-Wesley.

Goldberg, I. and Wagner, D. (1996). Randomness and the Netscape Browser. *Dr. Dobb's Journal* **9**, 66–70.

Goldfeld, S. and Quandt, R. (1972). *Nonlinear Methods in Econometrics.* Amsterdam: North-Holland.

Goldfeld, S., Quandt, R., and Trotter, H. (1966). Maximisation by Quadratic Hill-Climbing. *Econometrica* **34**, 541–51.

Good, I. J. (1957). On the Serial Test for Random Sequences. *Annals of Mathematical Statistics* **28**, 262–64.

Goodman, L. (1953). Ecological Regressions and the Behavior of Individuals. *American Sociological Review* **18**, 663–66.

Gorenstein, S. (1967). Testing a Random Number Generator. *Communications of the Association for Computing Machinery* **10**, 111–18.

Gould, W. and Sribney, W. (1999). *Maximum Likelihood Estimation with Stata.* College Station, TX: Stata Press.

Gove, W. R. and Hughes, M. (1980). Reexamining the Ecological Fallacy: A Study in Which Aggregate Data Are Critical in Investigating the Pathological Effects of Living Alone. *Social Forces* **58**, 1157–77.

Greene, W. (2003). *Econometric Analysis*, 5th ed. Upper Saddle River, NJ: Prentice Hall.

Grillenzoni, C. (1990). Modeling Time-Varying Dynamical Systems. *Journal of the American Statistical Association* **85**, 499–507.

Grochowski, J. (1995). *Winning Tips for Casino Games.* New York: Penguin.

Gujarati, D. N. (1995). *Basic Econometrics.* New York: McGraw-Hill.

Gupta, A. and Lam, M. S. (1996). Estimating Missing Values Using Neural Networks. *Journal of the Operational Research Society* **47**, 229–38.

Gurmu, S. (1991). Tests for Detecting Overdispersion in the Positive Poisson Regression Model. *Journal of Business and Economic Statistics* **9**, 1–12.

Gustafson, P. (2000). Local Robustness in Bayesian Analysis. In *Robust Bayesian Analysis*, D. R. Insua and F. Ruggeri (eds.). New York: Springer-Verlag, pp. 71–88.

Gustafson, P. and Wasserman, L. (1995). Local Sensitivity Diagnostics for Bayesian Inference. *Annals of Statistics* **23**, 2153–67.

Guttman, I., Dutter, R., and Freeman, P. R. (1978). Care and Handling of Univariate Outliers in the General Linear Model to Detect Spuriosity. *Technometrics* **20**, 187–93.

Haberman, S. J. (1989). Concavity and Estimation. *Annals of Statistics* **17**, 1631–61.

Häggström, O. (2002). *Finite Markov Chains and Algorithmic Applications.* Cambridge: Cambridge University Press.

Häggström, O. and Nelander, K. (1999). On Exact Simulation of Markov Random Fields Using Coupling from the Past. *Scandinavian Journal of Statistics* **26**, 395–411.

Hall, P. (1992). *The Bootstrap and Edgeworth Expansion.* New York: Springer-Verlag.

Hamilton, L. C. (1992). *Regression with Graphics: A Second Course in Applied Statistics.* Monterey, CA: Brooks/Cole.

Hampel, F. R. (1974). The Influence Curve and Its Role in Robust Estimation. *Journal of the American Statistical Association* **69**, 383–93.

Hampel, F. R., Rousseeuw, P. J., Ronchetti, E. M., and Stahel, W. A. (1986). *Robust Statistics: The Approach Based on Influence Functions.* New York: John Wiley & Sons.

Hanushek, E. A. and Jackson, J. E. (1977). *Statistical Methods for Social Scientists.* San Diego, CA: Academic Press.

Harrell, F. E. (2001). *Regression Modeling Strategies: With Applications to Linear Models, Logistic Regression, and Survival Analysis.* New York: Springer-Verlag.

Harris, T. E. (1956). The Existence of Stationary Measures for Certain Markov Processes. In *Proceedings of the 3rd Berkeley Symposium on Mathematical Statistics and Probability*, Volume II. Berkeley, CA: University of California Press, pp. 113–24.

Harville, D. A. (1997). *Matrix Algebra from a Statistician's Perspective.* New York: Springer-Verlag.

Hastings, W. K. (1970). Monte Carlo Sampling Methods Using Markov Chains and Their Applications. *Biometrika* **57**, 97–109.

Hathaway, R. J. (1985). A Constrained Formulation of Maximum-Likelihood Estimation for Normal Mixture Distributions. *Annals of Statistics* **13**, 795–800.

Haughton, D. (1997). Packages for Estimating Finite Mixtures: A Review. *American Statistician* **51**, 194–205.

Hausman, J. A. (1978). Specification Tests in Econometrics. *Econometrica* **46**, 1251–71.

Hausman, J., Hall, B., and Griliches, Z. (1984). Economic Models for Count Data with an Application to the Patents–R&D Relationship. *Econometrica* **52**, 909–38.

Hayakawa, M. and Kinoshita, T. (2001). *Comment on the Sign of the Pseudoscalar Pole Contribution to the Muon g-2.* Technical Report. Theory Division, KEK, Tsukuba Japan. <http://arxiv.org/abs/hep-ph/0112102>.

Heinze, G. (1999). *The Application of Firth's Procedure to Cox and Logistic Regression.* Technical Report 10/1999. Vienna: Department of Medical Computer Sciences, Section of Clinical Biometrics, Vienna University. <http://www.akh-wien.ac.at/imc/biometrie/programme/fl_en>.

Heinze, G. and Schemper, M. (2002). A Solution to the Problem of Separation in Logistic Regression. *Statistics in Medicine* **21**, 2409–19.

Hellekalek, P. (1998). Good Random Number Generators Are (Not So) Easy to Find. *Mathematics and Computers in Simulation* **46**, 485–505.

Herrnson, P. S. (1995). Replication, Verification, Secondary Analysis, and Data Collection in Political Science (with a response). *PS: Political Science & Politics* **28**, 452–5, 492–3.

Herron, M. C. and Shotts, K. W. (2003a). Logical Inconsistency in EI-Based Second Stage Regressions. *American Journal of Political Science*. Forthcoming.

Herron, M. C. and Shotts, K. W. (2003b). Using Ecological Inference Point Estimates As Dependent Variables in Second-Stage Linear Regressions. *Political Analysis* **11**(1), 44–64. With a response from Adolph and King (pp. 65–76), a reply (pp. 77–86), and a summary by Adolph, King, Herron and Shotts (pp. 86–94).

Higham, N. (2002). *Accuracy and Stability of Numerical Algorithms*. 2nd ed. Philadelphia: SIAM Press.

Hildreth, C. and Aborn, M. (1985). Report of the Committee on National Statistics. In *Sharing Research Data*, S. E. Fienberg, M. E. Martin, and M. L. Straf (eds.). Washington, DC: National Academy of Sciences.

Hill, T., O'Connor, M., and Remus, W. (1996). Neural Network Models for Time Series Forecasts. *Management Science* **42**, 1082–92.

Hinkley, D. V. (1988). Bootstrap Methods. *Journal of the Royal Statistical Society, Series B* **50**, 321–37.

Hirji, K. F., Tsiatis, A. A., and Mehta, C. R. (1989). Median Unbiased Estimation for Binary Data. *American Statistician* **43**, 7–11.

Hoaglin, D. C., Mosteller, F., and Tukey, J. W. (1983). *Understanding Robust and Exploratory Data Analysis*. New York: John Wiley & Sons.

Hoare, C. A. R. (1969). An Axiomatic Basis for Computer Programming. *Communications of the ACM* **12**, 576–580.

Hobert, J. P., Robert, C. P., and Titterington, D. M. (1999). On Perfect Simulation for Some Mixtures of Distributions. *Statistics and Computing* **9**, 287–98.

Hochbaum, D., Megiddo, N., Naor, J., and Tamir, A. (1993). Tight Bounds and Approximation Algorithms for Integer Programs with Two Variables per Inequality. *Mathematical Programming* **62**, 69–83.

Hodges, J. S. (1987). Uncertainty, Policy Analysis and Statistics. *Statistical Science* **2**, 259–75.

Hoel, P. G., Port, S. C., and Stone, C. J. (1987). *An Introduction to Stochastic Processes*. Prospect Heights, IL: Waveland Press.

Hoerl, A. E. and Kennard, R. W. (1970a). Ridge Regression: Biased Estimation for Nonorthogonal Problems. *Technometrics* **12**, 55–67.

Hoerl, A. E. and Kennard, R. W. (1970b). Ridge Regression: Applications to Nonorthogonal Problems. *Technometrics* **12**, 69–82.

Hoeting, J. A., Madigan, D., Raftery, D. E., and Volinsky, C. T. (1999). Bayesian Model Averaging: A Tutorial. *Statistical Science* **14**, 382–417.

Holland, J. H. (1975). *Adaptation in Natural and Artificial Systems*. Ann Arbor, MI: University of Michigan Press.

Hopfield, J. J. (1982). Neural Networks and Physical Systems with Emergent Collective Computational Abilities. *Proceedings of the National Academy of Sciences* **79**, 2554–58.

Hopfield, J. J. and Tank, D. W. (1985). "Neural" Computation of Decisions in Optimization Problems. *Biological Cybernetics* **52**, 141–52.

Hopfield, J. J. and Tank, D. W. (1986). Computing with Neural Circuits: A Model. *Science* **233**, 625–33.

Horowitz, J. L. (1997). Boostrap Performance in Econometrics. In *Advances in Econometrics*, D. M. Kreps and K. W. Wallis (eds.). Cambridge: Cambridge University Press.

Hosmer, D. W. (1973a). A Comparison of Iterative Maximum Likelihood Estimates of the Parameters of a Mixture of Two Normal Distributions under Three Different Types of Samples. *Biometrics* **29**, 761–70.

Hosmer, D. W. (1973b). On MLE of the Parameters of a Mixture of Two Normal Distributions. *Communications in Statistics* **1**, 217–27.

Hotelling, H. (1943). Some New Methods in Matrix Calculation. *Annals of Mathematical Statistics* **16**, 1–34.

Hsu, J. S. J. and Leonard, T. (1997). Hierarchical Bayesian Semiparametric Procedures for Logistic Regression. *Biometrika* **84**, 85–93.

Huber, P. J. (1972). Robust Statistics: A Review. *Annals of Mathematical Statistics* **43**, 1041–67.

Huber, P. J. (1973). Robust Regression: Asymptotics, Conjectures, and Monte Carlo. *Annals of Statistics* **1**, 799–821.

Huber, P. J. (1981). *Robust Statistics.* New York: John Wiley & Sons.

Hull, T. E. and Dobell, A. R. (1962). Random Number Generators. *SIAM Review* **4**, 230–54.

Hurst, R. L. and Knop, R. E. (1972). Generation of Random Normal Correlated Variables: Algorithm 425. *Communications of the Association for Computing Machinery* **15**, 355–57.

Hyndman, R. J. (1996). Computing and Graphing Highest Density Regions. *American Statistician* **50**, 120–26.

Ihaka, R. and Gentleman, R. (1996). R: A Language for Data Analysis and Graphics. *Journal of Computational and Graphical Statistics* **5**, 299–314.

Ingber, L. (1989). Very Fast Simulated Re–Annealing. *Mathematical Computer Modelling* **12**, 967–73.

Ingber, L. (1990). Statistical Mechanical Aids to Calculating Term Structure Models. *Physical Review A* **42**, 7057–64.

Ingber, L., Wehner, M. F., Jabbour, G. M., and Barnhill, T. M. (1991). Application of Statistical Mechanics Methodology to Term-Structure Bond-Pricing Models. *Mathematical Computer Modelling* **15**, 77–98.

Ingrassia, S. (1994). On the Rate of Convergence of the Metropolis Algorithm and Gibbs Sampler by Geometric Bounds. *Annals of Applied Probability* **4**, 347–89.

Intrator, O. and Intrator, N. (2001). Interpreting Neural-Network Results: A Simulation Study. *Computational Statistics and Data Analysis* **37**, 373–93.

Iturria, S. J., Carroll, R. J., and Firth, D. (1999). Polynomial Regression and Estimating Functions in the Presence of Multiplicative Measurement Error. *Journal of the Royal Statistical Society, Series B* **61**, 547–61.

Jaccard, J. and Wan, C. K. (1995). Measurement Error in the Analysis of Interaction Effects between Continuous Predictors Using Multiple Regression: Multiple Indicator and Structural Equation Approaches. *Psychological Bulletin* **117**, 348–57.

Jaeschke, G. (1993). On Strong Pseudoprimes to Several Bases. *Mathematics of Computation* **61**, 915–26.

Jagerman, D. L. (1965). Some Theorems Concerning Pseudo-random Numbers. *Mathematics of Computation* **19**, 418–26.

James, W. and Stein, C. (1961). Estimation with Quadratic Loss. In *Proceedings of the 4th Berkeley Symposium on Mathematical Statistics and Probability*, J. Neyman (ed.). Berkeley, CA: University of California Press.

Jansson, B. (1966). *Random Number Generators.* Stockholm: Victor Pettersons.

Johnson, V. E. (1996). Studying Convergence of Markov Chain Monte Carlo Algorithms Using Coupled Sample Paths. *Journal of the American Statistical Association* **91**, 154–66.

Johnson, G. (2001). Connoisseurs of Chaos Offer a Valuable Product: Randomness. *New York Times*, June 12, Late Edition-Final, Section F, Page 1, Column 3.

Johnson, G. (2002). At Lawrence Berkeley, Physicists Say a Colleague Took Them for a Ride. *New York Times*, October 15, Final Edition, Section F, Page 1, Column 3.

Joseph, L., Wolfson, D. B., and du Berger, R. (1995). Sample Size Calculations for Binomial Proportions via Highest Posterior Density Intervals. *Statistician* **44**, 143–54.

Judd, J. S. (1990). *Neural Network Design and the Complexity of Learning.* Cambridge, MA: MIT Press.

Judd, K. (1998). *Numerical Methods in Economics.* Cambridge, MA: MIT Press.

Judge, G. G., Miller, D. J., and Tam Cho, W. K. (2002). *An Information Theoretic Approach to Ecological Estimation and Inference.* <http://cho.pol.uiuc.edu/wendy/papers/jmc.pdf>.

Jun, B. and Kocher, P. (1999). *The Intel Random Number Generator Technical Report.* Cryptography Research, Inc. <http://developer.intel.com/design/security/CRIwp.pdf>.

Kadane, J. B. and Srinivasan, C. (1996). Bayesian Robustness and Stability. In *Bayesian Robustness*, J. O. Berger, B. Betró, E. Moreno, L. R. Pericchi, F. Ruggeri, G. Salinetti, and L. Wasserman (eds.). Monograph Series 29. Hayward, CA: Institute of Mathematical Statistics, pp. 139–56.

Kahn, M. J. and Raftery, A. E. (1996). Discharge Rates of Medicare Stroke Patients to Skilled Nursing Facilities: Bayesian Logistic Regression with Unobserved Heterogeneity. *Journal of the American Statistical Association* **91**, 29–41.

Karlin, S. and Taylor, H. M. (1981). *A Second Course in Stochastic Processes.* San Diego, CA: Academic Press.

Karlin, S. and Taylor, H. M. (1990). *A First Course in Stochastic Processes.* San Diego, CA: Academic Press.

Kass, R. E., Tierney, L., and Kadane, J. B. (1989). Approximate Methods for Assessing Influence and Sensitivity in Bayesian Analysis. *Biometrika* **76**, 663–74.

Kelley, C. T. (1999). *Iterative Methods for Optimization.* Philadelphia: SIAM Press.

Kelley, T. L. and McNemar, Q. (1929). Doolittle Versus the Kelley-Salisburg Iteration Method for Computing Multiple Regression Coefficients. *Journal of the American Statistical Association* **24**, 164–69.

Kendall, W. S. and Moller, J. (2000). Perfect Simulation Using Dominating Processes on Ordered Spaces, with Application to Locally Stable Point Processes. *Advances in Applied Probability* **32**, 844–65.

Kennedy, W. J. and Gentle, J. E. (1980). *Statistical Computing.* New York: Marcel Dekker.

King, G. (1989). *Unifying Political Methodology: The Likelihood Theory of Statistical Inference.* Ann Arbor, MI: University of Michigan Press.

King, G. (1997). *A Solution to the Ecological Inference Problem: Reconstructing Individual Behavior from Aggregate Data.* Princeton, NJ: Princeton University Press.

King, G. (1998). *EI: A Program for Ecological Inference, v. 1.61. Software Manual.* <http://gking.harvard.edu/stats.shtml/>.

King, G. (1999). The Future of Ecological Inference Research: A Comment on Freedman et al. *Journal of the American Statistical Association* **94**, 352–55.

King, G., Honaker, J., Joseph, A., and Scheve, K. (2001). Analyzing Incomplete Political Science Data: An Alternative Algorithm for Multiple Imputation. *American Political Science Review* **95**, 49–69.

King, G. and Signorino, C. S. (1996). The Generalization in the Generalized Event Count Model, with Comments on Achen, Amano, and Londregan. *Political Analysis* **6**, 225–252.

King, G. and Zeng, L. (2001a). Logistic Regression in Rare Events Data. *Political Analysis* **9**, 137–63.

King, G. and Zeng, L. (2001b). Improving Forecasts of State Failure. *World Politics* **53**, 623–58.

King, G., Rosen, O., and Tanner, M. (1999). Binomial-Beta Hierarchical Models for Ecological Inference. *Sociological Methods and Research* **28**, 61–90.

King, G., Tomz, M., and Wittenberg, J. (2000). Making the Most of Statistical Analyses: Improving Interpretation and Presentation. *American Journal of Political Science* **44**, 347–61.

Kirkpatrick, S., Gelatt, C. D., and Vecchi, M. P. (1983). Optimization by Simulated Annealing. *Science* **220**, 671–80.

Kit, E. and Finzi, S. (1995). *Software Testing in the Real World: Improving the Process.* New York: Addison-Wesley (ACM Press Books).

Kitagawa, G. and Akaike, H. (1982). A Quasi Bayesian Approach to Outlier Detection. *Annals of the Institute of Statistical Mathematics* **34B**, 389–98.

Klepper, S. and Leamer, E. E. (1984). Consistent Sets of Estimates for Regressions with Errors in All Variables. *Econometrica* **52**, 163–84.

Kleppner, D. and Ramsey, N. (1985). *Quick Calculus: A Self-Teaching Guide.* New York: John Wiley & Sons.

Knüsel, L. (1989). Computergestützte Berechnung statistischer Verteilungen. Munich: Oldenbourg. <http://www.stat.uni-muenchen.de/~knuesel/elv/elv.html>.

Knüsel, L. (1995). On the Accuracy of Statistical Distributions in GAUSS. *Computational Statistics and Data Analysis* **20**, 699–702.

Knüsel, L. (1998). On the Accuracy of the Statistical Distributions in Microsoft Excel. *Computational Statistics and Data Analysis* **26**, 375–77.

Knüsel, L. (2002). On the Reliability of Microsoft Excel XP for Statistical Purposes. *Computational Statistics and Data Analysis* **39**, 109–10.

Knuth, D. E. (1973). *The Art of Computing Programming: Fundamental Algorithms.* 2nd ed. Reading, MA: Addison-Wesley.

Knuth, D. E. (1998). *The Art of Computer Programming*, Volume 2: *Seminumerical Algorithms*, 3rd ed. Reading, MA: Addison-Wesley.

Kofler, M. (1997). *Maple: An Introduction and Reference.* Reading, MA: Addison-Wesley.

Krawczyk, H. (1992). How to Predict Congruential Generators. *Journal of Algorithms* **13**, 527–45.

Kronmal, R. (1964). The Evaluation of a Pseudorandom Normal Number Generator. *Journal of the Association for Computing Machinery* **11**, 357–63.

Krug, E. G., Kresnow, M., Peddicord, J. P., and Dahlberg, L. L. (1988). Suicide After Natural Disasters. *New England Journal of Medicine* **340**, 148–49.

Krzanowski, W. J. (1988). *Principles of Multivariate Analysis.* Oxford: Oxford University Press.

Kuan, C. and Liu, L. (1995). Forecasting Exchange Rates Using Feedforward and Recurrent Neural Networks. *Journal of Applied Econometrics* **10**, 347–64.

Kuan, C. M. and White, H. (1994). Artificial Neural Networks: An Econometric Perspective. *Econometric Reviews* **13**, 1–91.

Lange, K. (1999). *Numerical Analysis for Statisticians.* New York: Springer-Verlag.

Lange, K. L., Little, R. J. A., and Taylor, J. M. G. (1989). Robust Statistical Modeling Using the *t* Distribution. *Journal of the American Statistical Association* **84**, 881–96.

Lavine, M. (1991a). Sensitivity in Bayesian Statistics: The Prior and the Likelihood. *Journal of the American Statistical Association* **86**, 396–99.

Lavine, M. (1991b). An Approach to Robust Bayesian Analysis for Multidimensional Parameter Spaces. *Journal of the American Statistical Association* **86**, 400–403.

Lawrance, A. J. (1988). Regression Transformation Diagnostics Using Local Influence. *Journal of the American Statistical Association* **83**, 1067–72.

Lawson, C. L., Hanson, R. J., Kincaid, D., and Krogh, F. T. (1979). Basic Linear Algebra Subprograms for FORTRAN Usage. *ACM Transactions on Mathematical Software* **5**, 308–23.

Lax, P. D. (1997). *Linear Algebra.* New York: John Wiley & Sons.

Leamer, E. E. (1973). Multicollinearity: A Bayesian Interpretation. *Review of Economics and Statistics* **55** (3), 371–80.

Leamer, E. E. (1978). *Specification Searches: Ad Hoc Inference with Nonexperimental Data*. New York: John Wiley & Sons.

Leamer, E. E. (1983). Lets Take the Con Out of Econometrics. *American Economic Review* **73**, 31–43.

Leamer, E. E. (1984). Global Sensitivity Results for Generalized Least Squares Estimates. *Journal of the American Statistical Association* **79**, 867–70.

Leamer, E. E. (1985). Sensitivity Analysis Would Help. *American Economic Review* **75**, 308–13.

Leamer, E. E. and Leonard, H. (1983). Reporting the Fragility of Regression Estimates. *Review of Economics and Statistics* **65**, 306–17.

Learmonth, G. P. and Lewis, P. A. W. (1973). Some Widely Used and Recently Proposed Uniform Random Number Generators. *Proceedings of Computer Science and Statistics: 7th Annual Symposium on the Interface*, W. J. Kennedy (ed.). Ames, IA: Iowa State University, pp. 163–71.

L'Ecuyer, P. (1990). Random Numbers for Simulation. *Communications of the ACM* **33**, 85–97.

L'Ecuyer, P. (1992). Testing Random Number Generators. In *Proceedings of the 1992 Winter Simulation Conference*. Piscataway, NJ: IEEE Press.

L'Ecuyer, P. (1994). Uniform Random Number Generation. *Annals of Operations Research* **53**, 77–120.

L'Ecuyer, P. and Hellekalek, P. (1998). Random Number Generators: Selection Criteria and Testing. In *Random and Quasi-random Point Sets*, P. Hellekalek (ed.). New York: Springer-Verlag.

L'Ecuyer, P. and Panneton, F. (2000). A New Class of Linear Feedback Shift Register Generators. In *Proceedings of the 2000 Winter Simulation Conference*, J. A. Joines, R. R. Barton, K. Kang, and P. A. Fishwick (eds.). Piscataway, NJ: IEEE Press, pp. 690–96.

L'Ecuyer, P. and Simard, R. (2003). *TESTU01: A Software Library in ANSI C for Empirical Testing of Random Number Generators*. <http://www.iro.umontreal.ca/ simardr/>.

Lee, L. (1986). Specification Tests for Poisson Regression Models. *International Economic Review* **27**, 689–706.

Lehmann, E. L. (1999). *Elements of Large Sample Theory*. New York: Springer-Verlag.

Leimer, D. R. and Lesnoy, S. D. (1982). Social Security and Private Saving: New Time-Series Evidence. *Journal of Political Economy* **90**, 606–29.

LeSage, J. P. (1997). Bayesian Estimation of Spatial Autoregressive Models. *International Regional Science Review* **20**, 113–29.

LeSage, J. P. (2000). Bayesian Estimation of Limited Dependent Variable Spatial Autoregressive Models. *Geographical Analysis* **32**, 19–35.

LeSage, J. P. and Pace, R. K. (2001). Spatial Dependence in Data Mining. In *Data Mining for Scientific and Engineering Applications*, R. L. Grossman, C. Kamath, P. Kegelmeyer, V. Kumar, and R. R. Namburu (eds.). Boston: Kluwer Academic Publishers.

LeSage, J. P. and Simon, S. D. (1985). Numerical Accuracy of Statistical Algorithms for Microcomputers. *Computational Statistics and Data Analysis* **3**, 47–57.

Levine, R. and Renelt, D. (1992). A Sensitivity Analysis of Cross-Country Growth Regressions. *American Economic Review* **82**, 942–63.

Lewbel, A. (1997). Constructing Instruments for Regressions with Measurement Error When No Additional Data Are Available, with an Application to Patents and R&D. *Econometrica* **65**, 1201–13.

Lewis, J. and McCue, K. (2002). Comment on "The Statistical Foundations of the EI Method" (Lewis) and Reply (McCue) (an exchange in the letter's to the editor section). *American Statistician* **56**, 255–57.

Lewis, T. G. and Payne, W. H. (1973). Generalized Feedback Shift Register Pseudorandom Number Algorithm. *Journal of the Association for Computing Machinery* **20**, 456–68.

Lewis, P. A. W., Goodman, O. S., and Miller, J. W. (1969). A Pseudo-random Number Generator for the System 360. *IBM Systems Journal* **8**, 136–45.

Li, T. and Vuong, Q. (1998). Nonparametric Estimation of the Measurement Error Model Using Multiple Indicators. *Journal of Multivariate Analysis* **65**, 139–65.

Liseo, B., Petrella, L., and Salinetti, G. (1996). Bayesian Robustness: An Interactive Approach. In *Bayesian Statistics 5*, J. O. Berger, J. M. Bernardo, A. P. Dawid, and D. V. Lindley (eds.) Oxford: Oxford University Press, pp. 223–53.

Little, R. J. A. and Rubin, D. B. (1987). *Statistical Analysis with Missing Data*. New York: John Wiley & Sons.

Lo, A. Y. (1987). A Large Sample Study of the Bayesian Bootstrap. *Annals of Statistics* **15**, 360–75.

Locatelli, M. (2000). Simulated Annealing Algorithms for Continuous Global Optimization: Convergence Conditions. *Journal of Optimization Theory and Applications* **104**, 121–33.

Loh, W. Y. (1987). Calibrating Confidence Coefficients. *Journal of the American Statistical Association* **82**, 155–62.

Loh, W. Y. (1988). Discussion of "Theoretical Comparison of Bootstrap Confidence Intervals" by P. Hall. *Annals of Statistics* **16**, 972–76.

Loh, W. Y. (1991). Bootstrap Calibration for Confidence Interval Construction and Selection. *Statistica Sinica* **1**, 477–91.

Loh, W. Y. and Wu, C. F. J. (1991). Discussion of "Better Bootstrap Confidence Intervals" by B. Efron. *Journal of the American Statistical Association* **92**, 188–90.

Londregan, J. (1996). Some Remarks on the "Generalized Event Count" Distribution. *Political Analysis* **6**, 213–24.

Long, J. S. (1997). *Regression Models for Categorical and Limited Dependent Variables*. Thousand Oaks, CA: Sage Publications.

Longley, J. W. (1967). An Appraisal of Computer Programs for the Electronic Computer from the Point of View of the User. *Journal of the American Statistical Association* **62**, 819–41.

Lozier, D. W. and Olver, F. W. J. (1994). Numerical Evaluation of Special Functions. In *Mathematics of Computation, 1943–1993: A Half-Century of Computational Mathematics,* W. Gautschi (ed.), Proceedings of Symposia in Applied Mathematics. Providence, RI: American Mathematical Society (updated December 2000). <http://math.nist.gov/mcsd/Reports/2001/nesf/>.

Lundy, M. and Mees, A. (1986). Convergence of an Annealing Algorithm. *Mathematical Programming* **34**, 111–24.

MacKie-Mason, J. K. (1992). Econometric Software: A User's View. *Journal of Economic Perspectives* **6**, 165–87.

Maclaren, M. D. and Marsaglia, G. (1965). Uniform Random Number Generators. *Journal of the Association for Computing Machinery* **12**, 83–89.

Madanksy, A. (1959). The Fitting of Straight Lines When Both Variables Are Subject to Error. *Journal of the American Statistical Association* **54**, 173–205.

Malov, S. V. (1998). Random Variables Generated by Ranks in Dependent Schemes. *Metrika* **48**, 61–67.

Marais, M. L. and Wecker, W. E. (1998). Correcting for Omitted-Variables and Measurement-Error Bias in Regression with an Application to the Effect of Lead on IQ. *Journal of the American Statistical Association* **93**, 494–504.

Marín, J. M. (2000). A Robust Version of the Dynamic Linear Model with an Economic Application. In *Robust Bayesian Analysis,* D. R. Insua and F. Ruggeri (eds.). New York: Springer-Verlag, pp. 373–83.

Maros, I. (2002). *Computational Techniques of the Simplex Method.* Boston: Kluwer Academic Publishers.

Maros, I. and Khaliq, M. H. (2002). Advances in Design and Implementation of Optimization Software. *European Journal of Operational Research* **140**, 322–37.

Marquardt, D. W. (1970). Generalized Inverses, Ridge Regression, Biased Linear Estimation, and Nonlinear Estimation. *Technometrics* **12**, 591–612.

Marsaglia, G. (1968). Random Numbers Fall Mainly in the Planes. *Proceedings of the National Academy of Sciences* **61**, 25–28.

Marsaglia, G. (1984). A Current View of Random Number Generators. Paper presented at *Computer Science and Statistics: 16th Symposium on the Interface,* New York.

Marsaglia, G. (1985). A Current View of Random Number Generators. In *Computer Science and Statistics: 16th Symposium on the Interface,* L. Billard (ed.). Amsterdam: North-Holland.

Marsaglia, G. (1993). Monkey Tests for Random Number Generators. *Computers & Mathematics with Applications* **9**, 1–10.

Marsaglia, G. and Zaman, A. (1993). The KISS Generator. Technical Report, Department of Statistics, University of Florida.

Marsaglia, G. (1996). *DIEHARD: A Battery of Tests of Randomness* (software package). <http://stat.fsu.edu/geo/diehard.html>.

Marsaglia, G. and Tsang, W. W. (2002). Some Difficult-to-Pass Tests of Randomness. *Journal of Statistical Software* **7**, 1–8.

Mascagni, M. and Srinivasan, A. (2000). SPRNG: A Scalable Library for Pseudorandom Number Generation. *ACM Transactions on Mathematical Software* **26**, 436–61.

Matsumoto, M. and Nishimura, T. (1998). Mersenne Twister: A 623-Dimensionally Equidistributed Uniform Pseudorandom Number Generator. *ACM Transactions on Modeling and Computer Simulation* **8**, 3–30.

Matthews, A. and Davies, D. (1971). A Comparison of Modified Newton Methods for Unconstrained Optimization. *Computer Journal* **14**, 213–94.

McArdle, J. J. (1976). Empirical Test of Multivariate Generators. In *Proceedings of the 9th Annual Symposium on the Interface of Computer Science and Statistics*, D. C. Hoaglin and R. Welsch (eds.). Boston: Prindle, Weber & Schmidt, pp. 263–67.

McCue, K. (2001). The Statistical Foundations of the EI Method. *American Statistician* **55**, 106–11.

McCullagh, P. and Nelder, J. A. (1989). *Generalized Linear Models*, 2nd ed. New York: Chapman & Hall.

McCulloch, W. and Pitts, W. (1943). A Logical Calculus of the Ideas Immanent in Nervous Activity. *Bulletin of Mathematical Biophysics* **7**, 115–33.

McCulloch, C. E. and Searle, S. R. (2001). *Generalized, Linear, and Mixed Models.* New York: John Wiley & Sons.

McCullough, B. D. (1998). Assessing the Reliability of Statistical Software: Part I. *American Statistician* **52**, 358–66.

McCullough, B. D. (1999a). Econometric Software Reliability: Eviews, LIMDEP, SHAZAM, and TSP. *Journal of Applied Econometrics* **14**, 191–202.

McCullough, B. D. (1999b). Assessing the Reliability of Statistical Software: Part II. *American Statistician* **53**, 149–59.

McCullough, B. D. (2000). The Accuracy of Mathematica 4 as a Statistical Package. *Computational Statistics* **15**, 279–90.

McCullough, B. D. (2004). Review of TESTUO1. *Journal of Applied Econometrics*. Forthcoming.

McCullough, B. D. and Renfro, C. G. (1999). Benchmarks and Software Standards: A Case Study of GARCH Procedures. *Journal of Economic and Social Measurement* **25**, 59–71.

McCullough, B. D. and Vinod, H. D. (1999). The Numerical Reliability of Econometric Software. *Journal of Economic Literature* **37**, 633–65

McCullough, B. D. and Vinod, H. D. (2004). Verifying the Solution from a Nonlinear Solver: A Case Study. *American Economic Review*. Forthcoming.

McCullough, B. D. and Wilson, B. (1999). On the Accuracy of Statistical Procedures in Microsoft Excel 97. *Computational Statistics and Data Analysis* **31**, 27–37.

McCullough, B. D. and Wilson, B. (2002). On the Accuracy of Statistical Procedures in Microsoft Excel 2000 and Excel XP. *Computational Statistics and Data Analysis* **40**, 713–21.

McDaniel, W. L. (1989). Some Pseudoprimes and Related Numbers Having Special Forms. *Mathematics of Computation* **53**, 407–9.

McFadden, D. L. and Ruud, P. A. (1994). Estimation with Simulation. *Review of Economics and Statistics* **76**, 591–608.

Meier, K. J. and Smith, K. B. (1995). Representative Democracy and Representative Bureaucracy: Examining the Top Down and Bottom Up Linkages. *Social Science Quarterly* **75**, 790–803.

Meier, K. J., Polinard, J. L., and Wrinkle, R. (2000). Bureaucracy and Organizational Performance: Causality Arguments about Public Schools. *American Journal of Political Science* **44**, 590–602.

Mengersen, K. L. and Tweedie, R. L. (1996). Rates of Convergence of the Hastings and Metropolis Algorithms. *Annals of Statistics* **24**, 101–21.

Metropolis, N. and Ulam, S. (1949). The Monte Carlo Method. *Journal of the American Statistical Association* **44**, 335–41.

Metropolis, N., Rosenbluth, A. W., Rosenbluth, M. N., Teller, A. H., and Teller E. (1953). Equation of State Calculations by Fast Computing Machine. *Journal of Chemical Physics* **21**, 1087–91.

Meyn, S. P. and Tweedie, R. L. (1993). *Markov Chains and Stochastic Stability.* New York: Springer-Verlag.

Meyn, S. P. and Tweedie, R. L. (1994a). State-Dependent Criteria for Convergence of Markov Chains. *Annals of Applied Probability* **4**, 149–68.

Meyn, S. P. and Tweedie, R. L. (1994b). Computable Bounds for Convergence Rates of Markov Chains. *Annals of Applied Probability* **4**, 981–1011.

Michalewicz, Z. and Fogel, D. B. (1999). *How to Solve It: Modern Heuristics.* New York: Springer-Verlag.

Mihram, G. A. and Mihram, D. (1997). A Review and Update on Pseudo-random Number Generation, on Seeding, and on a Source of Seeds. *ASA Proceedings of the Statistical Computing Section.* Alexandria, VA: American Statistical Association, pp. 115–19.

Miller, W. E. and the American National Election Studies (1999). *American National Election Studies Cumulative Data File, 1948–1998* (data file), 10th ICPSR version. Ann Arbor, MI: University of Michigan, Center for Political Studies.

Mittelhammer, R. C., Judge, G. C., and Miller, D. J. (2000). *Econometric Foundations,* Cambridge: Cambridge University Press.

Mittelmann, H. D. and Spellucci, P. 2003. *Decision Tree for Optimization Software* (Web site). <http://plato.asu.edu/guide.html>.

Monahan, J. F. (2001). *Numerical Methods of Statistics.* New York: Cambridge University Press.

Mongeau, M., Karsenty, H., Rouze, V., and Hiriart-Urruty, J. B. (2000). Comparison of Public-Domain Software for Black-Box Global Optimization. *Optimization Methods and Software* **13**, 203–26.

Montgomery, D. C. C., Peck, E. A., and Vining, G. G. (2001). *Introduction to Linear Regression Analysis*, 3rd ed. New York: John Wiley & Sons.

Mooney, C. Z. (1997). *Monte Carlo Simulation.* Quantitative Applications in the Social Sciences 113. Thousand Oaks, CA: Sage Publications.

Moore, D. S. (1999). *The Basic Practice of Statistics*, 2nd ed. New York: W. H. Freeman.

Moore, D. S. and McCabe, G. P. (2002). *Introduction to the Practice of Statistics*, 4th ed. New York: W. H. Freeman.

Moore, E. H. (1920). On the Reciprocal of the General Algebraic Matrix. *Bulletin of the American Mathematical Society* **26**, 394–95.

Moré, J. J. (1978). The Levenberg–Marquardt Algorithm: Implementation and Theory. In *Numerical Analysis*, G. A. Watson (ed.). Lecture Notes in Mathematics 630. New York: Springer-Verlag, pp. 105–16.

Moré, J. and Wright, J. 1993. *Optimization Software Guide.* Philadelphia: Siam Press.

Moreno, E. (2000). Global Bayesian Robustness for Some Classes of Prior Distributions. In *Robust Bayesian Analysis*, D. R. Insua and F. Ruggeri (eds.). New York: Springer-Verlag, pp. 45–70.

Moreno, E. and Cano, J. A. (1991). Robust Bayesian Analysis with ϵ-Contaminations Partially Known. *Journal of the Royal Statistical Society, Series B* **53**, 143–55.

Moreno, E. and González, A. (1990). Empirical Bayes Analysis of ϵ-Contaminated Classes of Prior Distributions. *Brazilian Journal of Probability and Statistics* **4**, 177–200.

Moreno, E. and Pericchi, L. R. (1991). Robust Bayesian Analysis for ϵ-Contaminations with Shape and Quantile Restraints. In *Proceedings of the 5th International Symposium on Applied Stochastic Models*, R. Gutiéterrez, and M. Valderrama (eds.). Singapore: World Scientific, pp. 454–70.

Moreno, E. and Pericchi, L. R. (1993). Bayesian Robustness for Hierarchical ϵ-Contamination Models. *Journal of Statistical Planning and Inference* **37**, 159–68.

Moreno, E., Martínez, C., and Cano, J. A. (1996). Local Robustness and Influences for Contamination Classes of Prior Distributions. In *Bayesian Robustness*, J. O. Berger, B. Betró, E. Moreno, L. R. Pericchi, F. Ruggeri, G. Salinetti, and L. Wasserman (eds.). Monograph Series 29. Hayward, CA: Institute of Mathematical Statistics, pp. 139–56.

Morgan, B. J. T. (1984). *Elements of Simulation.* New York: Chapman & Hall.

Moshier, S. L. (1989). *Methods and Programs for Mathematical Functions.* Upper Saddle River, NJ: Prentice Hall. <http://www.netlib.org/cephes>.

Mroz, T. A. (1987). The Sensitivity of an Empirical Model of Married Women's Hours of Work to Economic and Statistical Assumptions. *Econometrica* **55**, 765–99.

Mukhopadhyay, P. (1997). Bayes Estimation of Small Area Totals under Measurement Error Models. *Journal of Applied Statistical Science* **5**, 105–11.

Myers, R. H. and Montgomery, D. C. (1997). A Tutorial on Generalized Linear Models. *Journal of Quality Technology* **29**, 274–91.

Myers, R. H., Montgomery, D. C., and Vining, G. G. (2002). *Generalized Linear Models with Applications in Engineering and the Sciences.* New York: John Wiley & Sons.

Nagler, J. (1994). Scobit: An Alternative Estimator to Logit and Probit. *American Journal of Political Science* **38**, 230–55.

Nair, K. R. and Banerjee, K. S. (1942). A Note on Fitting of Straight Lines If Both Variables Are Subject to Error. *Sankyā* **6**, 331.

National Research Council (NRC), Science, Technology, and Law Panel (2002). *Access to Research Data in the 21st Century: An Ongoing Dialogue among Interested Parties Report of a Workshop.* Washington, DC: National Academy Press.

Naylor, J. C. and Smith, A. F. M. (1982). Applications of a Method for the Efficient Computation of Posterior Distributions. *Applied Statistics* **31**, 214–25.

Naylor, J. C. and Smith, A. F. M. (1988a). Economic Illustrations of Novel Numerical Integration Strategies for Bayesian Inference. *Journal of Econometrics* **38**, 103–26.

Naylor, J. C. and Smith, A. F. M. (1988b). An Archeological Inference Problem. *Journal of the American Statistical Association* **83**, 588–95.

Nelder, J. A. and Mead, R. 1965. A Simplex Method for Function Minimization. *Computer Journal* **7**, 308–13.

Neter, J., Kutner, M. H., Nachtsheim, C., and Wasserman, W. (1996). *Applied Linear Regression Models.* Chicago: Richard D. Irwin.

Neumann, P. G. (1995). *Computer Related Risks.* New York: ACM Press Series (Addison-Wesley).

Newton, M. A. and Raftery, A. E. (1994). Approximate Bayesian Inference with the Weighted Likelihood Bootstrap. *Journal of the Royal Statistical Society, Series B* **56**, 3–48.

Nicole, S. (2000). Feedforward Neural Networks for Principal Components Extraction. *Computational Statistics and Data Analysis* **33**, 425–37.

Nocedal, J. and Wright, S. J. (1999). *Numerical Optimization.* New York: Springer-Verlag.

Nordbotten, S. (1996). Editing and Imputation by Means of Neural Networks. *Statistical Journal of the UN Economic Commission for Europe* **13**, 119–29.

Nordbotten, S. (1999). Small Area Statistics from Survey and Imputed Data. *Statistical Journal of the UN Economic Commission for Europe* **16**, 297–99.

Noreen, E. W. (1989). *Computer-Intensive Methods for Testing Hypotheses.* New York: John Wiley & Sons.

Norris, J. R. (1997). *Markov Chains.* Cambridge: Cambridge University Press.

Nummelin, E. (1984). *General Irreducible Markov Chains and Non-negative Operators.* Cambridge: Cambridge University Press.

Nummelin, E. and Tweedie, R. L. (1978). Geometric Ergodicity and R-Positivity for General Markov Chains. *Annals of Probability* **6**, 404–20.

Oh, M.-S. and Berger, J. O. (1993). Integration of Multimodal Functions by Monte Carlo Importance Sampling. *Journal of the American Statistical Association* **88**, 450–55.

O'Hagan, A. (1994). *Kendall's Advanced Theory of Statistics*: Volume 2B, *Bayesian Inference.* London: Edward Arnold.

O'Hagan, A. and Berger, J. O. (1988). Ranges of Posterior Probabilities for Quasiunimodal Priors with Specified Quantiles. *Journal of the American Statistical Association* **83**, 503–8.

O'Leary, D. P. and Rust, B. W. (1986). Confidence Intervals for Inequality-Constrained Least Squares Problems, with Applications to Ill-Posed Problems. *American Journal for Scientific and Statistical Computing* **7**, 473–89.

Orey, S. (1961). Strong Ratio Limit Property. *Bulletin of the American Mathematical Society* **67**, 571–74.

Ott, J. (1979). Maximum Likelihood Estimation by Counting Methods under Polygenic and Mixed Models in Human Pedigrees. *Journal of Human Genetics* **31**, 161–75.

Overton, M. L. (2001). *Numerical Computing with IEEE Floating Point Arithmetic.* Philadelphia: SIAM Press.

Pace, R. K. and Barry, R. (1997). Quick Computation of Spatial Autoregressive Estimators. *Geographical Analysis* **29**, 232–46.

Pace, R. K. and LeSage, J. P. (2003). Chebyshev Approximation of log Determinants Using Spatial Weight Matrices. *Computational Statistics & Data Analysis.* Forthcoming. Available at: <http://www.econ.utoledo.edu/faculty/lesage/workingp.html>.

Paik, H. (2000). Comments on Neural Networks. *Sociological Methods and Research* **28**, 425–53.

Panko, R. R. (1998). What We Know about Spreadsheet Errors. *Journal of End User Computing* **10**, 15–21.

Papadimitrious, C. (1994). *Computational Complexity.* Reading, MA: Addison-Wesley,

Park, S. K. and Miller, K. W. (1988). Random Number Generators: Good Ones Are Hard to Find. *Communications of the ACM* **31**, 1192–1201.

Parker, D. S. (1997). *Monte Carlo Arithmetic: Exploiting Randomness in Floating-Point Arithmetic.* Technical Report CSD-970002. Los Angeles: Computer Science Department, UCLA.

Parker, D. S., Pierce, B., and Eggert, P. R. (2000). Monte Carlo Arithmetic: A Framework for the Statistical Analysis of Roundoff Error. *IEEE Computation in Science and Engineering* **2**, 58–68.

Penrose, R. A. (1955). A Generalized Inverse for Matrices. *Proceedings of the Cambridge Philosophical Society* **51**, 406–13.

Peskun, P. H. (1973). Optimum Monte Carlo Sampling Using Markov Chains. *Biometrika* **60**, 607–12.

Peterson, C. and Soderberg, B. (1989). A New Method for Mapping Optimization Problems onto Neural Networks. *International Journal of Neural Systems* **1**, 3–22.

Pettit, L. I. and Smith, A. F. M. (1985). Outliers and Influential Observations in Linear Models. In *Bayesian Statistics 2*, J. M. Bernardo, M. H. DeGroot, D. V. Lindley, and A. F. M. Smith (eds.). Amsterdam: North-Holland, pp. 473–94.

Ploner, M. (2001). *An S-PLUS Library to Perform Logistic Regression without Convergence Problems.* Technical Report 2/2002. Vienna: Department of Medical Computer Sciences, Section of Clinical Biometrics, Vienna University. <http://www.akh-wien.ac.at/imc/biometrie/programme/ fl_en:>.

Poirer, D. J. (1988). Frequentist and Subjectivist Perspectives on the Problems of Model Building in Economics. *Journal of Economic Perspectives* **2**, 121–44.

Polasek, W. (1984). Regression Diagnostics for General Linear Regression Models. *Journal of the American Statistical Association* **79**, 336–40.

Polasek, W. (1987). Bounds on Rounding Errors in Linear Regression Models. *Statistician* **36**, 221–27.

Powell, K. E., Crosby, A. E., and Annest, J. L. (1999). Retraction of Krug E. G., M. Kresnow, J. P. Peddicord, and L. L. Dahlberg, "Suicide after Natural Disasters." *New England Journal of Medicine* **340**, 148–49.

Pratt, J. W. (1981). Concavity of the Log Likelihood. *Journal of the American Statistical Association* **76**, 103–6.

Pregibon, D. (1981). Logistic Regression Diagnostics. *Annals of Statistics* **9**, 705–24.

Press, W. H., Flannery, B. P., Teukolsky, S. A., and Vetterling, W. T. (1988). *Numerical Recipes: The Art of Scientific Computing.* Cambridge: Cambridge University Press.

Press, W. H., Teukolsky, S. A., Vetterling, W. T., and Flannery, B. P. (2002). *Numerical Recipes in C++: The Art of Scientific Computing*, 2nd ed. Cambridge: Cambridge University Press.

Propp, J. G. and Wilson, D. B. (1996). Exact Sampling with Coupled Markov Chains and Applications to Statistical Mechanics. *Random Structures and Algorithms* **9**, 223–52.

Rabin, M. O. (1980). Probabilistic Algorithms for Testing Primality. *Journal of Number Theory* **12**, 128–38.

Raftery, A. E. (1995). Bayesian Model Selection in Social Research. *Sociological Methodology* **25**, 111–63.

Raftery, A. E. (1996). Approximate Bayes Factors and Accounting for Model Uncertainty in Generalised Linear Models. *Biometrika* **83**, 251–66.

Raftery, A. E., Madigan, D., and Hoeting, J. A. (1997). Bayesian Model Averaging for Linear Regression Models. *Journal of the American Statistical Association* **92**, 179–91.

Rao, C. R. (1973). *Linear Statistical Inference and Its Applications.* New York: John Wiley & Sons.

Rao, C. R. and Mitra, S. K. (1971). *Generalized Inverse of Matrices and Its Applications.* New York: John Wiley & Sons.

R Development Core Team (2003). *R Language Definition.* Technical report. <http://cran.us.r-project.org/>.

Rechenberg, I. (1973). *Evolutionsstrategie: Optimierung technischer Systeme nach Prinzipien de biologischen Information.* Frommann. Freiburg, Germany.

Reiersol, O. (1950). Identifiability of a Linear Relation between Variables Which Are Subject to Errors. *Econometrica* **18**, 375–89.

Renfro, C. G. (1997). Normative Considerations in the Development of a Software Package for Econometric Estimation. *Journal of Economic and Social Measurement* **23**, 277–330.

Revkin, A. C. (2002). Data Revised on Soot in Air and Deaths. *New York Times* (National Edition), June 5, Section A, Page A23.

Revuz, D. (1975). *Markov Chains.* Amsterdam: North-Holland.

Riley, J. (1955). Solving Systems of Linear Equations with a Positive Definite, Symmetric but Possibly Ill-Conditioned Matrix. *Mathematical Tables and Other Aides to Computation* **9**, 96–101.

Ripley, B. D. (1987). *Stochastic Simulation.* New York: John Wiley & Sons.

Ripley, B. D. (1988). Uses and Abuses of Statistical Simulation. *Mathematical Programming* **42**, 53–68.

Ripley, B. D. (1990). Thoughts on Pseudorandom Number Generators. *Journal of Computational and Applied Math* **31**, 153–63.

Ritter, T. (1986). The Great CRC Mystery. *Dr. Dobb's Journal of Software Tools* **11**, 26–34, 76–83.

Rivest, R. (1992). *RFC 1321: The MD5 Message-Digest Algorithm.* Internet Activities Board.

Robbin, A. and Frost-Kumpf, L. (1997). Extending Theory for User-Centered Information Services: Diagnosing and Learning from Error in Complex Statistical Data. *Journal of the American Society of Information Science* **48**(2), 96–121.

Robert, C. P. (2001). *The Bayesian Choice: A Decision Theoretic Motivation*, 2nd ed. New York: Springer-Verlag.

Robert, C. P. and Casella, G. (1999). *Monte Carlo Statistical Methods.* New York: Springer-Verlag.

Roberts, G. O. and Polson, N. G. (1994). On the Geometric Convergence of the Gibbs Sampler. *Journal of the Royal Statistical Society, Series B* **56**, 377–84.

Roberts, G. O. and Smith, A. F. M. (1994). Simple Conditions for the Convergence of the Gibbs Sampler and Metropolis–Hastings Algorithms. *Stochastic Processes and Their Applications* **44**, 207–16.

Robinson, W. S. (1950). Ecological Correlation and the Behavior of Individuals. *American Sociological Review* **15**, 351–57.

Rogers, J., Filliben, J., Gill, L., Guthrie, W., Lagergren, E., and Vangel, M. (2000). *StRD: Statistical Reference Datasets for Testing the Numerical Accuracy of Statistical Software.* NIST 1396. Washington, DC: National Institute of Standards and Technology.

Rosenthal, J. S. (1993). Rates of Convergence for Data Augmentation on Finite Sample Spaces. *Annals of Applied Probability* **3**, 819–39.

Rosenthal, J. S. (1995a). Rates of Convergence for Gibbs Sampling for Variance Component Models. *Annals of Statistics* **23**, 740–61.

Rosenthal, J. S. (1995b). Minorization Conditions and Convergence Rates for Markov Chain Monte Carlo. *Journal of the American Statistical Association* **90**, 558–66.

Ross, S. (1996). *Stochastic Processes.* New York: John Wiley & Sons.

Rotkiewicz, A. (1972). W. Sierpinski's Works on the theory of numbers. *Rend. Circ. Mat. Palermo* **21**, 5–24.

Rousseeuw, P. J. and Leroy, A. M. (1987). *Robust Regression and Outlier Detection.* New York: John Wiley & Sons.

Rubin, D. B. (1981). The Bayesian Bootstrap. *Annals of Statistics* **9**, 130–34.

Rubin, D. B. (1987a). A Noniterative Sampling/Importance Resampling Alternative to the Data Augmentation Algorithm for Creating a Few Imputations When Fractions of Missing Information Are Modest: The SIR Algorithm. Discussion of Tanner & Wong (1987). *Journal of the American Statistical Society* **82**, 543–46.

Rubin, D. B. (1987b). *Multiple Imputation for Nonresponse in Surveys.* New York: John Wiley & Sons.

Rubinstein, R. Y. (1981). *Simulation and the Monte Carlo Method.* New York: John Wiley & Sons.

Ruggeri, F. and Wasserman, L. (1993). Infinitesimal Sensitivity of Posterior Distributions. *Canadian Journal of Statistics* **21**, 195–203.

Rukhin, A., Soto, J., Nechvatal, J., Smid, M., Barker, E., Leigh, S., Levenson, M., Vangel, M., Banks, D., Heckert, A., Dray, J., and Vo, S. (2000). *A Statistical Test Suite for Random and Pseudorandom Number Generators for Cryptographic Applications* (revised May 15, 2001). NIST SP 800-22. Washington, DC: National Institute of Standards and Technology. <http://csrc.nist.gov/rng/rng2.html>.

Runkle, D. E. (1987). Vector Autoregressions and Reality. *Journal of Business and Economic Statistics* **5**, 437–42.

Santner, T. J. and Duffy, D. E. (1986). A Note on A. Albert and J. A. Anderson's "Conditions for the Existence of Maximum Likelihood Estimates in Logistic Regression Models". *Biometrika* **73**, 755–58.

SAS Institute (1999). *SAS/STAT User's Guide*, Version 8, Volume 2. Carey, NC: SAS Institute.

Schmidt, F. L. and Hunter, J. E. (1996). Measurement Error in Psychological Research: Lessons from 26 Research Scenarios. *Psychological Methods* **1**, 199–223.

Schnabel, R. B. and Eskow, E. (1990). A New Modified Cholesky Factorization. *SIAM Journal of Scientific Statistical Computing* **11**, 1136–58.

Schneeweiss, H. (1976). Consistent Estimation of a Regression with Errors in the Variables. *Metrika* **23**, 101–17.

Schneier, B. (1994). *Applied Cryptography: Protocols, Algorithms, and Source Code in C.* New York: John Wiley & Sons.

Schumacher, M., Robner, R., and Vach, W. (1996). Neural Networks and Logistic Regression. *Computational Statistics and Data Analysis* **21**, 661–701.

Schwefel, H. P. (1977). *Numerische Optimierung von Computer-Modellen mittels der Evolutionsstrategie.* Basel, Switzerland: Birkhauser.

Searle, S. R. (1971). *Linear Models.* New York: Wiley & Sons.

Seber, G. A. F. and Wild, C. J. (1989). *Nonlinear Regression.* New York: John Wiley & Sons.

Sekhon, J. S. and Mebane, W., Jr. (1998). Genetic Optimization Using Derivatives: Theory and Application to Nonlinear Models. *Political Analysis* **7**, 187–210.

Selke, W., Talapov, A. L., and Schur, L. N. (1993). Cluster-Flipping Monte Carlo Algorithm and Correlations in "Good" Random Number Generators. *JETP Letters* **58**, 665–68.

Semmler, W. and Gong, G. (1996). Estimating Parameters of Real Business Cycle Models. *Journal of Economic Behavior and Organization* **30**, 301–25.

Serfling, R. J. (1980). *Approximation Theorems of Mathematical Statistics*. New York: John Wiley & Sons.

Shao, J. and Tu, D. (1995). *The Jackknife and Bootstrap*. New York: Springer-Verlag.

Sieber, J. E. (1991). Sharing Social Science Data. In *Sharing Social Science Data*, J. E. Sieber (ed.). Newbury Park, CA: Sage Publications, pp. 1–19.

Sierpinski, W. (1960). Sur un problème concernant les nombres. *Elemente der Mathematik* **15**, 73–74.

Simon, S. D. and LeSage, J. P. (1988). Benchmarking Numerical Accuracy of Statistical Algorithms. *Computational Statistics and Data Analysis* **7**, 197–209.

Simon, S. D. and LeSage, J. P. (1990). Assessing the Accuracy of ANOVA Calculations in Statistical Software. *Computational Statistics and Data Analysis* **8**, 325–32.

Sirkin, R. M. (1999). *Statistics for the Social Sciences*, 2nd ed. Thousand Oaks, CA: Sage Publications.

Sivaganesan, S. (1993). Robust Bayesian Diagnostics. *Journal of Statistical Planning and Inference* **35**, 171–88.

Sivaganesan, S. (2000). Global and Local Robustness Approaches: Uses and Limitations. In *Robust Bayesian Analysis*, D. R. Insua and F. Ruggeri (eds.). New York: Springer-Verlag, pp. 89–108.

Skinner, C. J. (1998). Logistic Modelling of Longitudinal Survey Data with Measurement Error. *Statistica Sinica* **8**, 1045–58.

Smirnov, O. and Anselin, L. (2001). Fast Maximum Likelihood Estimation of Very Large Spatial Autoregressive Models: A Characteristic Polynomial Approach. *Computational Statistics and Data Analysis* **35**, 301–19.

Smith, P. H. (1972). The Social Base of Peronism. *Hispanic American Historical Review* **52**, 55–73.

Smith, D. M. (1988). Algorithm 693; A FORTRAN Package for Floating-Point Multiple-Precision Arithmetic. *ACM Transactions on Mathematical Software (TOMS)* **17**, 273–83.

Smith, D. M. (2001). Algorithm 814: Fortran 90 Software for Floating-Point Multiple Precision Arithmetic, Gamma and Related Functions. *ACM Transactions on Mathematical Software (TOMS)* **27**, 377–87.

Smith, K. B. (2003). *The Ideology of Education: The Commonwealth, the Market, and America's Schools*. Albany, NY: SUNY Press.

Smith, G. and Campbell, F. (1980). A Critique of Some Ridge Regression Methods (with comments). *Journal of the American Statistical Association* **75**, 74–103.

Smith, A. F. M. and Gelfand, A. E. (1992). Bayesian Statistics without the Tears. *American Statistician* **46**, 84–88.

Smith, A. F. M. and Makov, U. E. (1978). A Quasi-Bayes Sequential Procedure for Mixtures. *Journal of the Royal Statistical Society, Series B* **40**, 106–12.

Smith, K. B. and Meier, K. J. (1995). *The Case against School Choice: Politics, Markets, and Fools*. Armonk, NY: M.E. Sharpe.

Smith, A. F. M., Skene, A. M., Shaw, J. E. H., and Naylor, J. C. (1985). The Implementation of the Bayesian Paradigm. *Communications in Statistics* **14**, 1079–1102.

Smith, A. F. M., Skene, A. M., Shaw, J. E. H., and Naylor, J. C. (1987). Progress with Numerical and Graphical Methods for Practical Bayesian Statistics. *Statistician* **36**, 75–82.

Srinivasan, A., Ceperley, D. M., and Mascagni, M. (1999). Random Number Generators for Parallel Applications. In *Advances in Chemical Physics* **105**, *Monte Carlo Methods in Chemical Physics*, D. Ferguson, J. I. Siepmann, and D. G. Truhlar (eds.). New York: John Wiley & Sons.

Starr, N. (1979). Linear Estimation of the Probability of Discovering a New Species. *Annals of Statistics* **7**, 644–52.

Stata Corporation (1999). *Stata Statistical Software Release 6.0.* College Station, TX: Stata Corporation.

Stefanski, L. A. (1989). Unbiased Estimation of a Nonlinear Function of a Normal Mean with an Application to Measurement Error Models. *Communications in Statistics A* **18**, 4335–58.

Stern, S. (1997). Simulation-Based Estimation. *Journal of Economic Literature* **35**, 2006–39.

Stewart, T. J. (1986). Experience with a Bayesian Bootstrap Method Incorporating Proper Prior Information. *Communications in Statistics A* **15**, 3205–25.

St. Laurent, R. T. and Cook, R. D. (1993). Leverage, Local Influence and Curvature in Nonlinear Regression. *Biometrika* **80**, 99–106.

Stokes, H. H. (2003). On the Advantage of Using More Than One Package to Solve a Problem. *Journal of Economic and Social Measurement*. Forthcoming.

Strawderman, W. E. (1978). Minimax Adaptive Generalized Ridge Regression Estimators. *Journal of the American Statistical Association* **72**, 890–91.

Sun, D., Tsutakawa, R. K., and Speckman, P. L. (1999). Posterior Distribution of Hierarchical Models Using car(1) Distributions. *Biometrika* **86**, 341–50.

Sutradhar, B. C. and Rao, J. N. K. (1996). Estimation of Regression Parameters in Generalized Linear Models for Cluster Correlated Data with Measurement Error. *Canadian Journal of Statistics* **24**, 177–92.

Symons, M. J., Grimson, R. C., and Yuan, Y. C. (1983). Clustering of Rare Events. *Biometrics* **39**, 193–205.

Tam, K. Y. and Kiang, M. Y. (1992). Managerial Applications of Neural Networks: The Case of Bank Failure Predictions. *Management Science* **38**, 926–47.

Tam Cho, W. K. (1998). Iff the Assumption Fits. . . : A Comment on the King Ecological Inference Model. *Political Analysis* **7**, 143–64.

Tam Cho, W. K. and Gaines, B. (2001). *Reassessing the Study of Split-Ticket Voting.* Unpublished manuscript, archived as ICPSR publication related archive item 1264. <ftp://anonymous@ftp.icpsr.umich.edu/ pub/PRA/outgoing/ s1264/>.

Tanner, M. A. (1996). *Tools for Statistic Inference: Methods for the Exploration of Posterior Distributions and Likelihood Functions.* New York: Springer-Verlag.

Thistead, R. A. (1988). *Elements of Statistical Computing: Numerical Computation.* New York: Chapman & Hall/CRC Press.

Thompson, S. P. and Gardner, M. (1998). *Calculus Made Easy*, rev. ed. New York: St. Martins Press.

Thursby, J. G. (1985). The Relationship among the Specification Tests of Hausman, Ramsey, and Chow. *Journal of the American Statistical Association* **80**, 926–28.

Tierney, L. (1994). Markov Chains for Exploring Posterior Distributions. *Annals of Statistics* **22**, 1701–28.

Titterington, D. M., Smith, A. F. M., and Makov, U. E. (1985). *Statistical Analysis of Finite Mixture Distributions.* New York: John Wiley & Sons.

Toothill, J. P. R., Robinson, W. D., and Adams, A. G. (1971). The Runs Up and Down Performance of Tausworthe Pseudo-random Number Generators. *Journal of the Association for Computing Machinery* **18**, 381–99.

Traub, J. F. and Wozniakowsi, H. (1992). The Monte Carlo Algorithm with a Pseudo-random Generator. *Mathematics of Computation* **58**, 303–39.

Turing, A. M. (1948). Rounding-off Errors in Matrix Processes. *Quarterly Journal of Mechanics and Applied Mathematics* **1**, 287–308.

Tweedie, R. L. (1975). Sufficient Conditions for Ergodicity and Recurrence of Markov Chains on a General State-Space. *Stochastic Processes and Applications* **3**, 385–403.

Vattulainen, L., Nissila, T. A., and Kankalla, K. (1994). Physical Tests for Random Numbers in Simulations. *Physical Review Letters* **73**, 1513–16.

Veall, M. R. (1990). Testing for a Global Maximum in an Econometric Context. *Econometrica* **58**, 1459–65.

Venables, W. N. and Ripley, B. D. (1999). *Modern Applied Statistics Using S-Plus*, 3rd ed. New York: Springer-Verlag.

Viega, J. and McGraw, G. (2001). *Building Secure Software: How to Avoid Security Problems the Right Way.* Reading, MA: Addison-Wesley.

Vinod, H. D. (2000). Review of GAUSS for Windows Including Its Numerical Accuracy. *Journal of Applied Econometrics* **15**, 211–20.

Volinsky, C. T., Madigan, D., Raftery, A. E., and Kronmal, R. A. (1997). Bayesian Model Averaging in Proportional Hazard Models: Assessing the Risk of a Stroke. *Applied Statistics* **46**, 433–48.

Von Neumann, J. (1951). Various Techniques Used in Connection with Random Digits, "Monte Carlo Method." *U.S. National Bureau of Standards Applied Mathematics Series* **12**, 36–38.

von zur Gathen, J. and Gerhard, J. (1999). *Modern Computer Algebra.* Cambridge: Cambridge University Press.

Wagner, K. and Gill, J. (2004). Bayesian Inference in Public Administration Research: Substantive Differences from Somewhat Different Assumptions. *International Journal of Public Administration* (forthcoming).

Wald, A. (1940). The Fitting of Straight Lines if Both Variables are Subject to Error. *Annals of Mathematical Statistics* **11**, 284–300.

Wall, L., Christiansen, T., and Orwant, J. (2000). *Programming Perl*, 3rd ed. Sebastapol, CA: O'Reilly & Associates.

Wampler, R. H. (1980). Test Procedures and Test Problems for Least Squares Algorithms. *Journal of Econometrics* **12**, 3–22.

Wang, C. Y. and Wang, S. (1997). Semiparametric Methods in Logistic Regression with Measurement Error. *Statistica Sinica* **7**, 1103–20.

Wasserman, L. (1992). Recent Methodological Advances in Robust Bayesian Inference. In *Bayesian Statistics 4*, J. O. Berger, J. M. Bernardo, A. P. Dawid, and A. F. M. Smith (eds.). Oxford: Oxford University Press, pp. 763–73.

Wei, G. C. G. and Tanner, M. A. (1990). A Monte Carlo Implementation of the EM Algorithm and the Poor Man's Data Augmentation Algorithm. *Journal of the American Statistical Association* **85**, 699–704.

Weiss, R. E. (1996). An Approach to Bayesian Sensitivity Analysis. *Journal of the Royal Statistical Society, Series B* **58**, 739–50.

Weng, C. S. (1989). On a Second Order Property of the Bayesian Bootstrap. *Annals of Statistics* **17**, 705–10.

West, M. (1984). Outlier Models and Prior Distributions in Bayesian Linear Regression. *Journal of the Royal Statistical Society, Series B* **46**, 431–39.

West, P. M., Brockett, P. L., and Golden, L. L. (1997). A Comparative Analysis of Neural Networks and Statistical Methods for Predicting Consumer Choice. *Marketing Science* **16**, 370–91.

Western, B. (1995). A Comparative Study of Working-Class Disorganization: Union Decline in Eighteen Advanced Capitalist Countries. *American Sociological Review* **60**, 179–201.

Western, B. (1998). Causal Heterogeneity in Comparative Research: A Bayesian Hierarchical Modeling Approach. *American Journal of Political Science* **42**, 1233–59.

Western, B. (1999). Bayesian Methods for Sociologists: An Introduction. *Sociological Methods & Research* **28**, 7–34.

White, H. (1981). Consequences and Detection of Misspecified Nonlinear Regression Models. *Journal of the American Statistical Association* **76**, 419–33.

White, H. (1982). Maximum Likelihood Estimation of Misspecified Models. *Econometrica* **50**, 1–26.

Whittlesey, J. R. B. (1969). On the Multidimensional Uniformity of Pseudo-random Generators. *Communications of the Association for Computing Machinery* **12**, 247.

Wichmann, B. A. and Hill, I. D. (1982). An Efficient and Portable Pseudorandom Number Generator. *Applied Statistics* **31**, 188–90.

Wilkinson, L. (1994). Practical Guidelines for Testing Statistical Software. In *Computational Statistics*, P. Dirschedl and R. Ostermann (eds.). Heidelberg: Physica-Verlag.

Wilkinson, L. and Dallal, G. E. (1977). Accuracy of Sample Moments Calculations among Widely Used Statistical Programs. *American Statistician* **21**, 128–31.

Wolak, F. (1991). The Local Nature of Hypothesis Tests Involving Inequality Constraints in Nonlinear Models. *Econometrica* **59**, 981–95.

Wolfinger, R. and Rosenstone, S. J. (1980). *Who Votes?* New Haven, CT: Yale University Press.

Wolfram, S. (1999). *The Mathematica Book*, 4th ed. Cambridge: Cambridge University Press.

Wolfram, S. (2001). *Mathematica, Version 4.1.* Champaign, IL: Wolfram Research.

Wolpert, D. H. and Macready, W. G. (1997). No Free Lunch Theorems for Optimization. *IEEE Transactions on Evolutionary Computation* **1**, 67–82.

Wong, M. Y. (1989). Likelihood Estimation of a Simple Linear Regression Model When Both Variables Have Error. *Biometrika* **76**, 141–48.

Wooldridge, J. M. (2000). *Introductory Econometrics: A Modern Approach.* Cincinnati, OH: South-Western Publishing.

Wu, B. and Chang, C. (2002). Using Genetic Algorithms to Parameters $(d; r)$ Estimation for Threshold Autoregressive Models. *Computational Statistics and Data Analysis* **38**, 315–30.

Xianga, A., Lapuertab, P., Ryutova, A., Buckleya, J., and Azena, S. (2000). Comparison of the Performance of Neural Network Methods and Cox Regression for Censored Survival Data *Computational Statistics and Data Analysis* **34**, 243–57.

Yeo, G.-K. (1984). A Note of Caution on Using Statistical Software. *Statistician* **33**, 181–84.

Young, M. R., DeSarbo, W. S., and Morwitz, V. G. (1998). The Stochastic Modeling of Purchase Intentions and Behavior. *Management Science* **44**, 188–202.

Zellner, A. and Moulton, B. R. (1985). Bayesian Regression Diagnostics with Applications to International Consumption and Income Data. *Journal of Econometrics* **29**, 187–211.

Zhao, Y. and Lee, A. H. (1996). A Simulation Study of Estimators for Generalized Linear Measurement Error Models. *Journal of Statistical Computation and Simulation* **54**, 55–74.

Zhenting, H. and Qingfeng, G. (1978). *Homogeneous Denumerable Markov Chains.* Berlin: Spring-Verlag Science Press.

Author Index

Boldface page number indicates that the author is the primary topic of discussion.

Numerical Issues in Statistical Computing for the Social Scientist, by Micah Altman, Jeff Gill, and Michael P. McDonald
ISBN 0-471-23633-0   Copyright © 2004 John Wiley & Sons, Inc.

Subject Index

Numerical Issues in Statistical Computing for the Social Scientist, by Micah Altman, Jeff Gill,
and Michael P. McDonald
ISBN 0-471-23633-0   Copyright © 2004 John Wiley & Sons, Inc.

WILEY SERIES IN PROBABILITY AND STATISTICS
ESTABLISHED BY WALTER A. SHEWHART AND SAMUEL S. WILKS

Editors *David J. Balding, Noel A. C. Cressie, Nicholas I. Fisher,*
Iain M. Johnstone, J. B. Kadane, Louise M. Ryan, David W. Scott,
Adrian F. M. Smith, Jozef L. Teugels
Editors Emeriti: *Vic Barnett, J. Stuart Hunter, David G. Kendall*

The *Wiley Series in Probability and Statistics* is well established and authoritative. It covers many topics of current research interest in both pure and applied statistics and probability theory. Written by leading statisticians and institutions, the titles span both state-of-the-art developments in the field and classical methods.

Reflecting the wide range of current research in statistics, the series encompasses applied, methodological and theoretical statistics, ranging from applications and new techniques made possible by advances in computerized practice to rigorous treatment of theoretical approaches.

This series provides essential and invaluable reading for all statisticians, whether in academia, industry, government, or research.

ABRAHAM and LEDOLTER · Statistical Methods for Forecasting
AGRESTI · Analysis of Ordinal Categorical Data
AGRESTI · An Introduction to Categorical Data Analysis
AGRESTI · Categorical Data Analysis, *Second Edition*
ALTMAN, GILL, and McDONALD · Numerical Issues in Statistical Computing
    for the Social Scientist
ANDĚL · Mathematics of Chance
ANDERSON · An Introduction to Multivariate Statistical Analysis, *Third Edition*
 *ANDERSON · The Statistical Analysis of Time Series
ANDERSON, AUQUIER, HAUCK, OAKES, VANDAELE, and WEISBERG ·
    Statistical Methods for Comparative Studies
ANDERSON and LOYNES · The Teaching of Practical Statistics
ARMITAGE and DAVID (editors) · Advances in Biometry
ARNOLD, BALAKRISHNAN, and NAGARAJA · Records
 *ARTHANARI and DODGE · Mathematical Programming in Statistics
 *BAILEY · The Elements of Stochastic Processes with Applications to the Natural
    Sciences
BALAKRISHNAN and KOUTRAS · Runs and Scans with Applications
BARNETT · Comparative Statistical Inference, *Third Edition*
BARNETT and LEWIS · Outliers in Statistical Data, *Third Edition*
BARTOSZYNSKI and NIEWIADOMSKA-BUGAJ · Probability and Statistical Inference
BASILEVSKY · Statistical Factor Analysis and Related Methods: Theory and
    Applications
BASU and RIGDON · Statistical Methods for the Reliability of Repairable Systems
BATES and WATTS · Nonlinear Regression Analysis and Its Applications
BECHHOFER, SANTNER, and GOLDSMAN · Design and Analysis of Experiments for
    Statistical Selection, Screening, and Multiple Comparisons
BELSLEY · Conditioning Diagnostics: Collinearity and Weak Data in Regression
BELSLEY, KUH, and WELSCH · Regression Diagnostics: Identifying Influential
    Data and Sources of Collinearity
BENDAT and PIERSOL · Random Data: Analysis and Measurement Procedures,
    Third Edition
BERRY, CHALONER, and GEWEKE · Bayesian Analysis in Statistics and
    Econometrics: Essays in Honor of Arnold Zellner
BERNARDO and SMITH · Bayesian Theory

*Now available in a lower priced paperback edition in the Wiley Classics Library.

DAVID and NAGARAJA · Order Statistics, *Third Edition*
*DEGROOT, FIENBERG, and KADANE · Statistics and the Law
DEL CASTILLO · Statistical Process Adjustment for Quality Control
DETTE and STUDDEN · The Theory of Canonical Moments with Applications in
    Statistics, Probability, and Analysis
DEY and MUKERJEE · Fractional Factorial Plans
DILLON and GOLDSTEIN · Multivariate Analysis: Methods and Applications
DODGE · Alternative Methods of Regression
*DODGE and ROMIG · Sampling Inspection Tables, *Second Edition*
*DOOB · Stochastic Processes
DOWDY and WEARDEN · Statistics for Research, *Second Edition*
DRAPER and SMITH · Applied Regression Analysis, *Third Edition*
DRYDEN and MARDIA · Statistical Shape Analysis
DUDEWICZ and MISHRA · Modern Mathematical Statistics
DUNN and CLARK · Applied Statistics: Analysis of Variance and Regression, *Second Edition*
DUNN and CLARK · Basic Statistics: A Primer for the Biomedical Sciences,
    Third Edition
DUPUIS and ELLIS · A Weak Convergence Approach to the Theory of Large Deviations
*ELANDT-JOHNSON and JOHNSON · Survival Models and Data Analysis
ENDERS · Applied Econometric Time Series
ETHIER and KURTZ · Markov Processes: Characterization and Convergence
EVANS, HASTINGS, and PEACOCK · Statistical Distributions, *Third Edition*
FELLER · An Introduction to Probability Theory and Its Applications, Volume I,
    Third Edition, Revised; Volume II, *Second Edition*
FISHER and VAN BELLE · Biostatistics: A Methodology for the Health Sciences
*FLEISS · The Design and Analysis of Clinical Experiments
FLEISS · Statistical Methods for Rates and Proportions, *Second Edition*
FLEMING and HARRINGTON · Counting Processes and Survival Analysis
FULLER · Introduction to Statistical Time Series, *Second Edition*
FULLER · Measurement Error Models
GALLANT · Nonlinear Statistical Models
GHOSH, MUKHOPADHYAY, and SEN · Sequential Estimation
GIFI · Nonlinear Multivariate Analysis
GLASSERMAN and YAO · Monotone Structure in Discrete-Event Systems
GNANADESIKAN · Methods for Statistical Data Analysis of Multivariate Observations,
    Second Edition
GOLDSTEIN and LEWIS · Assessment: Problems, Development, and Statistical Issues
GREENWOOD and NIKULIN · A Guide to Chi-Squared Testing
GROSS and HARRIS · Fundamentals of Queueing Theory, *Third Edition*
*HAHN and SHAPIRO · Statistical Models in Engineering
HAHN and MEEKER · Statistical Intervals: A Guide for Practitioners
HALD · A History of Probability and Statistics and their Applications Before 1750
HALD · A History of Mathematical Statistics from 1750 to 1930
HAMPEL · Robust Statistics: The Approach Based on Influence Functions
HANNAN and DEISTLER · The Statistical Theory of Linear Systems
HEIBERGER · Computation for the Analysis of Designed Experiments
HEDAYAT and SINHA · Design and Inference in Finite Population Sampling
HELLER · MACSYMA for Statisticians
HINKELMAN and KEMPTHORNE: · Design and Analysis of Experiments, Volume 1:
    Introduction to Experimental Design
HOAGLIN, MOSTELLER, and TUKEY · Exploratory Approach to Analysis
    of Variance
HOAGLIN, MOSTELLER, and TUKEY · Exploring Data Tables, Trends and Shapes
*HOAGLIN, MOSTELLER, and TUKEY · Understanding Robust and Exploratory Data Analysis

*Now available in a lower priced paperback edition in the Wiley Classics Library.

*Now available in a lower priced paperback edition in the Wiley Classics Library.

*Now available in a lower priced paperback edition in the Wiley Classics Library.